Multi-Phase Fueling and Feedback Processes in Jetted AGN

Multi-Phase Fueling and Feedback Processes in Jetted AGN

Guest Editors

Isabella Prandoni
Ilaria Ruffa

MDPI

Basel • Beijing • Wuhan • Barcelona • Belgrade • Novi Sad • Cluj • Manchester

Guest Editors
Isabella Prandoni
Institute of Radioastronomy
INAF
Bologna
Italy

Ilaria Ruffa
Arcetri Observatory
INAF
Florence
Italy

Editorial Office
MDPI AG
Grosspeteranlage 5
4052 Basel, Switzerland

This is a reprint of the Special Issue, published open access by the journal *Galaxies* (ISSN 2075-4434), freely accessible at: https://www.mdpi.com/journal/galaxies/special_issues/A034OQPK44.

For citation purposes, cite each article independently as indicated on the article page online and as indicated below:

Lastname, A.A.; Lastname, B.B. Article Title. *Journal Name* **Year**, *Volume Number*, Page Range.

ISBN 978-3-7258-7026-4 (Hbk)
ISBN 978-3-7258-7027-1 (PDF)
https://doi.org/10.3390/books978-3-7258-7027-1

Contents

About the Editors

Isabella Prandoni

Isabella Prandoni is a Science Director at the Institute of Radioastronomy (IRA) of the National Institute for Astrophysics (INAF) in Bologna, Italy. Her research primarily focuses on galaxy formation and evolution, with a specific emphasis on the role of Active Galactic Nuclei (AGN) in shaping galaxies across cosmic time. As an internationally recognized expert in radio interferometry and radio surveys, she holds leading roles in several major projects conducted with new-generation radio telescopes, including LOFAR, MeerKAT, and ASKAP. Dr. Prandoni coordinates national activities in preparation for the forthcoming SKA Observatory (SKAO), the world's largest radio telescope, and serves as a member of the Italian delegation to the SKAO Council. She has previously served as the Project Scientist for the Sardinia Radio Telescope and as a member of the European ALMA Science Advisory Committee.

Ilaria Ruffa

Ilaria Ruffa is a Postdoctoral Researcher at the INAF, Arcetri Astrophysical Observatory in Florence, Italy. Her research primarily investigates the physics of galaxy formation and evolution, with a special focus on the co-evolution of Supermassive Black Holes (SMBHs) and their host galaxies. An expert in radio interferometry and cold gas dynamics, she utilizes high-resolution data from world-class facilities, such as the Atacama Large Millimeter/submillimeter Array (ALMA) and the SKA precursor MeerKAT, to probe the feeding and feedback mechanisms of Active Galactic Nuclei (AGN). Ruffa is a key member of various international collaborations, such as the WISDOM (mm-Wave Interferometric Survey of Dark Object Masses) and the KILOGAS (Kiloparsec Investigations of Local Objects' Gas And Star-formation), where she leads efforts involving accurate studies of the cold gas physics and kinematics. Her work has provided significant new insights into the interplay between black hole accretion and the physical properties of the cold interstellar medium.

Preface

Relativistic radio jets associated with active galactic nuclei (AGN) are widely recognized as key agents in the complex interplay between supermassive black holes (SMBHs) and their surrounding environments. Over the past decades, substantial observational and theoretical progress has demonstrated that jets can profoundly affect the thermodynamic and dynamic states of gas across a wide range of spatial scales, from the immediate vicinity of the central engine to the galactic, circumgalactic and intergalactic medium. However, despite this progress, fundamental questions remain, placing the study of jetted AGN and their impact on galaxy evolution at the forefront of the current observational and theoretical debate.

A major challenge in this field arises from the intrinsically multi-scale and multi-physics nature of the problem. Achieving a comprehensive understanding of jetted AGN requires addressing a broad and interconnected set of processes. These processes range from the mechanisms responsible for fueling the SMBH and triggering jet activity, as well as those governing radio jet lifetimes and duty cycles, to the detailed physics of jet-driven feedback and the quantitative assessment of its impact on the host galaxy and its large-scale environment. On sub-galactic scales, jets interact with the dense, multiphase interstellar medium, driving shocks, turbulence, and outflows whose net effect on star formation is still debated. On the galactic, circumgalactic and intergalactic scales, jets can redistribute gas reservoirs and regulate cooling, creating conditions that can influence the long-term evolutionary paths of galaxies. These processes are, in turn, tightly linked to the physics of accretion and jet launching, which depend on angular momentum transport, magnetic fields, and the properties of the central SMBH. Capturing this full picture requires bridging traditionally separate research areas, which remains one of the central challenges of modern AGN studies.

Thanks to the advent of a new generation of observational facilities and numerical tools, this topic represents one of the most active and rapidly evolving fields in extragalactic astrophysics. For instance, integral-field spectroscopy with instruments such as VLT/MUSE, high-resolution millimeter observations with ALMA, and the growing capabilities of SKA precursors (such as MeerKAT) have begun to reveal the multiphase structure and kinematics of gas affected by radio jets with unprecedented detail. When combined with state-of-the-art numerical simulations, these observations provide new insights into how relativistic outflows couple their energy and momentum to the surrounding medium. As a result, the field is currently witnessing a rapid succession of new findings, both complementary and apparently contradictory, reflecting the richness and complexity of the underlying physics.

The aim of this Special Issue, *Multi Phase Fuelling and Feedback Processes in Jetted AGN*, is to provide a snapshot of recent research in this broad, hot topic, as well as to identify key open questions that will shape future research. Given the pace at which new observational and theoretical developments are emerging, a focused collection of review and research articles could be particularly useful to the community, offering both an overview of recent progress and a framework for interpreting forthcoming results.

This reprint originates from the international meeting *AGN on the Beach*, held in Tropea (Italy) in September 2023, which was organized by the editors of this volume together with other colleagues active in this field. The meeting brought together experts working on different aspects of AGN fueling and feedback, fostering lively discussions across observational and theoretical perspectives. Invited review speakers were encouraged to contribute overview articles based on the topics presented at the meeting, and additional contributions were solicited on a voluntary basis from other participants.

Many of the articles included in this volume were directly inspired by the discussions that took place in Tropea, reflecting the collaborative and interdisciplinary spirit of the meeting.

The reprint is structured around eight contributions that collectively address key aspects of jet fueling and feedback, spanning topics such as the interaction of radio jets with multiphase gas, the impact of jets on star formation, and the role of feedback across different spatial scales and environments. An additional review on jet feedback on sub-parsec scales by Dr. Dipanjan Mukherjee, closely aligned with the themes of this volume and also triggered by the meeting in Tropea, is available separately online and provides further context to the broader picture explored here.

In light of the success of the workshop in Tropea, we decided to establish a new series of *AGN on the Beach* meetings to facilitate discussion of jetted AGN in a relaxed yet scientifically stimulating environment. The next conference in this series will take place from 21 to 25 September 2026 at Diani Beach, Kenya.

We hope that this reprint will serve as a useful reference for both researchers entering the field and experts seeking a consolidated overview of recent progress. By bringing together diverse perspectives on fueling and feedback in jetted AGN, we aim to stimulate further discussion and inspire future studies in the era of next-generation facilities.

Isabella Prandoni and Ilaria Ruffa
Guest Editors

galaxies

MDPI

Review

Fueling Processes on (Sub-)kpc Scales

Francoise Combes

Observatoire de Paris, LERMA, Collège de France, CNRS, PSL University, Sorbonne University, 75014 Paris, France; francoise.combes@obspm.fr

Abstract: Since the 1970s, astronomers have struggled with the issue of how matter can be accreted to promote black-hole growth. While low-angular-momentum stars may be devoured by a black hole, they are not a sustainable source of fuel. Gas, which could potentially provide an abundant fuel source, presents another challenge due to its enormous angular momentum. While viscous torques are not significant, gas is subject to gravity torques from non-axisymmetric potentials such as bars and spirals. Primary bars can exchange angular momentum with the gas within corotation, causing it to spiral inwards until reaching the inner Lindblad resonance. An embedded nuclear bar can then take over. As the gas reaches the black hole's sphere of influence, the torque becomes negative, fueling the center. Dynamical friction also accelerates the infall of gas clouds closer to the nucleus. However, because of the Eddington limit, growing a black hole from a stellar-mass seed is a slow process. The existence of very massive black holes in the early universe remains a puzzle that could potentially be solved through direct collapse of massive clouds into black holes or super-Eddington accretion.

Keywords: galaxies: active nuclei; galaxies: bars; spirals; black holes; angular momentum; molecular torus; fueling; feedback; warps

1. Introduction

One of the main issues when attempting to explain the fueling of active galactic nuclei (AGN) is the mechanism to get rid of the angular momentum (AM). A black hole may swallow neighboring stars, which will create a depletion among those stars with a low AM. A long relaxation time will be required to replenish this loss cone, so the fueling will rely on the gas infall. The latter requires gravity torques, tangential forces, and therefore non-axisymmetric features, such as bars or spirals. Large-scale features are not sufficient and should be supported by embedded bars to prolong their action towards the center.

The high spatial resolution provided by the ALMA interferometer can reveal these embedded structures, in particular nuclear spirals inside the nuclear rings, corresponding to the inner Lindblad resonance (ILR) of the bars. Examples of these structures have been unveiled in nearby Seyfert galaxies.

Inside these nuclear spirals, at a scale of 10 pc, molecular tori have been revealed as circumnuclear disks that are kinematically decoupled from the large-scale disk. The origin of the decoupling could be as a result of several causes, like the accretion of gas with different AM, and/or precession and warping of the very central disk due to relativistic effects, coupled with the supermassive black-hole spin.

2. Angular Momentum Problem

This problem is as a result of the large contrast between the angular momentum of the gas in the last stable orbit, $L = 2 \times 10^{24}$ $(M/10^8 M_\odot)$ cm^2/s, for a typical black-hole mass of $10^8 M_\odot$, and the gas AM at 3 kpc, $L = 10^{29}$ cm^2/s. The ratio of these two values means that the gas has to lose 5 orders of magnitude in AM within a relatively short dynamical time. The illustration of the AM increase with radius is shown in Figure 1. For a typical luminosity of 10^{46} erg/s, the central engine has to swallow 2 $M_{\odot}$/yr during a duty cycle of 100 Myr.

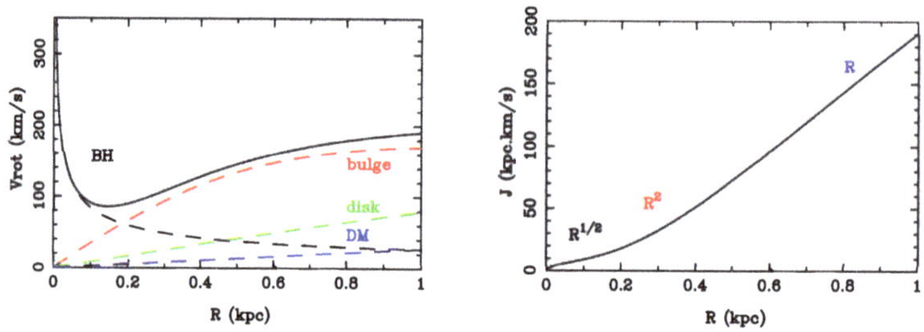

Figure 1. How does the angular momentum increase in radius around a super-massive black hole(BH)?: (**left**) the typical rotation curve in a spiral galaxy is dominated by the BH mass in the center (Keplerian potential), then rises due to the bulge, and is finally flat due to the conspiracy of the disk and dark matter halo (DM). These various contributions are marked in different colors; (**right**) the corresponding angular momentum per unit mass increases with the power laws of the radius, with the first slopes being a half (Keplerian), then two and one.

Stars with a low AM in the neighborhood of the black hole can be tidally sheared, and their gas is accreted in a rotating disk. These TDEs (tidal disruption events) can occur at a frequency of once every 10,000 years in the Milky Way. Some TDEs have been detected in external galaxies, signified by a characteristic light curve decreasing with time as a $-5/3$ power law [1]. However, after a brief period, the depletion and loss cone effect deplete this fueling source unless galaxy interactions re-shuffle the stellar distribution, creating nuclear star clusters in a nuclear starburst.

Gas is, however, the main fuel, and is driven inwards through non-axisymmetries in the galaxy potential. Several steps can be distinguished in this process. First, primary bars, typically with diameters of 10 kpc, drive the gas from their corotation to R∼100 pc, where the gas is stalled in a ring.

Then, embedded nuclear bars prolong their action from 100 pc to 10 pc. Non-axisymmetries usually have morphologies of m = 2, but m = 1 (lopsidedness), or tidal forces from companions can also play a role.

At smaller scales of 1–10 pc, other processes also have to be considered, such as turbulence, viscosity, warps, bends, dynamical friction, and the formation of thick disks, as long as a sufficient amount of gas remains.

2.1. *Dynamics of Bars*

Let us recall the main features of barred galaxies: the stellar orbits are classified through a skeleton of periodic orbits. The latter are orbits that close on themselves after one or more turns in the bar rotating frame. These are the building blocks that determine the stellar distribution function, as they define families of trapped orbits around them. Trapped orbits are non-periodic, but oscillate about one periodic orbit, with a similar shape. The periodic orbits are numerous (see the review by Contopoulos and Grosbol [2]); we will discuss the most important ones for bar support in the following. Inside corotation, the x1 family is the main family supporting the bar. Orbits are elongated parallel to the bar, within the corotation. The x2 family also exists, but only between the two inner Lindblad resonances (ILRs), if they exist. They are more round and are elongated perpendicular to the bar. The existence of two ILRs in the axisymmetric sense might not be sufficient for the x2 family to appear. When the bar is strong enough, the x2 orbits disappear. The bar strength necessary to eliminate the x2 family depends on the pattern speed; the lower the speed, the stronger the bar must be. Outside the corotation, the 2/1 orbits that are run in the retrograde sense in the rotating frame are perpendicular to the bar inside the outer

Lindblad resonance (OLR), and are parallel to the bar slightly outside (see discussion by Kalnajs [3]). Their shape is a characteristic figure eight, which is very similar to the dimpled shape of some outer rings in barred galaxies.

In summary, periodic orbits are aligned parallel or perpendicular to the bar, and their orientation changes by 90° at each resonance [4,5]. Gas tends to follow periodic orbits, but its dissipative character, due to cloud collisions, means that orbits cannot cross. Instead, their orbits are tilted, and they change gradually by 90° at each resonance. The crowding of these stream lines produces a spiral morphology [6]. The spiral is open, and at its maximum can rotate by 180–360°.

2.2. Embedded Structures

Because the gas and stars are not in phase, stellar bars exert torque on the gas, except at resonances where the gas piles up and stalls in rings aligned with the bar in some way. When mass has accumulated in the center, all frequencies of Ω and $\Omega - \kappa/2$, the orbit precessing rate, increase significantly in value, implying the existence of two ILRs. Between these ILRs, the periodic orbits (x2) are perpendicular to the bar.Stars cannot sustain the bar anymore and weaken it. The z-resonance, creating peanut-shape bulges, also weakens the bar. This triggers the decoupling of a second bar, an embedded nuclear bar [7]. This second bar rotates faster than the primary bar, and both are misaligned [8].

2.3. Fueling AGN: Removing Angular Momentum (AM)

The torques exerted on the gas by the bar change sign at each resonance (cf Figure 2). Outside the corotation radius (CR), the gas is driven outwards and accumulates at the outer Lindblad resonance (OLR). Inside CR, the gas is driven inwards, at least down to ILR. To quantify the phenomenon, the bar gravity torque can be computed from the red images by tracing old stars, and using the potential. When the gas distribution is overlaid on the torque map (cf Figure 3), the sign and amount of the gas infall can be estimated, as in NGC3627 [9]. The correlation between the bars and AGN is still debated [10–13], as the time-scales corresponding to the primary and secondary bars are quite different, as well as being different from the AGN duty cycle.

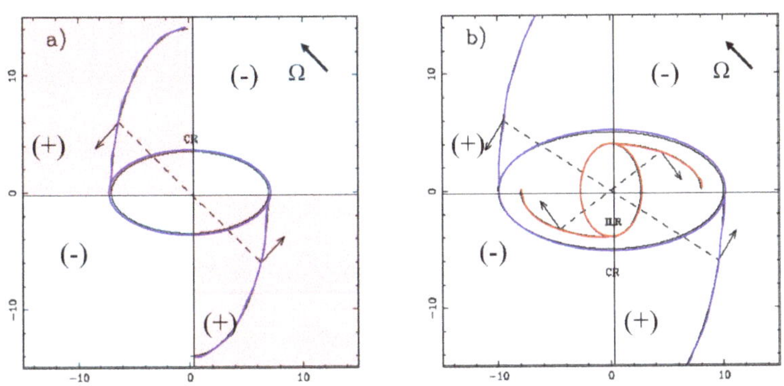

Figure 2. The sign of the torques exerted by the bar on the gas can be obtained geometrically through these schematic diagrams: (**a**) outside corotation, or roughly outside the bar, the torque is positive (red quadrant) with respect to the sense of rotation (Ω), and the gas is driven out toward the outer Lindblad resonance (OLR); (**b**) inside the corotation, the torque is negative (blue quadrant), and the gas is driven inward, down to the inner resonance (ILR).

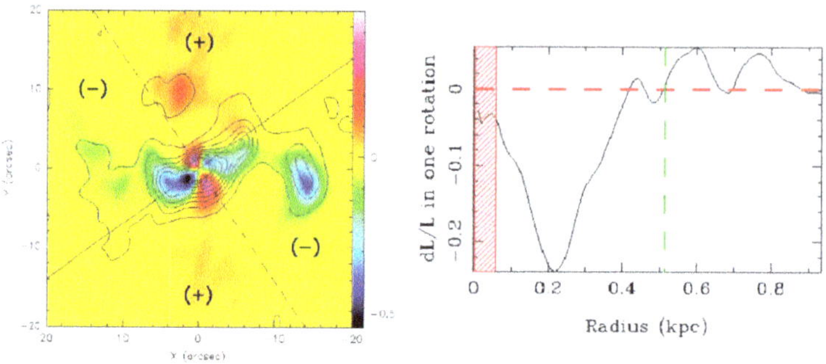

Figure 3. The gravity torques can be measured on real galaxies by computing the potential in each pixel and the forces from a red image (old stars), and comparing the results with the observed gas density. The (**left**) image is the torque map in color, with the bar splitting the plane in four quadrants, for NGC 3627. The gas density is overlaid in contours. The (**right**) plot quantifies the relative angular momentum lost in one rotation, while averaging the torque over the azimuth, weighted by the gas density. Adapted from Casasola et al. [9].

During a survey of about 20 galaxies using the IRAM interferometer, statistics of fueling at the 10–100 pc scale were obtained [14]. Only 35% of negative torques were measured in the center; the remainder of the measurements were positive torque or gas was stalled in a ring. For the latter case, future fueling has to wait for the decoupling of a secondary bar, as shown in the simulation of Figure 4. This means that the fueling phases are short, a few 10^7 years, and may be due to feedback. Star formation is also fueled by torque, and is always associated with AGN activity as well as longer time scales.

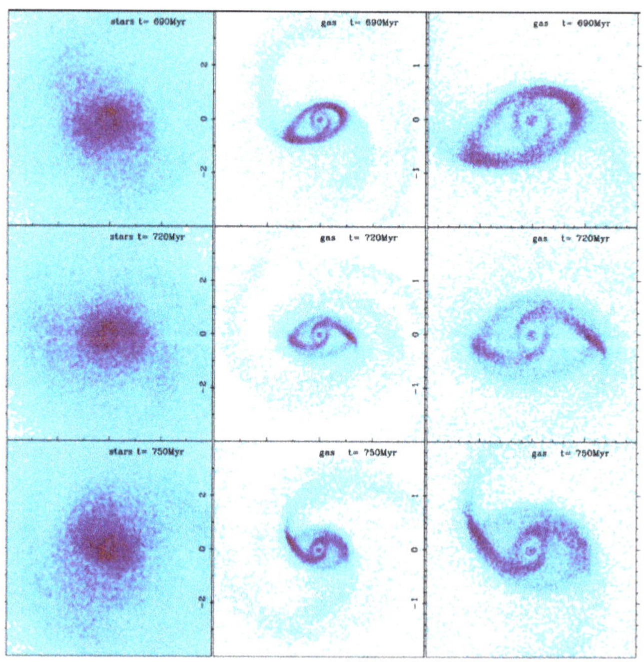

Figure 4. Hydro-N-body simulation of a double bar in a spiral galaxy. The (**left**) panel shows the stellar component surface density (linear scale, axes in kpc), the (**middle**) panel shows the gas

component (log scale), while the (**right**) panel is also the gas but in a zoomed spatial scale. The simulation shows clearly that the gas first piles up in the external ring (ILR of the primary bar), then progressively infalls inside the ILR to gather at the nuclear bar resonance. Adapted from Hunt et al. [15].

3. Small-Scale Fueling with ALMA

With the advent of ALMA, higher spatial resolution is possible, and nearby galaxies can be explored using 10 pc scales. One of the first barred spiral observed was NGC 1433 (cf Figure 5). While only a star-forming ring was observed at ILR with HST, a second ring was detected in the molecular gas with ALMA, corresponding to the second ILR. The computation of the torques showed that AGN is not presently fueled, but positive torques bring the gas from the center to the second ring. Negative torques outside this second ring contribute to the accumulation of gas there [16]. On the minor axis, an outflow with small velocity has been detected. It is one of the smallest outflows that has been detected, with $M(H_2) = 3.6 \times 10^6$ $M_\odot$ and a flow rate of 7 $M_\odot$/yr [17].

Figure 5. HST image of NGC1433 bar, with characteristic leading dust lanes, resulting in a blue ring at the inner Lindblad resonance (ILR). The insert displays the blue ring from HST overlaid with the orange-red image from ALMA of the molecular gas, traced by its CO(3-2) emission. The latter reveals a second ring inside the first ILR. Adapted from Combes et al. [17].

Other barred Seyfert galaxies, like NGC1566, were found in the feeding phase (cf Figure 6). Inside the ILR ring, the CO(3-2) emission map from ALMA revealed the existence of a nuclear disk, with a trailing nuclear bar. The existence of the trailing spiral was a surprise, as without the gravitational influence of the central black hole, a leading spiral was expected. This was due to the shape of the $\Omega - \kappa/2$ curve as a function of the radius. When the gas would infall, it precessed more rapidly if the curve increased when the radius was reduced, which means a trailing arm; the inverse was true if the curve declined (cf Figure 7).

If the spiral was trailing, this means that the gas experienced a climbing curve, due to BH, and it entered the sphere of influence of BH. The computation of the torques in NGC1566 confirmed the evidence of fueling [18].

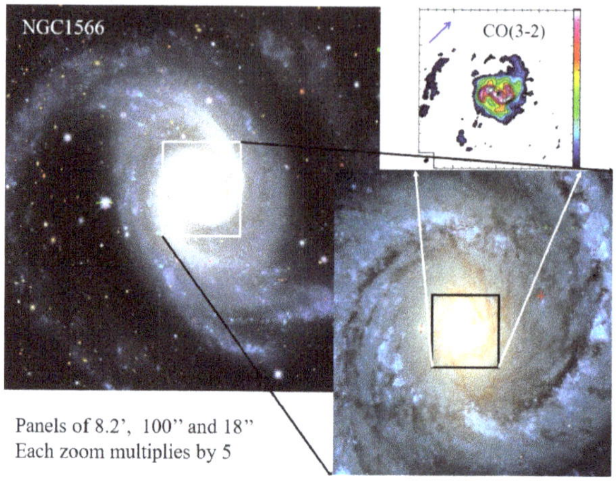

Figure 6. The barred spiral galaxy NGC 1566 reveals several embedded structures, as seen in these three progressively zoomed images. The last one reveals a trailing nuclear spiral in the molecular gas, traced by its CO(3-2) emission observed with ALMA. Adapted from Combes et al. [18].

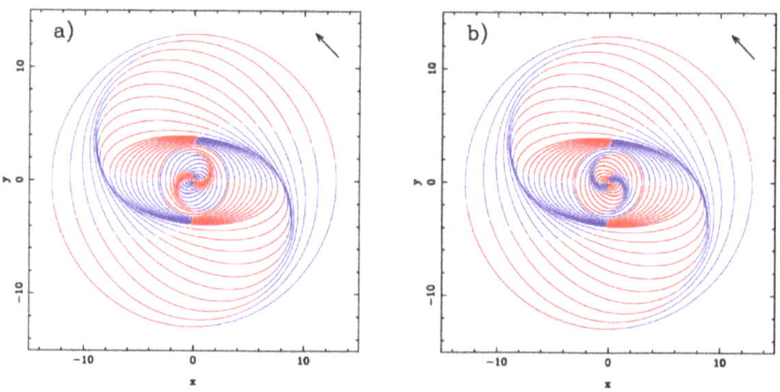

Figure 7. Gas streamlines schematically represented by elliptical orbits, precessed to form logarithmic spirals. The streamlines are initially aligned on the horizontal axis (parallel to the bar) and colored according to the four quadrants. The pattern speed is such that an ILR exists inside the bar, delineated by a ring. (**a**) without a super-massive black hole (BH), the precessing rate decreases towards the center and the gas forms a leading nuclear spiral. (**b**) with a BH dominating the potential inside the ILR, the precessing rate increases towards the center and the gas forms a trailing spiral. This changes the sign of the torque exerted on the gas.

In several other barred galaxies, a trailing nuclear spiral has been revealed in the molecular component with ALMA, for instance in NGC 1808 or NGC 613. This indicates that the gas has entered the sphere of influence of the black hole, making the torque negative, ensuring the AGN is fueled. This trailing spiral always develops inside the ILR ring of the bar [19].

NGC 613 is the academic case of a strong barred galaxy, with a star-forming gas ring at ILR. The first ALMA observations, with a moderate spatial resolution, could see only the ring, without resolving the internal structure [20]. With a beam of 60mas, it was possible to clearly see an internal trailing nuclear spiral and a molecular torus inside the spiral (cf Figure 8). The computation of torques indicated that, when inside a 50 pc radius, the gas could lose all its angular momentum in only one rotation [21]. In addition, a molecular outflow was detected along the minor axis, parallel to the cm wave detected by the radio jet.

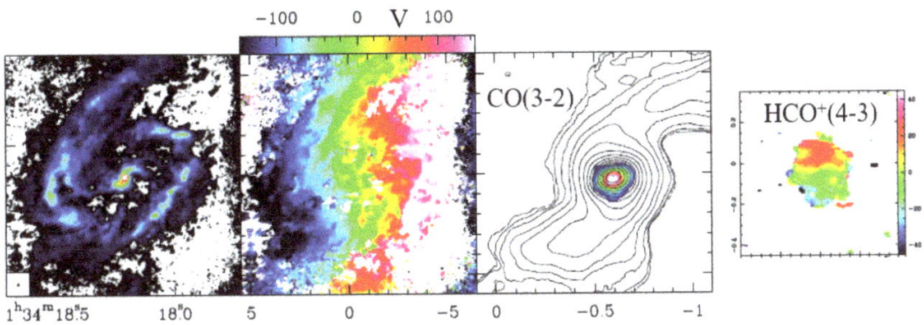

Figure 8. The barred galaxy NGC 613 shows a contrasted ring at its ILR, in the molecular gas traced by CO(3-2) with ALMA. From left to right is the gas surface density, then the velocity field, and a 10-fold zoom of the gas density (contours), with the radio continuum (color-scale) at the center. The right-most panel is the velocity field of the HCO^+(4-3) emission, revealing a misalignment with the large-scale gas. Adapted from Combes et al. [22] and Audibert et al. [21].

4. Molecular Tori

Our vision of the central regions of AGN surrounding the black hole have changed significantly in recent years. For a long time, a dusty torus with a donut shape was assumed to exist around the accretion disk and the broad line region (BLR). It was thought to obscure the BLR for observers observing the accretion disk almost edge-on. But infrared interferometers have shown that the dust is frequently detected in the polar direction, instead of being aligned along the putative torus. This polar dust must be ejected through the AGN wind, starting from the dust sublimation radius (pc size), and forming the border of a hollow cone [23]. The gas motion is then both infall in a thin disk, forming a molecular torus, and then outflows in the perpendicular direction.

Figure 8 shows that the molecular torus, inserted in the nuclear spiral, has a decoupled kinematics, i.e., the kinematic major axis is not aligned with that of the large-scale disk [21,22].

This misalignment is frequent in all nearby galaxies observed using ALMA with 10 pc resolution, such as NGC 1672 or NGC 1326 in the NUGA sample [22], or NGC 5643 or NGC 6300 in the GATOS sample [24]. It is not yet possible to determine whether the central molecular torus is simply continuously warped or tilted, with discontinuous disks torn into a few pieces. Circinus is one of the nearest galaxies, observed down to 2 pc resolution, and shows a nuclear spiral of 20–30 pc and another circum-nuclear disk inside, which can be interpreted as the molecular torus. It is self-absorbed in CO(3-2) but not in CO(6-5), nor in the dense gas tracers, such as HCO^+. In addition to this cold thin disk fueling the AGN, there is an outflow in the polar direction, composed of warm dust and ionized gas (cf Figure 9) [25,26].

The prototypical Seyfert 2 NGC 1068 has been intensively observed at a high resolution at many wavelengths. While the near-infrared clearly reveals a hollow polar cone in warm dust [27], the cold component in the millimeter domain reveals a molecular torus of 7–10 pc in diameter [28,29]. The torus is quasi edge-on, misaligned with the large-scale disk, warped, more inclined than the water maser disk, and might suffer from counter rotation [29,30]. In contrast, the barred Seyfert 1 galaxy NGC 1097, using ALMA at 10 pc resolution, does not show any dense and compact torus. There is a nuclear spiral inside the star-forming ring at the ILR of the bar [31,32]. The absence of torus concerns about 10–20% of low-luminosity AGN. This percentage increases with the Eddington ratio, and could be related to the AGN feedback [24].

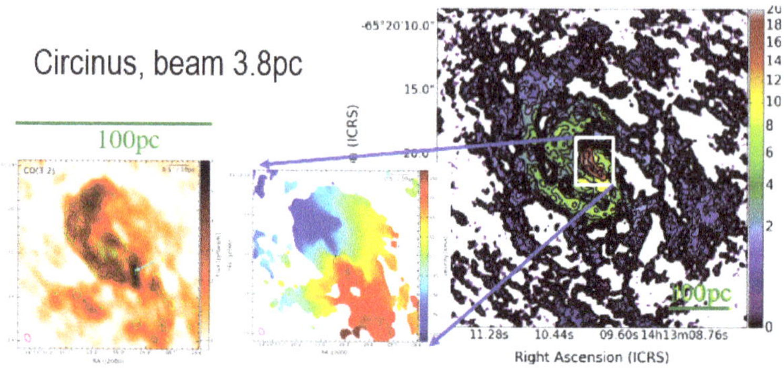

Figure 9. Image of the molecular gas, obtained in CO(3-2) with ALMA, of the Circinus galaxy, with a resolution of 3.8 pc. There is a nuclear spiral, of 20–30 pc, and another circum-nuclear disk inside, which could be the molecular torus (edge-on, for this type 2 AGN). Adapted from Tristram et al. [26].

The influence of the AGN activity is indeed visible on the molecular gas concentration. The latter can be quantified by the surface density ratio of the molecular gas inside 50 pc and inside 200 pc. The most active AGNs have less H_2 concentration. The latter drops for X-ray luminosities Lx $> 10^{42.5}$ erg/s, or for Eddington ratios $> 10^{-3}$, see Figure 10, [24].

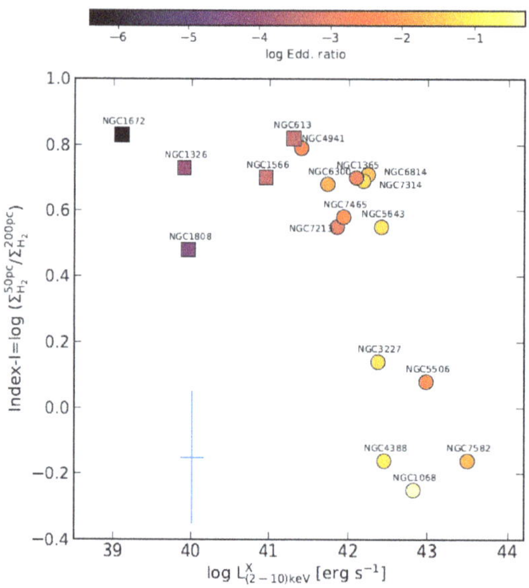

Figure 10. Concentration of molecular gas in the central regions of galaxies versus their AGN luminosities measured in the 2–10 keV X-ray band. The concentration is measured using the ratio of the average H_2 surface densities at two spatial scales, r $<$ 50 pc and r $<$ 200 pc, characteristic of the nuclear and circumnuclear regions. Symbols are color-coded as a function of the Eddington ratios. The sample galaxies: NUGA (square markers) and GATOS (circle markers) can be separated both by their AGN activity and molecular gas concentration. Adapted from García-Burillo et al. [24].

5. Early Black Holes

Along their lifetime, galaxies grow both their bulge and their black hole, in synergy, so that a tight relation can be observed between both masses [33]. Secular evolution, together with some tidal interaction with companions, can provide the fuel for AGN and BH growth. However, AGN feedback makes the BH growth intermittent, as described above. Some different processes should occur at a high redshift, to account for the observations of high black-hole masses (M $>$ 10^9 $M_\odot$) already at z $>$ 6 [34]. Hundreds of quasars have been detected at a high redshift, z $>$ 6, with black hole masses between 10^8 and 10^{10} M⊙, in the first billion years of the universe (e.g., [35–37]). The black-hole growth rate through accretion is proportional to its mass, and is thus very slow at the beginning, starting with stellar mass seeds. At any time, the Eddington luminosity imposes an upper limit in the accretion rate, which is proportional to the BH mass. In contrast, at a high redshift, black holes appear to grow faster than their host bulges, and the proportionality relation breaks. Several solutions have been proposed. One is to assume that the BH in the initial phase accrete exceptionally faster than their Eddington rate, without being stopped by the AGN feedback.

Simulations of black-hole growth with super-Eddington accretion, even in a massive over-density environment, progenitor of a cluster, in general, do not reproduce the observed massive black holes at z = 6 [38], nor explain why black-hole growth is more rapid than the stellar mass growth, contrary to what is observed at z = 0. However, with some modifications to the usual scenario, taking into account less efficient AGN feedback, more efficient accretion, and starting earlier with a more massive seed, it is possible to account for the observations [39].

An alternative solution is to assume that BH collapses directly from a massive cloud, due to primordial metal abundance and a lack of fragmentation. The seeds forming the first black holes would then be much more massive than stellar, and the growth rate would be strong from the start. This, however, assumes contrived suppression of the H_2 molecule formation, and remains debated.

The existence of intermediate mass black holes (IMBHs) is part of the puzzle, as they have not yet been detected unambiguously. The direct collapse scenario could explain their low abundance. The major challenge for observations of IMBHs is their small mass and small impact on their surroundings. As the radius of influence of a black-hole is scaled linearly with its mass, it requires a very high spatial resolution in order to detect their impact on the gas or stellar component. Also, the dynamical friction time-scale for a black hole to decay towards the nuclei is inversely proportional to the mass, and IMBH might not have time to infall within a Hubble time (e.g., [40]).

An promising way to detect IMBH is to search for AGN signatures, and candidates have been observed in several low-mass galaxies [41]. Candidates for IMBH exist in several dwarf active galaxy nuclei (e.g., [42]), however, in the core of globular clusters, where they were expected, only a collection of stellar-mass black holes have been detected instead [43,44].

6. Summary

AGN fueling requires non-axisymmetries in the galaxy potential to drive the gas inwards. The primary bar in spiral galaxies can drive the gas from its corotation to the inner Linblad resonance, where the gas accumulates and forms a starbursting ring, at the scale of 100 pc. Further fueling has to await the formation of a nuclear bar, embedded inside the ILR ring. ALMA has revealed 10 pc scale structures inside the ring, in particular, trailing nuclear spirals, indicating that the gas has entered the sphere of influence of the central black hole. The stellar potential then prolongs the action of the primary bar to fuel the AGN. Inside the nuclear spiral, a molecular disk, kinematically decoupled, is often found, which can be interpreted as the molecular torus.

The misalignment of the torus can be explained by gas accretion with a different angular momentum. This occurs naturally through star formation/supernovae feedback, which ejects gas above the plane. When this gas comes back, by the fountain effect, it can arrive in any direction with a random angular momentum. Numerical simulations have shown that even polar gas rings or disks can be formed [45,46].

The decoupling of molecular tori is not too surprising, given the very different dynamical time scales between the 10 pc and 100 pc scales. In addition, the material at much less than 1 pc from the black hole is certainly influenced by relativistic effects, related to the black-hole spin: the Bardeen–Petterson effect, where the torque exerted by the BH tends to align the material perpendicular to its spin, and produces warps. Warping might also be radiation-driven [47,48].

Contrary to the existence of a tight relation between bulge and black hole masses in nearby galaxies, super-massive black holes grow relatively faster than the stellar component in early galaxies, in the first billion years of the universe. To reproduce this behavior, simulations have assumed super-Eddington accretion and low-efficiency AGN feedback at these epochs. Alternatively, direct collapse of super-massive stars into black holes might contribute to the solution. Future observations in the early universe with JWST might enlighten the issue.

Funding: This research received no external funding.

Informed Consent Statement: Not applicable.

Acknowledgments: Thanks to Ilaria Ruffa and the organisers of the conference "AGN on the beach" in Tropea, Italy, where this material was discussed. FC has benefited from the support of the Programme National Cosmologie et Galaxies.

Conflicts of Interest: The author declares no conflict of interest.

References

1. Miller, J.M.; Kaastra, J.S.; Miller, M.C.; Reynolds, M.T.; Brown, G.; Cenko, S.B.; Drake, J.J.; Gezari, S.; Guillochon, J.; Gultekin, K.; et al. Flows of X-ray gas reveal the disruption of a star by a massive black hole. *Nature* **2015**, *526*, 542–545. [CrossRef]
2. Contopoulos, G.; Grosbol, P. Orbits in barred galaxies. *A&A Rev.* **1989**, *1*, 261–289.
3. Kalnajs, A.J. Pattern Speeds of Density Waves. In *Dynamics of Disc Galaxies*; Sundelius, B., Ed.; Göteborgs University and Chalmers University of Technology: Göteborgs, Sweden, 1991; p. 323.
4. Athanassoula, E. Morphology of bar orbits. *MNRAS* **1992**, *259*, 328–344. [CrossRef]
5. Contopoulos, G.; Papayannopoulos, T. Orbits in weak and strong bars. *A&A* **1980**, *92*, 33–46.
6. Sanders, R.H.; Huntley, J.M. Gas response to oval distortions in disk galaxies. *ApJ* **1976**, *209*, 53–65. [CrossRef]
7. Friedli, D.; Martinet, L. Bars within bars in lenticular and spiral galaxies: A step in secular evolution? *A&A* **1993**, *277*, 27–41.
8. Buta, R.; Combes, F. Galactic Rings. *Fund. Cosmic Phys.* **1996**, *17*, 95–281.
9. Casasola, V.; Hunt, L.K.; Combes, F.; García-Burillo, S.; Neri, R. Molecular Gas in NUclei of GAlaxies (NUGA). XIV. The barred LINER/Seyfert 2 galaxy NGC 3627. *A&A* **2011**, *527*, A92.
10. Alonso, S.; Coldwell, G.; Lambas, D.G. AGN spiral galaxies in groups: Effects of bars. *A&A* **2014**, *572*, A86.
11. Cardamone, C.N.; Schawinski, K.; Masters, K.; Lintott, C.; Fortson, L. Galaxy Zoo: The Connection between AGN Activity and Bars in Late Type Galaxies. In *American Astronomical Society Meeting Abstracts #218*; American Astronomical Society: Boston, MA, USA, 2011; Volume 218.
12. Masters, K.L.; Nichol, R.C.; Hoyle, B.; Lintott, C.; Bamford, S.P.; Edmondson, E.M.; Fortson, L.; Keel, W.C.; Schawinski, K.; Smith, A.M.; et al. Galaxy Zoo: Bars in disc galaxies. *MNRAS* **2011**, *411*, 2026–2034. [CrossRef]
13. Schawinski, K.; Dowlin, N.; Thomas, D.; Urry, C.M.; Edmondson, E. The Role of Mergers in Early-type Galaxy Evolution and Black Hole Growth. *ApJ* **2010**, *714*, L108–L112. [CrossRef]
14. García-Burillo, S.; Combes, F. The feeding of activity in galaxies: A molecular line perspective. *J. Phys. Conf. Ser.* **2012**, *372*, 012050. [CrossRef]
15. Hunt, L.K.; Combes, F.; García-Burillo, S.; Schinnerer, E.; Krips, M.; Baker, A.J.; Boone, F.; Eckart, A.; Léon, S.; Neri, R.; et al. Molecular Gas in NUclei of GAlaxies (NUGA). IX. The decoupled bars and gas inflow in NGC 2782. *A&A* **2018**, *482*, 133–150.
16. Smajić, S.; Moser, L.; Eckart, A.; Valencia-S., M.; Combes, F.; Horrobin, M.; García-Burillo, S.; García-Marín, M.; Fischer, S.; Zuther, J. ALMA-backed NIR high resolution integral field spectroscopy of the NUGA galaxy NGC 1433. *A&A* **2014**, *567*, A119.
17. Combes, F.; García-Burillo, S.; Casasola, V.; Hunt, L.; Krips, M.; Baker, A.J.; Boone, F.; Eckart, A.; Marquez, I.; Neri, R.; et al. ALMA observations of feeding and feedback in nearby Seyfert galaxies: An AGN-driven outflow in NGC 1433. *A&A* **2013**, *558*, A124.
18. Combes, F.; García-Burillo, S.; Casasola, V.; Hunt, L.K.; Krips, M.; Baker, A.J.; Boone, F.; Eckart, A.; Marquez, I.; Neri, R.; et al. ALMA reveals the feeding of the Seyfert 1 nucleus in NGC 1566. *A&A* **2014**, *565*, A97.
19. Audibert, A.; Combes, F.; García-Burillo, S.; Hunt, L.; Eckart, A.; Aalto, S.; Casasola, V.; Boone, F.; Krips, M.; Viti, S.; et al. Black hole feeding and star formation in NGC 1808. *A&A* **2021**, *656*, A60.
20. Miyamoto, Y.; Nakai, N.; Seta, M.; Salak, D.; Nagai, M.; Kaneko, H. ALMA multiline observations toward the central region of NGC 613. *PASJ* **2017**, *69*, 83. [CrossRef]
21. Audibert, A.; Combes, F.; García-Burillo, S.; Hunt, L.; Eckart, A.; Aalto, S.; Casasola, V.; Boone, F.; Krips, M.; Viti, S.; et al. ALMA captures feeding and feedback from the active galactic nucleus in NGC 613. *A&A* **2019**, *632*, A33.

22. Combes, F.; García-Burillo, S.; Audibert, A.; Hunt, L.; Eckart, A.; Aalto, S.; Casasola, V.; Boone, F.; Krips, M.; Viti, S.; et al. ALMA observations of molecular tori around massive black holes. *A&A* **2019**, *623*, A79.
23. Hönig, S.F. Redefining the Torus: A Unifying View of AGNs in the Infrared and Submillimeter. *ApJ* **2019**, *884*, 171. [CrossRef]
24. García-Burillo, S.; Alonso-Herrero, A.; Ramos Almeida, C.; González-Martín, O.; Combes, F.; Usero, A.; Hönig, S.; Querejeta, M.; Hicks, E.K.S.; Hunt, L.K.; et al. The Galaxy Activity, Torus, and Outflow Survey (GATOS). I. ALMA images of dusty molecular tori in Seyfert galaxies. *A&A* **2021**, *652*, A98.
25. Izumi, T.; Wada, K.; Fukushige, R.; Hamamura, S.; Kohno, K. Circumnuclear Multiphase Gas in the Circinus Galaxy. II. The Molecular and Atomic Obscuring Structures Revealed with ALMA. *ApJ* **2018**, *867*, 8. [CrossRef]
26. Tristram, K.R.W.; Impellizzeri, C.M.V.; Zhang, Z.-Y.; Villard, E.; Henkel, C.; Viti, S.; Burtscher, L.; Combes, F.; García-Burillo, S.; Martín, S.; et al. ALMA imaging of the cold molecular and dusty disk in the type 2 active nucleus of the Circinus galaxy. *A&A* **2022**, *664*, A142.
27. Gratadour, D.; Rouan, D.; Grosset, L.; Boccaletti, A.; Clénet, Y. Polarimetric imaging of NGC 1068 at high angular resolution in the near infrared. Direct evidence of an extended nuclear torus. *A&A* **2015**, *581*, L8.
28. García-Burillo, S.; Combes, F.; Ramos Almeida, C.; Usero, A.; Krips, M.; Alonso-Herrero, A.; Aalto, S.; Casasola, V.; Hunt, L.K.; Martín, S.; et al. ALMA Resolves the Torus of NGC 1068: Continuum and Molecular Line Emission. *ApJ* **2016**, *823*, L12. [CrossRef]
29. Imanishi, M.; Nguyen, D.D.; Wada, K.; Hagiwara, Y.; Iguchi, S.; Izumi, T.; Kawakatu, N.; Nakanishi, K.; Onishi, K. ALMA 0″02 Resolution Observations Reveal HCN-abundance-enhanced Counter-rotating and Outflowing Dense Molecular Gas at the NGC 1068 Nucleus. *ApJ* **2020**, *902*, 99. [CrossRef]
30. Impellizzeri, C.M.V.; Gallimore, J.F.; Baum, S.A.; Elitzur, M.; Davies, R.; Lutz, D.; Maiolino, R.; Marconi, A.; Nikutta, R.; O'Dea, C.P.; et al. Counter-rotation and High-velocity Outflow in the Parsec-scale Molecular Torus of NGC 1068. *ApJ* **2019**, *884*, L28. [CrossRef]
31. Fathi, K.; Storchi-Bergmann, T.; Riffel, R.A.; Winge, C.; Axon, D.J.; Robinson, A.; Capetti, A.; Marconi, A. Streaming Motions toward the Supermassive Black Hole in NGC 1097. *ApJ* **2006**, *641*, L25–L28. [CrossRef]
32. Izumi, T.; Kohno, K.; Fathi, K.; Hatziminaoglou, E.; Davies, R.I.; Martín, S.; Matsushita, S.; Schinnerer, E.; Espada, D.; Aalto, S.; et al. On the Disappearance of a Cold Molecular Torus around the Low-luminosity Active Galactic Nucleus of NGC 1097. *ApJ* **2017**, *845*, L5. [CrossRef]
33. Kormendy, J.; Ho, L.C. Coevolution (Or Not) of Supermassive Black Holes and Host Galaxies. *ARA&A* **2013**, *51*, 511–653.
34. Venemans, B.P.; Walter, F.; Zschaechner, L.; Decarli, R.; De Rosa, G.; Findlay, J.R.; McMahon, R.G.; Sutherland, W.J. Bright [C II] and Dust Emission in Three z > 6.6 Quasar Host Galaxies Observed by ALMA. *ApJ* **2016**, *816*, 37. [CrossRef]
35. Farina, E.P.; Schindler, J.-T.; Walter, F.; Bañados, E.; Davies, F.B.; Decarli, R.; Eilers, A.-C.; Fan, X.; Hennawi, J.F.; Mazzucchelli, C.; et al. The X-shooter/ALMA Sample of Quasars in the Epoch of Reionization. II. Black Hole Masses, Eddington Ratios, and the Formation of the First Quasars. *ApJ* **2022**, *941*, 106. [CrossRef]
36. Mortlock, D.J.; Warren, S.J.; Venemans, B.P.; Patel, M.; Hewett, P.C.; McMahon, R.G.; Simpson, C.; Theuns, T.; Gonzáles-Solares, E.A.; Adamson, A.; et al. A luminous quasar at a redshift of z = 7.085. *Nature* **2011**, *474*, 616–619. [CrossRef]
37. Yang, J.; Wang, F.; Fan, X.; Hennawi, J.F.; Davies, F.B.; Yue, M.; Banados, E.; Wu, X.-B.; Venemans, B.; Barth, A.J.; et al. Pōniuā'ena: A Luminous z = 7.5 Quasar Hosting a 1.5 Billion Solar Mass Black Hole. *ApJ* **2020**, *897*, L14. [CrossRef]
38. Sassano, F.; Capelo, P.R.; Mayer, L.; Schneider, R.; Valiante, R. Super-critical accretion of medium-weight seed black holes in gaseous proto-galactic nuclei. *MNRAS* **2023**, *519*, 1837–1855. [CrossRef]
39. Bennett, J.S.; Sijacki, D.; Costa, T.; Laporte, N.; Witten, C. The growth of the gargantuan black holes powering high-redshift quasars and their impact on the formation of early galaxies and protoclusters. *MNRAS* **2024** , *527*, 1033–1054. [CrossRef]
40. Di Matteo, T.; Ni, Y.; Chen, N.; Croft, R.; Bird, S.; Pacucci, F.; Ricarte, A.; Tremmel, M. A vast population of wandering and merging IMBHs at cosmic noon. *MNRAS* **2023**, *525*, 1479–1497. [CrossRef]
41. Chilingarian, I.V.; Katkov, I.Y.; Zolotukhin, I.Y.; Grishin, K.A.; Beletsky, Y.; Boutsia, K.; Osip, D.J. A Population of Bona Fide Intermediate-mass Black Holes Identified as Low-luminosity Active Galactic Nuclei. *ApJ* **2018**, *863*, 1. [CrossRef]
42. Mezcua, M. Observational evidence for intermediate-mass black holes. *Int. J. Mod. Phys. D* **2017**, *26*, 1730021. [CrossRef]
43. Vitral, E.; Libralato, M.; Kremer, K.; Mamon, G.A.; Bellini, A.; Bedin, L.R.; Anderson, J. An elusive dark central mass in the globular cluster M4. *MNRAS* **2023**, *522*, 5740–5757. [CrossRef]
44. Vitral, E.; Mamon, G.A. Does NGC 6397 contain an intermediate-mass black hole or a more diffuse inner subcluster? *A&A* **2021**, *646*, A63.
45. Emsellem, E.; Renaud, F.; Bournaud, F.; Elmegreen, B.; Combes, F.; Gabor, J.M. The interplay between a galactic bar and a supermassive black hole: Nuclear fueling in a subparsec resolution galaxy simulation. *MNRAS* **2015**, *446*, 2468–2482. [CrossRef]
46. Renaud, F.; Bournaud, F.; Emsellem, E.; Agertz, O.; Athanassoula, E.; Combes, F.; Elmegreen, B.; Kraljic, K.; Motte, F.; Teyssier, R. Environmental regulation of cloud and star formation in galactic bars. *MNRAS* **2015**, *454*, 3299–3310. [CrossRef]
47. Maloney, P.R.; Begelman, M.C. The Origin of Warped, Precessing Accretions Disks in X-Ray Binaries. *ApJ* **1997**, *491*, L43–L46. [CrossRef]
48. Pringle, J.E. Self-induced warping of accretion discs. *MNRAS* **1996**, *281*, 357–361. [CrossRef]

galaxies

MDPI

Review

Observational Tests of Active Galactic Nuclei Feedback: An Overview of Approaches and Interpretation

Chris M. Harrison [1,*] and Cristina Ramos Almeida [2,3,*]

1 School of Mathematics, Statistics and Physics, Newcastle University, Newcastle upon Tyne NE1 7RU, UK
2 Instituto de Astrofísica de Canarias, Calle Vía Láctea, s/n, 38205 La Laguna, Tenerife, Spain
3 Departamento de Astrofísica, Universidad de La Laguna, 38206 La Laguna, Tenerife, Spain
* Correspondence: christopher.harrison@newcastle.ac.uk (C.M.H.); cra@iac.es (C.R.A.)

Abstract: Growing supermassive black holes (Active Galactic Nuclei; AGN) release energy with the potential to alter their host galaxies and larger-scale environment; a process named "AGN feedback". Feedback is a required component of galaxy formation models and simulations to explain the observed properties of galaxy populations. We provide a broad overview of observational approaches that are designed to establish the physical processes that couple AGN energy to the multi-phase gas, or to find evidence that AGN impact upon galaxy evolution. The orders-of-magnitude range in spatial, temporal, and temperature scales, requires a diverse set of observational studies. For example, studying individual targets in detail sheds light on coupling mechanisms; however, evidence for the long-term impact of AGN is better established within galaxy populations that are not necessarily currently active. We emphasise how modern surveys have revealed the importance of radio emission for identifying and characterising feedback mechanisms. At the achieved sensitivities, the detected radio emission can trace a range of processes, including a shocked interstellar medium caused by AGN outflows (driven by various mechanisms including radiation pressure, accretion disc winds, and jets). We also describe how interpreting observations in the context of theoretical work can be challenging, in part, due to some of the adopted terminology.

Keywords: galaxies; active galactic nuclei; feedback; jets; outflows

1. Introduction

The presence of supermassive black holes (SMBHs) at the centre of galaxies, including our own Milky Way, is confirmed by the measurements of the motions of stars and gas in galaxy centres, and by the gravitational lensing signatures within multi-wavelength images (e.g., [1–5]). These black holes (BHs) grow through periods of gas accretion, which are known as Active Galactic Nuclei (AGN), and are identified using observations across the electromagnetic spectrum [6–8]. The energy released by AGN was discussed as potentially important for influencing galaxy evolution as early as the 1980s; however, it is over the last three decades that there has been prolific research into this so-called "AGN feedback" process (see discussion in [9]).

The explosion of interest in AGN feedback was largely triggered by: (1) the observed correlation between black hole masses and stellar bulge properties [10–12], which early analytical models sought to explain with the effect of AGN driving gaseous outflows into their hosts [13–15]; (2) the lower-than-expected rate of gas cooling identified around the most massive galaxies, which implies a heating source that is attributed to AGN [16–18]; and (3) semi-analytical models and hydrodynamic simulations, which invoked AGN to regulate black hole growth and to reduce the efficiency of star formation (SF) at the highest stellar masses, as a required process to successfully reproduce the observed local stellar mass functions, and colour bi-modality, of galaxies [19–23].

Whilst energetic feedback from stellar winds and supernovae are typically considered sufficient for low mass galaxies (with stellar masses $\lesssim 10^{10}$ solar masses), AGN feedback remains a crucial component of galaxy evolution theory to explain the massive end of the galaxy population. This is demonstrated by its ubiquitous role within the current generation of cosmological simulations (e.g., [24–30]). Across these simulations, AGN remain the only known solution to regulate the cooling in dense galaxy environments, and to sufficiently control the efficiency of star formation for massive galaxies. Furthermore, within modern models and simulations, AGN are considered important for establishing other observed properties of galaxies and their environments. This includes galaxy structures, mass profiles and dynamics, and many of the properties of the gas observed in and around galaxies (e.g., [28,31–37]).

Observations are required to test any theoretical ideas and implementations of AGN feedback within simulations whilst carefully keeping in mind the difference between true predictions and observations that were used to calibrate the models (e.g., [24]—see Sections 4.3 and 6). Furthermore, observations should be used to constrain the physical mechanisms by which the energy released by AGN is able to couple to the gas. Indeed, there is currently no theoretical consensus on this, with an ongoing discussion on the relative importance of different mechanisms including radiation pressure on dusty gas; accretion disc winds; radio jets; and cosmic rays (e.g., [38–47]). Consequently, the literature is full of observational studies exploring these topics. They range from multi-wavelength, highly detailed observations of individual targets through to statistical tests on large galaxy samples. Due to the diversity of these studies, it can be challenging to interpret the results within the context of one another, and within the context of the theoretical ideas around AGN feedback. We aim to address this challenge in this article by providing a broad overview of the different observational approaches to test AGN feedback, and discuss how they can (or cannot) be related to one another and to theoretical predictions.

Observational work on AGN feedback has been reviewed multiple times over the last decade, including by [2,8,9,48–57]. These reviews typically focus in depth on one aspect of the feedback processes (e.g., the role of jets or the properties of the multi-phase gaseous outflows), and/or they are dedicated to one particular sub-set of the AGN population (e.g., radio-detected AGN, those in the densest galactic environments, or the most luminous AGN). This article does not aim to supersede these previously focused reviews. Instead, we aim to complement them by providing a more holistic overview of the breadth of observational approaches taken to search for, and to characterise, AGN feedback. The goal is to place the different observational approaches within the context of one another, and to comment upon how they can collectively be used to test or refine different theoretical ideas of feedback.

2. A Multi-Faceted Approach to a Complex Problem

AGN feedback is a problem of many scales. Both black hole growth (AGN) and stellar growth (star formation) are fed by gas. Therefore, we are interested in understanding a balance between feeding from and feedback upon a gaseous fuel supply. As discussed below, the relationship between these processes is a multi-scale problem over many orders of magnitude in spatial, temperature, and temporal scales. This range of scales is represented in the schematic diagram shown in Figure 1.

2.1. Multiple Spatial Scales, Temperatures, Gas Phases, and Timescales

To assess the relationship between AGN and galaxy formation, we need to understand the physical processes that can be occurring over more than six orders of magnitude in spatial scales. The spatial scales of accretion flows onto SMBHs are sub-parsec. Indeed, the physics that determine both the final accretion on to black holes and the corresponding energy output can take place on extremely small scales, down to the innermost stable circular orbit. These accretion flows are fed from the surrounding circumnuclear region (CNR), which is considered to be on scales of $\sim$1–10 pc, and is also responsible for the significant obscuration of the central AGN (i.e., the torus [58]). Therefore, the interaction between the AGN and the material in the CNR can determine both the rate of black hole growth and the ability to observe AGN emission. Galaxies are typically considered to have their interstellar medium (ISM) located on scales from a few parsecs up to tens of kiloparsecs, and star formation primarily occurs within this ISM [59]. If AGN impact directly on the ISM (e.g., by entraining/expelling it, or by heating/exciting/ionising it; see Section 4), they have the potential to impact directly upon the efficiency of star formation within the host galaxy (see Section 5).

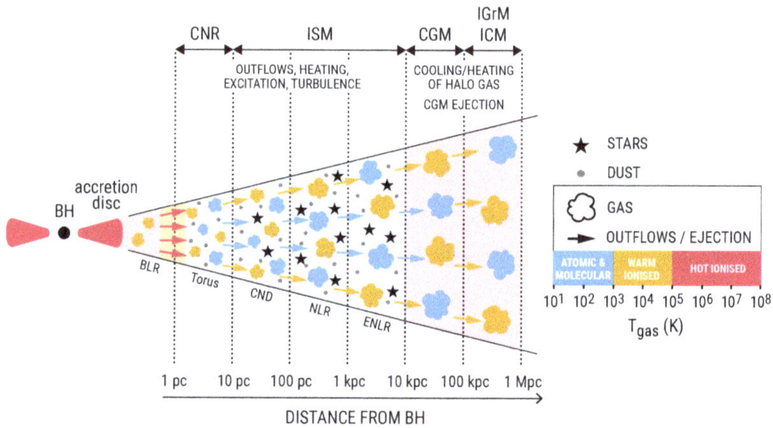

Figure 1. A schematic diagram to highlight the orders-of-magnitude range in temperature (colour-coding of gas clouds, intra-cloud medium, and outflows) and spatial scales of AGN feedback. Gas in a range of phases, with a huge range in temperatures, can be affected by the energy released by AGN. This can occur from nuclear scales through to galaxy cluster scales. This large range in spatial scales naturally corresponds to an equivalently large range in temporal scales. See discussion in Section 2.

The gas surrounding galaxies, outside of the their discs or ISM, but within their virial radii, are known as the circumgalactic region (CGM; [60]). Broadly speaking, this gas extends from ~10 kpc up to ~100 kpc. The CGM is a reservoir that can fuel a galaxy's ISM. It can be enriched by material that has been expelled from the host galaxy (e.g., by AGN outflows; Section 4). If an AGN can expel gas from the CGM, this could ultimately prevent or regulate future star formation in the host (e.g., [61]). The medium between galaxies that are located within groups or clusters can extend up to hundreds of kiloparsecs. This is called the intragroup medium (IGrM) or intracluster medium (ICM), respectively. AGN can act as a source of heating on this gas, and the ability for this gas to cool and flow onto galaxies plays a role in establishing the rate of star formation (see Section 6).

As summarised in Figure 1, the CNR, ISM, CGM, and IGrM/ICM can all contain gas at a range of temperatures, and in a range of phases (i.e., molecular, atomic, or ionised; see, e.g., [58,59,62]). For example, on spatial scales ranging from tens to hundreds of parsecs, the cold and dense ISM (i.e., the molecular gas [63]) is generally referred to as the circumnuclear disc (CND). The ionised gas on similar scales and up to ~1 kpc is known as the narrow-line region (NLR), and up to ~10 kpc as the extended-NLR (ENLR; see Figure 1). It is the coldest and densest gas that is directly associated with star formation, relating to temperatures of 10–100 s of Kelvin. Therefore, observations that trace atomic and molecular gas are usually considered necessary for measuring the properties of the available fuel supply for star formation (see Section 5). However, much of this gas will have cooled from hotter phases. Furthermore, AGN are expected to heat and expel considerable gas in hot/ionised phases (e.g., [42,56,64]), which may also cool or help trigger the formation of colder phases (e.g., [65]). Therefore, it is necessary to consider a wide range of gas phases. The corresponding temperature range may span >6 orders of magnitude. This creates a huge observational challenge. It has only recently become feasible to obtain multi-phase measurements for representative galaxy samples, yet the properties of some gas phases still remain largely unconstrained (see Section 4).

Due to the orders-of-magnitude range of *spatial* scales expected for the AGN feedback process (see Figure 1), we naturally have to consider a huge range of *time* scales. Furthermore, we observe the signatures of historic (not currently active) AGN events and variability in AGN luminosity (e.g., [66–68]). Such observations confirm that black hole growth, and the resulting energy output, is a variable and transient process. Indeed, it is better to consider AGN as *events* in the life-cycle of a galaxy and not to consider them as *objects* that persist in time.

The relative amount of time that a black hole is accreting (active), compared to not accreting (inactive), is sometimes referred to as the 'duty cycle'; although its exact definition varies in the literature. There are potentially multiple relevant time scales (or 'duty cycles') for the AGN feedback process, which may also depend on the 'accretion mode' (as can be seen in Section 3). These can range from variability on timescales of days–months for the observed high-energy AGN emissions [66], a proposed typical high-accretion rate period of growth lasting $\sim 10^5$ years [69,70], through to the idea that AGN could nearly always be active at low levels for the most massive galaxies [71]. An important example study in the context of understanding AGN feedback, Hickox et al. [72], use the simulations of Novak et al. [73] to demonstrate that black holes may switch between active (i.e., luminous AGN) and non-active states, on timescales of $\lesssim 1$ Myr. This is far shorter than the typical timescales for the star formation of $\gtrsim 100$ Myr. This severely complicates how we can connect a specific observed AGN event to any impact that it may have on the star formation in its host galaxy (see further discussion in Sections 5 and 6).

2.2. The Multi-Faceted Approach to Observational Work on AGN Feedback

The challenges of scales outlined above, and in Figure 1, necessitate a multi-faceted approach to observationally test the ideas around AGN feedback. Figure 2 provides a schematic overview of the primary regimes involved in addressing this AGN feedback problem.

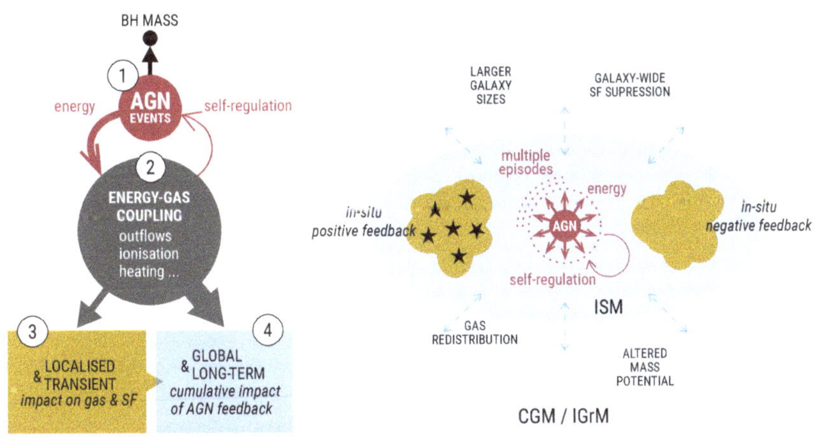

Figure 2. A schematic overview of the different regimes needed to tackle the overall AGN feedback problem with observations (see Section 2.2). This is presented in the form of a flow chart (**left**) and a schematic representation of an AGN-host galaxy embedded in a halo (**right**). Different colours and numbers within the flow chart correspond to different regions and processes labelled on the right panel and with the same colours.

Firstly, it is crucial to establish the demographics of AGN across different galaxy populations to assess: (a) the energy budget and duty cycles of AGN events; and (b) the prevalence of different mechanisms of energy injection into the host galaxies (e.g., jets, winds and radiation). This is labelled with a "1" in Figure 2, and is discussed further in Section 3. The energy released by AGN events can have an impact upon the kinematics, distribution, and excitation/ionisation state of the multi-phase gas (labelled with "2" in Figure 2). The properties and level of these effects will depend on how efficiently the energy can couple to the gas on various spatial scales. This is important to establish with observations, and is discussed in Section 4. An energy injection into the ISM may have an in situ, localised affect, on the ability for gas to form stars. This may have a negative or a positive impact on the star formation efficiency (labelled "3" in Figure 2) and it is important to understand the details of this process using high angular resolution observations (see Section 5). Finally, the cumulative impact of multiple AGN events could have many long-term effects on the global properties of the host galaxies and their larger-scale environment (labelled "4" in Figure 2). Assessing the "imprint" of AGN feedback on these properties is an important observational test for refining the different AGN feedback models invoked in cosmological simulations, and this is further discussed in Section 6.

3. AGN Accretion Modes and Methods of Energy Injection into the Host

Understanding the physical process of AGN feedback requires an understanding of the prevalence and power of different mechanisms of energy output during AGN events (see region "1" in Figure 2; discussed in this section) and constraining how efficiently this energy can couple to the gas (discussed in Section 4). It is beyond the scope of this article to review AGN demographics, but we note the importance of establishing a complete census of AGN events across different galaxy populations and cosmic epochs (see instead, e.g., [8,54,74,75]). Here, we present a broad overview of the mechanisms by which AGN may inject energy into their host galaxies (Section 3.1), reflect on what can be learnt on this topic from modern-day radio surveys (Section 3.2), and briefly comment on these in the context of the theoretical 'modes' of AGN feedback implemented in some simulations (Section 3.3).

3.1. AGN Accretion Modes and Mechanisms of Energy Output

We will consider two main classes of AGN events: those which are "radiatively efficient", where the accreting mass is efficiently converted to radiation, and those which are "radiatively inefficient", where the converse is true[1]. Although a simplification, and a matter of ongoing research, we can broadly relate this dichotomy to the accretion rates and the resulting properties of the accretion flow. Sources with the highest accretion rates, i.e., those with luminosities $\gtrsim 1\%$ of their Eddington luminosity are expected to have a radiatively efficient accretion disc that is geometrically thin and optically thick. In contrast, black holes with lower accretion rates are expected to have radiatively inefficient flows that are geometrically thick and optically thin [79–81]. The spin of the black holes may also be crucial in determining the mechanical power output available through jets [82,83]. Potential methods of energy injection from these two AGN classes are summarised in Figure 3.

For radiatively inefficient AGN events (left side of Figure 3; sometimes called low-luminosity AGN; LLAGN), the mechanical energy output greatly exceeds the radiative output. This is likely to be dominated by the collimated jets of charged particles, although there can be a non-negligible contribution of magnetically or thermally driven isotropic (wide-angle) winds from the accretion flow and from Compton (X-ray) heating (e.g., [54,84–88]). For radiatively efficient AGN[2] (right side of Figure 3), the energy injected into the host can be from multiple mechanisms, including (1) the photo-ionisation of gas up to CGM scales (e.g., [90,91]); (2) in situ radiation pressure on gas and dust (located on CNR/torus scales to ISM scales; e.g., [39,92]); (3) powerful wide-angle winds associated with the accretion discs, which are likely radiatively driven through electron scattering or through ultra-violet line driving (e.g., [42,93–95]); (4) collimated jets, which could efficiently couple to the gas on various spatial scales depending on power, inclination, ISM properties, etc. (e.g., [64,96]; see Section 4.2) and; (5) Compton (X-ray) heating in the region of the accretion disc, or even further out [87,97][3]. Mechanisms (2)–(5) could all result in the propagation of mechanical energy into the host galaxies by shocking gas and/or by sweeping up material into an outflow (see Section 4).

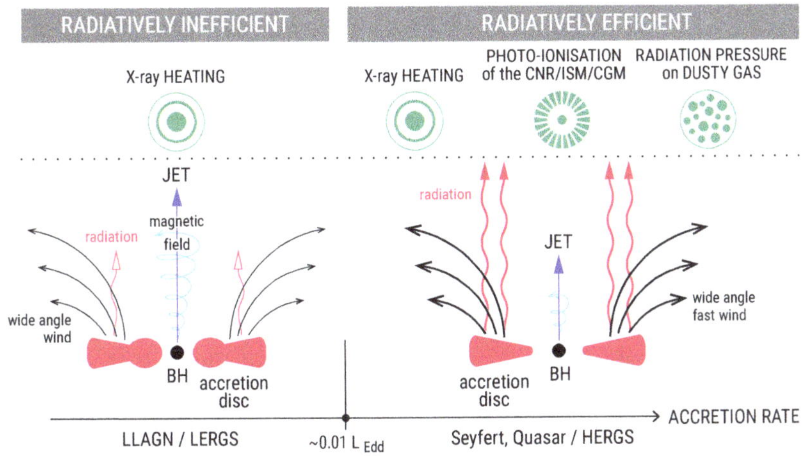

Figure 3. A schematic overview of different mechanisms by which AGN can inject energy into the gas in and around their host galaxies. This is shown for two regimes of accretion: 'radiatively inefficient' for lower accretion rates (**left**) and 'radiatively efficient' for higher accretion rates (**right**). Multiple mechanisms co-exist in both regimes. This is described in Section 3.1.

In the literature, the terms "wind", "outflow", and "jet" can be used interchangeably in a way that can cause confusion for interpretation. For AGN, we encourage using the term "wind" to describe the launching mechanism associated with accretion discs and "outflow" to describe ISM/CGM material that has been swept up/entrained by the various driving mechanisms. The term "jet" may refer to the initial physical process (relativistic, collimated, beam of charged particles), or any other collimated outflowing structure, but this should be carefully defined in each study.

3.2. Radio-Identified AGN

We consider it important to provide a discussion on radio emission from AGN in the context of AGN feedback studies. The related terminology can cause some of the greatest confusion across both observational and theoretical works.

The earliest radio observations were only sensitive to the most powerful extra-galactic sources which are clearly dominated by jets. The term 'radio galaxy' is sometimes still used to define any galaxies considered to host radio-detected jets. However, in the modern era of radio astronomy, the origin of the detected radio emission can trace a wider range of processes (see below). The terms 'radio loud' and 'radio quiet' are also commonly used to classify AGN. Traditionally, radiatively efficient AGN are separated into these classes based upon the ratio of radio luminosity to an accretion luminosity (e.g., [99–101]). AGN that are dominated by radio emission above some threshold are classified as 'radio loud'. However, 'radio loud' is sometimes used more loosely in the literature to refer to galaxies where any radio emission associated with an AGN is identified, or to simply define whether a galaxy is radio-detected. It is crucial to carefully define what is meant by a 'radio galaxy' or 'radio loud' AGN in each individual study, and to take caution in determining whether the radio emission can or cannot be unambiguously associated with radio jets.

Due to the relationship between the star formation rate (SFR) and associated radio luminosity in star forming galaxies, it is possible to estimate the expected contribution of star formation to the total observed radio luminosity (e.g., [102,103]). Therefore, any galaxy with significant radio emission above this is expected to have radio emission associated with an AGN, and are sometimes labelled as a 'radio excess' galaxy/AGN or as a 'radio AGN' (e.g., [104–106]). In Figure 4, we provide a schematic representation of the difference between the traditional 'radio loud' criteria and a radio excess criteria for radiatively efficient AGN. This figure is motivated by the data presented in Figure 12 in MacFarlane et al. [106], who study SDSS quasars with LOFAR observations[4].

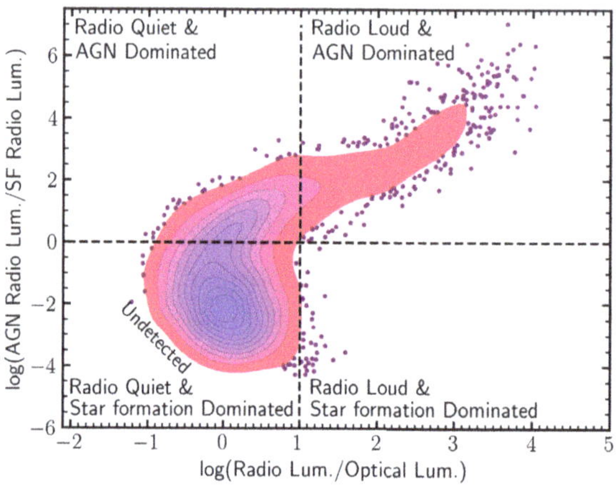

Figure 4. Schematic representation of the data from Figure 12 in [106], showing a comparison between the 'radio loud' criteria for quasars (x axis) (i.e., $R = L_{5\mathrm{GHz}}/L_{4400\text{Å}}$ with a cut-off value at $R = 10$) and the ratio of radio luminosity attributed to AGN to that attributed to star formation (y axis). Sources below the detection threshold of the surveys would fill out the bottom-left of this diagram.

The traditional definition of 'radio loud' effectively identifies galaxies with powerful jets and it is commonly accepted that ∼10–20% of quasars are 'radio loud' (Figure 4, top-right quadrant; e.g., [99,106–108]). Nonetheless, *it does not necessarily mean that the radiative energy output is not energetically important for feedback.* A 'radio loud' quasar is expected to host powerful accretion disc winds and have significant radiation pressure (e.g., [39,53]; Figure 3). Furthermore, the 'radio loud' definition misses the *majority* of the radiatively efficient AGN where the radio emission is AGN dominated (Figure 4, top-left quadrant).

A crucial point is that populations of radio-detected AGN contain a mixture of radiatively efficient and radiatively inefficient sources (see Figure 3). Radio-detected AGN, which exhibit weak or no optical emission lines, are known as "low-excitation radio galaxies" (LERGs) and appear to have little-to-no evidence of radiative emission associated with an accretion flow; this makes LERGs one candidate class of radiatively inefficient AGN (see review in [54])[5]. Radio-detected AGN that also have the signatures of radiatively efficient accretion known as "high-excitation radio galaxies" (HERGs).

Whilst it is sometimes assumed that excess radio emission (above star formation) should be attributed to radio jets, this is not necessarily the case. For example, if AGN-driven outflows cause shocks in the ISM, this can be another significant contribution to the observed radio luminosity (e.g., [38,112,113]). Whilst diagnostics such as the 'radio excess' criteria and high brightness temperatures (e.g., [114]) can be used to distinguish between AGN and star formation related emission, it remains observationally challenging to directly disentangle emission associated with a radio jet and that from outflow-induced shocks (e.g., [115,116]). Theoretical predictions will help guide the appropriate diagnostics (e.g., [117,118]).

In summary, radio emission associated with AGN events (both radiatively efficient and radiatively inefficient) is extremely prevalent and could even be ubiquitous (at least at low levels) for galaxies with the highest stellar masses (i.e., $\sim 10^{11}$ solar masses; [71]). The origin of this radio emission is often ambiguous and detailed work is required to establish the underlying physical processes and the connection to AGN feedback.

3.3. Connecting Observed AGN Populations to Traditional 'Feedback Modes'

We do not attempt to review AGN feedback models in this article. However, we make some brief notes about interpreting the observed AGN populations in the context of some feedback models. Early semi-analytical models and hydrodynamic simulations invoked simple prescriptions for AGN feedback (e.g., [20–23,119–121]), which have been influential on the community thinking about different 'modes' of black hole growth and corresponding 'modes' of AGN feedback (see discussion in, e.g., [122]). For example, the term 'quasar mode' ('radiative mode', or 'starburst mode') was considered important for rapid periods of growth (sometimes associated with mergers), with the idea that ISM could be expelled from the host galaxies. The term 'radio mode' (or 'hot-halo mode', 'maintenance mode', or 'kinetic mode') was considered important for lower accretion rate systems and acts more through a regulation of cooling of the gas on large scales. Whilst appropriately simple at the time, the direct connection to *observed* populations of radio-detected AGN and quasars is not fully applicable with a modern understanding. Furthermore, two distinct modes of feedback are unlikely to be the complete story.

We have established that, during radiatively efficient AGN events, multiple mechanisms of energy input can result in significant *mechanical* forms of energy being turned into the gas through outflows driven by winds, jets etc. (Figure 3; Section 3.1). Furthermore, as discussed in Section 3.2, radio emission associated with AGN is prevalent for both radiatively efficient (high accretion rate) and radiatively inefficient accretion (low accretion rate) regimes. Therefore, we should avoid assuming that radio-detected AGN are universally associated with the traditional 'radio mode' of feedback. One should also avoid assuming that quasars only provide an 'ejective' type of feedback on the ISM, and are only associated with accretion disc winds or radiation (cf. the traditional implements of a 'quasar mode'). Quasars, and all radiatively efficient AGN, can also regulate gas cooling ('regulative' feedback), could have an important impact on the ejection of gas out to CGM scales, and can have jets as an important energy injection mechanism (e.g., [41,42,123–125]).

4. Energy–Gas Coupling: Outflow Kinematics and Energetics

In Section 3, we discussed the potential mechanisms by which AGN can inject energy into their host galaxies. However, it is important to remember that the relevance for AGN feedback requires that this energy couple with the multi-phase gas. Here, we summarise the observational results on multi-phase gas outflows based on samples of AGN (Section 4.1), discuss different AGN and galaxy properties that might have a key impact on the energy–gas coupling (Section 4.2), and describe how to constrain the outflow energetics and potential for impact from the observational point of view (Section 4.3).

4.1. Multi-Phase Gas Outflows

Gas outflows, one of the clearest and hence most studied forms of AGN feedback, impact the ISM by removing, heating, and/or mixing the gas (see Figure 1). Hundreds of observational studies report evidence of outflows observed in different gas phases, including the (1) hot ionised ($T_{gas}\sim10^{5-8}$ K and $n_{gas}\sim10^{6-8}$ cm^{-3}) using X-ray absorption lines [126–128]; (2) warm ionised ($T_{gas}\sim10^{3-5}$ K and $n_{gas}\sim10^{2-5}$ cm^{-3}), mainly through emission lines detected in the rest-frame UV [129–132], optical [133–139] and near-infrared [140–143]; (3) neutral atomic ($T_{gas}\sim10^{2-3}$ K and $n_{gas}\sim10^{1-2}$ cm^{-3}) by means of the H I absorption line [144,145], the sodium doublet absorption (NaID; [146–148] and the [C II] fine-structure emission [149,150]; and (4) molecular ($T_{gas}\sim10^{1-3}$ K and $n_{gas} > 10^3$ cm^{-3}). The latter phase can be probed, for example, using the rotational and vibrational H_2 lines in the near-infrared (hot molecular; [140,141,143]), the rotational lines in the mid-infrared (warm molecular; [151,152]), the hydroxyl (OH) transitions in the far-infrared [153–156] and the carbon monoxide (CO) emission lines in the (sub)millimetre [157–163]. These spectral features trace the outflows on different spatial scales, that go from sub-parsec in the case of the hot ionised phase (i.e., BLR scales), and from hundreds of parsecs to $\sim$10 kpc for the ionised and molecular gas phases (see Figure 1).

Despite the wealth of observational effort invested in characterising AGN-driven outflows, the overall majority of studies report single-phase estimates of their properties that provide an incomplete view of the AGN feedback phenomenon [164]. Accurate outflow properties in different gas phases need to be determined in representative AGN samples to estimate how much each phase contributes to the overall mass and energy budget. From the few studies reporting multi-phase outflow properties for samples of AGN (e.g., [143,165–167]), the cold molecular outflows would carry the bulk of the mass, followed by the neutral and the warm ionised. In the local universe, the molecular outflows are slower (v $\lesssim$ 300–500 km s^{-1}) and more compact (r $\lesssim$ 1–2 kpc) than the ionised outflows (~500–1000 km s^{-1}; r~1–10 kpc). In the high redshift Universe (at cosmic noon, i.e., z~1–3, and earlier), most of the works reporting outflow measurements focus on the ionised gas phase, since the bright optical emission lines are shifted to the infrared. The integral field spectroscopy of quasars with different radio luminosities revealed the evidence of fast ionised outflows on galaxy scales [139,168–174]. The neutral atomic and molecular gas phases of the outflows have also been studied at high redshift, and although some works report the presence of massive, galaxy-scale high-velocity outflows in luminous quasars [149,150,175,176], overall there is no evidence for them being a widespread phenomenon, at least those probed by the [C II] emission line [177,178]. A handful of studies on individual galaxies reported multi-phase outflow measurements, mainly targeting the cold molecular and ionised phases [179] but also the ionised and neutral atomic [180]. Vayner et al. [181] characterised the ionised and cold molecular outflows in four quasars at z~2 with powerful jets. They found that, although most of the outflow mass is in the molecular phase (which is clumpier and more compact), the kinetic energies are higher in the ionised phase.

With a few exceptions, the multi-phase outflow masses derived from observations are modest, but observational limitations might be preventing us from detecting the most extended and diffuse component of the multi-phase outflows, which might carry the bulk of their mass and/or energy. For example, in the case of the cold molecular gas, this diffuse component might be missed by interferometric observations that do not include short baselines [164]. Deep X-ray observations have revealed the presence of hot X-ray-emitting gas (T_{gas}~10^{6-7} K) extending to CGM scales in quasars (e.g., [182–184]), and diffuse, extended gas on scales of up to ~50 kpc have been reported for neutral hydrogen around radio galaxies (e.g., [185]). Some of this diffuse emissions might be participating in, or being entrained by, the outflows. Therefore, it would largely increase the outflow mass, and in the case of the hot ionised gas, its energy. Whilst we wait for new a generation of X-ray observatory, that will be required to fully characterise the hottest gaseous phases [186], indirect approaches will need to be taken [187–190].

From an observational point of view, it is not clear yet whether the different outflow phases are the different faces of the same phenomenon, with certain phases transitioning into others [92], or just unrelated events. Deep, spatially resolved multi-phase outflow measurements of AGN of different luminosities, hosted in galaxies with different properties and environments, are necessary for advancing our understanding.

4.2. Coupling between Energy and Gas

According to observational scaling relations found from the outflow measurements of AGN in the local Universe, the more luminous the AGN, the more massive and faster the gas outflows that it drives (e.g., [158,191]). However, these scaling relations were derived from observations of CO-bright targets known to have strong multi-phase outflows (e.g., Mrk 231 [192], NGC 1068 [193,194], IC 5063 [195,196]), some of them hosted in ultra-luminous galaxies (ULIRGs). More recent works have now started to populate the AGN luminosity-outflow mass rate plane with less biased AGN and ULIRG samples [161,162,167], and the linear relation vanishes (see left panel of Figure 5). This indicates that having a high AGN luminosity does not guarantee the presence of a massive outflow, as the coupling between energy and gas depends on other factors that we mention below.

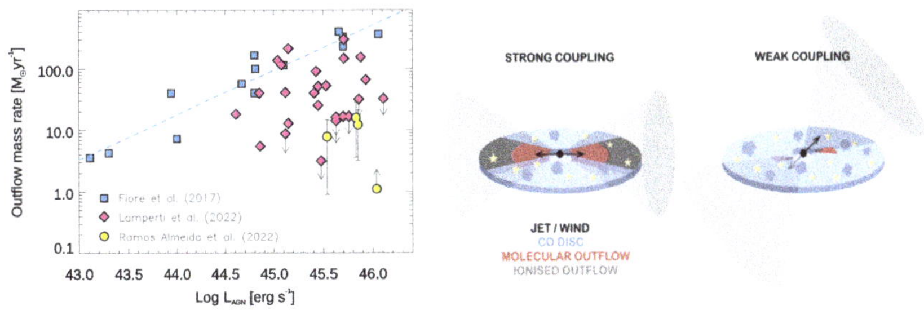

Figure 5. Molecular mass outflow rate versus AGN luminosity (**left panel**). The dashed line is the fit to the data compiled by Fiore et al. [191]. Diamonds are the ULIRGs from Lamperti et al. [161] and circles the type-2 quasars from Ramos Almeida et al. [162]. The plots on the right are examples of a strong (weak) coupling scenario, where the jet/wind/ionised outflow subtends a small (large) angle with the molecular gas disc, launching a massive (modest) molecular outflow, shown in red. Credit (**right panel**): Ramos Almeida, A&A, 658, A155 (2022) [162], reproduced with permission ©ESO.

Aside from AGN luminosity, a key property that determines the level of coupling between energy and gas is orientation. When a jet and/or disc wind subtend a small angle relative to the gas disc, the latter is entrained and the jet/wind energy is transmitted more efficiently (a strong coupling scenario in Figure 5) when the angle is large (weak coupling scenario). Something similar can happen when an ionised outflow is coplanar or quasi-coplanar with a galaxy disc: the molecular gas is then entrained by the ionised outflow, becoming a molecular outflow, normally directed along the kinematic minor axis [63,162]. The importance of orientation has been shown by the hydrodynamic simulations of jet–ISM interactions: jets almost coplanar with the galaxy discs (strong coupling scenario in Figure 5) entrain the gas more efficiently than jets subtending a larger angle (weak coupling scenario) inducing higher outflow velocities and disturbance [64,118,125,197,198]. Jets have been proposed to be responsible for the enhancement in velocity dispersion in the direction perpendicular to the jet observed in the AGN of different bolometric luminosities with modest jet powers [44,162,199–203]. Another factor with potential influence on the coupling between jets and ISM is jet power. Simulations show that low-power jets ($P_{jet} \lesssim 10^{45}$ erg s^{-1}) become trapped for a longer time by the ISM and disrupt the surrounding gas over a larger volume, and for a longer period, than jets of higher powers ($P_{jet} \gtrsim 10^{45}$ erg s^{-1}), which are capable of "drilling" through the ISM [204]. Therefore, the low-power jets often detected in radiatively efficient AGN such as Seyferts and quasars can be more disruptive and induce more massive outflows than high-power jets [52,196,205–208]. Finally, the distribution of molecular gas is another factor leading to more or less efficient coupling. Some of the most powerful molecular outflows are found in ULIRGs [157,158], associated with galaxy mergers and high nuclear column densities. The latter might favour the launch of massive molecular outflows by increasing the coupling between the jets/winds and the ISM [42,209,210]. It has been proposed that this dust-enshrouded phase is followed by an active feedback phase (blow-out phase) [8,211,212] that modifies the nuclear molecular gas distribution, creating cavities that have been predicted by hydrodynamic simulations of massive galaxies with disc winds [213] and observed in nearby AGN using ALMA [63,214,215] (see Section 5).

The observations of representative samples of AGN of different bolometric and radio luminosities, hosted in galaxies of different masses and morphological types, are needed to understand the interplay between the different outflow gas phases and the impact of the different factors listed above on the coupling between the outflows and the ambient gas. This requires observing proposals in different telescopes that are expensive in terms of observing time and that can be considered "incremental" in terms of science goals, but we argue that they are necessary to move forward in AGN feedback studies.

4.3. Outflow Energetics and Potential for Impact

Here, we consider gas outflows as any gas that is not just photoionised, excited, and/or shocked by the AGN, but also kinematically disturbed by it, as in [61]. Identifying outflows through emission line profiles is challenging because different gas components across the host galaxy having different kinematics contribute to the shape of the line profiles: rotation, stripped material, infalling gas, companion objects, etc. (see, e.g., [172]). The most widely used method consists of assuming that each kinematic component can be described by a Gaussian (i.e., parametric methods). In the simplest case scenario, the emission lines can be described by the sum of a narrow component ($\lesssim$400 km s^{-1}) tracing virial motions, and another component(s), generally but not necessarily broader and shifted relative to the systemic velocity, associated with outflowing gas. Using parametric methods, it is possible to measure the properties of each kinematic component, but in galaxies with complex kinematics, it is difficult to ascribe a physical meaning to each of them [216,217]. In the latter case, non-parametric methods might be more appropriate [133,218,219] because they make it possible to identify high-velocity gas by measuring the emission line velocities at fixed fractions of either the peak or integrated line flux (see Figure 4 in [133]). We refer the reader to Hervealla-Seoane et al. [217] for an analysis of the influence that using different parametric and non-parametric methods have on the outflow kinematics and energetics.

In the case of the molecular gas, often studied using the CO emission lines detected in the (sub)millimetre regime, outflows usually have lower velocities than their ionised counterparts [162,167,181], making it even more difficult to disentangle them from the virial motions around rotation. Therefore, unless clear emission line wings are detected (e.g., [158,220]), it is not possible to identify any detected non-circular motion with radial outflows without a careful analysis of the gas kinematics. The observed moment maps and position velocity diagrams can be compared to simple rotating disc models in order to identify molecular outflows [162,193,221]. Once identified, the outflow flux can be calculated using more or less conservative approaches [125]. Regardless of the method used to characterise the gas kinematics, several observational effects including the spectral response and resolution of the instrument, beam smearing, projection, etc., need to be considered and accounted for in order to reliably characterise the outflow properties. These effects, their implications and ways of mitigation have been largely discussed in the literature [9,169,222].

To infer the potential impact of the outflows on the host galaxies, we need to measure the outflow physical properties such as mass, mass rate, and kinetic power (M, $\dot{M}$, $\dot{E}_{kin}$). This requires the estimates of other quantities such as the outflow density (see [9] for a discussion of the different methods and caveats) and metallicity, and to assume an outflow geometry. The latter can be spherical, (multi-)conical, or a slab/shell of certain radial thickness, and it also has an impact on the derived outflow mass rates [223]. Other assumptions that are often made are that the outflow clouds have the same density and metallicity and no acceleration or deceleration of the outflowing gas. In the case of the cold molecular outflows estimated from CO emission lines, an α_{CO} factor needs to be assumed to convert from measured carbon monoxide luminosities to estimated molecular hydrogen (H_2) masses. The values of $\alpha_{CO} \leq 1$ are usually measured/assumed for molecular outflows [162,193,196], which is significantly lower than the Galactic factor (α_{CO} = 4.35; [224]).

The outflow mass rates can be compared with the SFR to derive the mass loading factor (η = $\dot{M}$/SFR) and hence have an estimate of the gas mass that is being evacuated by the outflow compared to that being consumed by star formation. The trouble is that, often, the spatial scales corresponding to the outflow (few kiloparsecs at most) and to integrated SFRs (galaxy scales) are different, and therefore, the mass loading factors might be misleading [162]. The outflow momentum rate ($\dot{M} \times v$) divided by the AGN radiation momentum rate (L_{bol}/c) can provide information about whether the outflows are energy- or momentum-driven [53,191], and the kinetic power ($\dot{E}_{kin}$) is usually compared with either the jet power or the AGN bolometric luminosity to have an idea of whether the kinetic power of jets and/or AGN radiation can drive the outflow [125,133,167,203]. However, many factors compound to make it observationally challenging to obtain reliable measurements of bolometric luminosity. The required bolometric correction factors, to convert from a luminosity in one wavelength regime to the total AGN luminosity remain challenging and uncertain [225]. Furthermore, a dominant fraction of AGN may be heavily obscured and can even be completely missed in many surveys [74]. Finally, as previously discussed in Section 2, accretion rates and the resulting energy output can be highly variable on temporal scales much shorter than the outflow dynamical timescales. A significant challenge also exists when attempting to determine the available energy from jets. Scaling relationships have been created between observed radio luminosities and jet powers, using the measurements of the mechanical work inferred from the jet-induced cavities inside the X-ray-emitting atmospheres of massive galaxy environments (e.g., [226–228]). These relationships are used extensively in the literature to convert a radio luminosity into the available power from jets as a potential feedback mechanism. However, there are many systematic uncertainties in these relationships, even for the regimes (i.e., jet powers, galaxy environments) from which they were calibrated (e.g., [229,230]). Crucially, it is not at all clear whether they can be extended to different galactic environments, or to different regimes of jets. For example, lower power jets, which are orientated into a gas-rich ISM, can be quickly frustrated/disrupted, and the observed radio luminosity may be a poor tracer of jet power (e.g., [64,96]). Furthermore, the observed radio luminosity may be the result of many processes, and not simply a direct tracer of jets ([113,116]; Section 3.2). Therefore, when using the observed AGN luminosities and jet powers to test theoretical predictions, or to establish the dominant energy driving mechanisms, it is crucial to consider all of these systematic uncertainties.

The ratio between the outflow kinetic power and AGN luminosity is often referred to as the kinetic coupling efficiency and it has been widely used by the community in an attempt to compare the observational results with the predictions from cosmological simulations. In the seminal work of Di Matteo et al. [231], the feedback efficiency (ϵ_f; the fraction of AGN luminosity that couples to the gas in the vicinity of the black hole) was calibrated to 5%, so the local M_{BH}-σ relation was reproduced by the simulation. As widely discussed in [9], the comparison between the observed kinetic coupling efficiencies ($\dot{E}_{kin}/L_{bol}$) and the ϵ_f values adopted in cosmological simulations is not straightforward. First, in the case of radiation-driven gas outflows, the feedback efficiencies are calibrated (to ~0.005–0.15 in more recent simulations; see [9] and references therein) and hence are not a prediction themselves [24,98,232]. Second, only a fraction of the injected energy will be transformed into outflow kinetic energy depending on the ISM properties, the gravitational potential, etc. Therefore, only a small kinetic coupling efficiency may be actually measured from observations. Finally, as several observational studies show, outflows can be launched/accelerated by compact jets, even in radiatively efficient AGN, challenging how useful a comparison to the bolometric AGN output would be in these cases (see Sections 3.3 and 4.1). As discussed in Sections 5 and 6, there are better ways to compare observations with the predictions from simulations than just taking a fiducial theoretical value as a reference to determine whether kiloparsec scale outflows are relevant from an energetic point of view.

5. Localised and Transient Impact on Gas and Star Formation

According to cosmological simulations, it is the cumulative output of several AGN events that are most relevant for the global and long-term impact on their host galaxies (see Section 6). The current AGN state of a galaxy is therefore not a useful proxy for studying the global and long-term impact of AGN feedback on gas and star formation, at least in the local universe (see [233] for a recent discussion of the potential impact of AGN feedback across cosmic time). However, by studying currently active AGN, using spatially resolved observations, we can obtain crucial information on the localised impact of AGN feedback (labelled as '3' in Figure 2), which is needed to understand how the energy couples with the gas, and under which circumstances it enhances or reduces the star formation efficiency.

As already mentioned in Section 2, AGN should be treated as events or episodes that are much shorter than the typical timescales of star formation. In the case of the outflows, the dynamical timescales are usually ∼1–10 Myr depending on their velocity and extent (for example, for an unimpeded outflow expanding at an average velocity of 1000 km s^{-1} it takes one million years to reach a radius of one kiloparsec). Therefore, if we aim at investigating the impact, either positive or negative, that the present outflows have on star formation, we should be looking at young stellar populations (YSPs), dominated by O and B stars. The latter can be traced using the stellar population modelling of the rest-frame UV/optical spectra [234] as well as the polycyclic aromatic hydrocarbon (PAH) features detected in the rest-frame mid-infrared [235]. In the case of other star formation tracers such as Hα, [Ne II], and the 24 μm flux, it is difficult to disentangle the contribution from star formation and AGN (e.g., [236–238]), although it has been argued that PAHs could also be excited by AGN radiation [239,240]. Stellar synthesis modelling is challenging in AGN because, in the case of type-1 AGN, the strong continuum dilutes the stellar absorption features. Type-2 AGN are better suited for this type of analysis, although scattered AGN light from the obscured nucleus can significantly contribute to the UV/optical continuum, and can therefore be interpreted as a YSP [241]. This degeneracy can be broken if high-order Balmer lines (e.g., H9, H10, H11, and H12; 3835–3750 Å) are detected, since they are tracers of young stars. It is also important to account for the nebular continuum coming from the free–free, free–bound, and two-photon decay continua associated with the gas nebula [242] to accurately characterise the YSPs.

Evidence for recent star formation in AGN based on the detection of PAH features has been reported from kiloparsec scales [38,243–245] to as close as tens of parsecs from the AGN [246–250]. However, it is important to remember that AGN radiation and shocks might modify the structure of the aromatic molecules and/or destroy the smallest grains, hence suppressing the short-wavelength features (6.2, 7.7, and 8.6 μm; [251,252]). Luckily, the larger and neutral molecules that produce the 11.3 and 17 μm features are more resilient and they have been found to be enhanced in AGN relative to the short-wavelength PAHs, at least in Seyfert galaxies [253]. At higher AGN luminosities, it is not clear whether even the largest PAH molecules can survive the AGN radiation field [254–256], but upcoming JWST datasets will be crucial to study the viability of PAH features as the tracers of star formation in AGN of different luminosities, at different distances from the AGN, and in and out of the outflow/jet regions [257–259].

AGN-driven outflows and jets can trigger star formation locally by compressing the surrounding gas, as it has been shown theoretically (e.g., [213,260–262]) and observationally using integral field spectroscopy and/or interferometry (e.g., [216,263–266]). Mercedes-Feliz et al. [213] used the FIRE simulations incorporating the hyper-refined disc winds to study the impact of quasar outflows with kinetic powers of 10^{46} erg s^{-1} on the circumnuclear disc (CND; scales of ∼100 pc; see Figure 1) of massive galaxies at z∼2. These outflows drive the formation of a central gas cavity and increase the local star formation efficiency at its inner edges, where gas is being compressed. Despite this localised positive feedback, the simulations show that the outflows reduce the surface density of star formation across the galaxy (i.e., negative feedback) by limiting the availability of gas rather than evacuating it [213,267]. Recent observational work using data from legacy integral field spectroscopic surveys report evidence for suppressed gas fractions and SFRs in the central kiloparsec of galaxies hosting AGN as compared to matched controls (e.g., [268–270]).

The creation of molecular gas cavities of tens to hundreds of parsecs by AGN-driven outflows and/or jets is one of the most clear observational evidences of the direct impact of AGN feedback on the host galaxies [63,214,215,271]. Using ALMA CO observations at ∼10 pc resolution of a sample of nearby AGN with X-ray luminosities in the range $\sim10^{39-43}$ erg s^{-1}, García-Burillo et al. [63] found that the most luminous AGN within the sample show evidence for molecular outflows and smaller molecular gas concentrations in the inner r∼50 pc (torus scales) of the galaxies relative to the inner r∼200 pc (CND scales) than the less luminous AGN in the sample. They interpreted these results as the imprint of feedback on the nuclear distribution of molecular gas: at low AGN luminosities, galaxies build their nuclear molecular gas reservoirs via the accretion of gas, and as accretion and subsequent feedback become more efficient, jets and winds push molecular gas outwards, producing the cavities predicted by hydrodynamic simulations (e.g., [213,262, 267]). These molecular gas cavities, which have been detected in low-J CO transitions such as CO(1-0) [272], CO(2-1) [162,214,215,273], and CO(3-2) [63,271], are filled with ionised and warm molecular (H_2) gas [214,215,271], indicating that the AGN is modifying the properties of the cold molecular gas on CND scales. Indeed, for AGN observed in more than one CO transition, such as NGC 1068 [271], NGC 3100 [272] and the Teacup [125], higher gas excitation is found in the central regions of the galaxies. All these galaxies have relatively low jet powers ($P_{jet}{\sim}10^{43}$ erg s^{-1}) with different inclinations relative to the molecular gas discs, which might be responsible, together with AGN radiation and outflows, for the high gas excitation in specific regions of the galaxy centres.

In summary, spatially resolved studies of nearby AGN are revealing the complex interplay between AGN and the surrounding gas. Obtaining the accurate measurements of this interplay, including outflow energetics and ambient gas and recent star formation properties, is key to revise the recipes used in simulations. The high angular resolution and sensitivity of facilities such as ALMA and the JWST now allow us to obtain spatially resolved constraints at high redshift (e.g., [180,274,275]).

6. Global and Long-Term Cumulative Impact of AGN Feedback

To assess the relevance of AGN feedback in a cosmological context requires statistical studies that search for the imprint of feedback on the overall galaxy population (labelled '4' in Figure 2).

Some of the best evidence that AGN have a global and long-term impact on galaxy evolution comes from the observation that AGN jets can offset cooling in dense galactic environments. Hence, they can regulate the rate at which gas can cool into galaxies and form stars. X-ray cavities excavated by radio jets are well studied in local cluster and group environments, and provide a direct indication of the work performed as these jet-induced lobes expand into the ICM/IGrM [49,227,276,277]. This process appears to provide the required amount of heating to balance cooling and may apply for massive early-type galaxies in a wider range of environments and out to $z \sim 1$ [278–283]. In broad terms, there appears to be convincing evidence that the cumulative effect of AGN (see Figure 2) is able to regulate cooling in massive systems, at least in the densest environments and the latest cosmic times. Nonetheless, some of these calculations remain uncertain, and it is still an open question as to how the energy is communicated to the ICM/IGrM, and the universal applicability across different environments and accretion states remains a matter of ongoing study [54,56,284,285]. Furthermore, whilst this could be an effective mechanism for maintaining low star formation levels, it may be insufficient to shut down, i.e., 'quench', high star formation levels, which may require a more ejective form of feedback (e.g., [61,121,286,287]).

One common type of study has been to measure the SFRs, gas fractions, and star formation efficiencies in samples of AGN host galaxies (see [51,288], and the references therein). Some of these studies look for the instantaneous impact of the observed AGN events on either the global (galaxy-wide) fuel for star formation (the molecular gas content) or directly on the SFRs. The two broad approaches are to compare these measured properties to the matched samples of inactive galaxies (e.g., [138,162,289–292]), or to explore these properties as a function of AGN luminosity or the prevalence of AGN-driven outflows (e.g., [293–301]). There has been some evidence (although not a consensus) that luminous AGN at cosmic noon have moderately depleted levels of molecular gas (e.g., [290,302]). This potentially indicates that this gas has been excited to higher transitions (see Section 5) or has been ejected. However, despite the wide range of approaches, and the different AGN samples, the general consensus is that radiatively efficient AGN typically reside in gas-rich, star-forming galaxies. This general picture is also consistent with statistical studies that look at the galaxy population as a whole; although not *all* radiatively efficient AGN live in star-forming galaxies, this is more likely to be the case for the most luminous AGN (e.g., [303,304]). In contrast, radiatively inefficient AGN are more likely to be found in quenched (non-star-forming) galaxies (e.g., [305]).

A comparison with cosmological simulations shows broad agreement with the above observational result that the high accretion rate AGN typically resides in gas-rich, star-forming galaxies [138,288,306,307]. This likely tells us something about the shared fuel supply and feeding processes needed to invoke a luminous AGN event. Furthermore, it shows that these types of observations do not rule out AGN feedback models that are invoked in simulations. Indeed, a measured instantaneous radiative luminosity of an AGN is likely unconnected to the eventual global, cumulative, feedback on the host galaxy because an instantaneous AGN luminosity does not constrain the total (cumulative) energy injection that could occur during multiple periods of black hole growth (see Figure 2). This all suggests that the molecular global gas content and SFRs of statistical samples of galaxies hosting radiatively efficient AGN are not expected to reveal the strong signatures of galaxy-wide (global) AGN feedback (also see [138,288,306–308]). Instead, the global evidence for AGN feedback may be found in the wider galaxy population, which do not necessarily currently host an observable AGN event (e.g., [309,310]).

AGN feedback is largely required in galaxy evolution theory to explain the build-up of quenched galaxies. Therefore, another approach to investigate the role of AGN feedback is to consider the galaxy population as a whole, and ask what is the most important factor/property for a galaxy to be quenched. The total black hole mass can be considered a tracer of the historic record of the total radiative energy output by AGN (see Figure 2), because the majority of black hole growth occurs during radiatively efficient phases. Work has suggested a strong connection between black hole mass and quiescence in galaxies, for stellar-mass matched samples, with broad agreement between both observations and simulations (at least for massive central galaxies, as opposed to satellite galaxies; [306–308,311]). These works imply that it is the integrated (cumulative) history of energy injection from AGN that is the most important for suppressing star formation, and not the instantaneous AGN luminosity. Nonetheless, caution must be taken when comparing these observations to simulations that invoke effective feedback prescriptions that are only turned on above a certain black hole mass (also see [288]). Furthermore, significant energy output may occur during AGN events that do not significantly contribute to black hole growth, for example, from jets during radiatively inefficient AGN events.

Indeed, an important difference between some feedback models in cosmological simulations is that of whether the feedback is most efficient during periods of rapid black hole growth (cf. a 'radiatively efficient' regime), or periods of low accretion (cf. a 'radiatively inefficient' regime). In the former case, the total energy output will be related to the black hole mass, whilst for the latter, this is not necessarily the case. Voit et al. [312] show that the failure of one particular cosmological simulation to reproduce the observed black hole mass-to-halo mass relationship can be attributed to the fact that the implemented *efficient mode of feedback* for lifting baryons (which results in the star-formation quenching) only occurs during periods of *low accretion rates*. This appears to be in contrast to observations that suggest that the majority of black hole growth occurs during the quenching process [312,313]. Similar comparisons between the observations of the overall galaxy population and simulations will be important to further test and refine the different prescriptions of feedback models in simulations (e.g., [314,315]). The spatially resolved measurements of star formation, and stellar populations, on large samples of galaxies in different environments, can shed further insight into the quenching process and the potential role of AGN across different populations (e.g., [270,316]).

Finally, we note that the effects of AGN feedback may be imprinted on CGM's surface brightness profiles of emission-line regions and on the distribution of metals (e.g., [37,91]). Different prescriptions of feedback result in the different predicted properties of the CGM (e.g., [61,91,317,318]). This remains a promising route to further test the feedback prescriptions and bridge the gap between theoretical predictions and observational studies.

7. Concluding Remarks

After decades of work, we draw the conclusion that there are no observations which are *inconsistent* with the broadest theoretical idea that AGN feedback is a crucial component of galaxy formation theory for explaining the properties of massive galaxies. However, obtaining a complete physical picture of how this process works in the real Universe is an ongoing challenge. It is a multi-scale problem over many orders of magnitude in spatial, temperature, and time scale (see Figure 1 and Section 2). Therefore, it is inescapable that multiple observational approaches are required to tackle this problem, using a range of facilities and a breadth of galaxy samples, for addressing the many components of the overall process (see Figure 2). Some key takeaway points from our holistic review are as follows:

- AGN should be considered *events* and not *objects* that persist in time. AGN are self-regulatory and variable; therefore, it may be difficult to directly relate a single accretion episode to a significant, global impact on galaxy properties. Ultimately, the properties of a galaxy will be influenced by the cumulative output of multiple accretion episodes/feedback events (see Section 2).
- Both the high accretion rate ('radiatively efficient') and low accretion rate ('radiatively inefficient') AGN can have multiple, overlapping mechanisms for injecting energy into their hosts (see Figure 3). These can contribute to both ejective and regulative channels of feedback (see Figure 2). Radio emissions can trace a range of feedback mechanisms over a range accretion rates (see Figures 3 and 4). Therefore, care should be taken when comparing the simplified theoretical feedback modes with the observed AGN populations (see Section 3).
- An AGN-driven outflow is gas that has been kinematically disturbed by a variety of possible driving mechanisms including accretion disc winds, radiation pressure, and jets. For the outflows to be relevant for feedback, their energy has to couple to the multiphase gas, and this coupling depends on several factors including AGN luminosity or jet-power, jet/wind orientation, and ISM properties. More efficient coupling will result in a more significant impact on the properties/distribution of the gas, both localised and globally (see Figure 2). The observations of the representative samples of AGN of different luminosities, hosted in galaxies with diverse properties, are needed to quantify the relevance of the previously mentioned factors on the coupling (see Section 4).
- Whilst the current AGN state of a galaxy is not a priori a useful proxy for assessing the impact on *global* galaxy properties, by studying currently active AGN with spatially resolved observations, we can obtain crucial information on the physics of *localised* impact. This is essential to determine how the energy couples with the gas, and under which circumstances it enhances or reduces star formation efficiency (see Figure 2 and Section 5).
- Evidence of the cumulative impact of AGN episodes on global galaxy properties is likely found in the galaxy population as a whole (not necessarily currently active). The distributions of galaxy properties are important for testing and ruling out different AGN feedback prescriptions implemented in cosmological simulations (see Section 6).

Over the next decade, we expect an explosion of data that will build-up an increasingly complete observational picture of how AGN couple to multi-phase gas, and the resulting impact of this interaction. With the latest observatories and instrumentation, this will be possible for ever-more representative galaxy samples, including for the relatively unexplored lowest mass galaxies [319–322] and those in the earliest Universe [274,323]. High-resolution simulations, that incorporate more physics, and that can be meaningfully compared to observations (e.g., [56,91,96,118,125,213,324]), provide a promising future for understanding the physics of the *localised and transient* processes of interactions between AGN and their host galaxies, and the potential *cumulative, global, and long-term impacts* of these interactions. All aspects of the multi-faceted observational approaches, summarised in Figure 2, remain important as we strive for a comprehensive understanding of the role of AGN feedback in galaxy evolution.

Author Contributions: The conceptualisation, research, and writing were carried out jointly by C.M.H. and C.R.A. All authors have read and agreed to the published version of the manuscript.

Funding: C.M.H. acknowledges funding from a United Kingdom Research and Innovation grant (code: MR/V022830/1). C.R.A. thanks support from the EU H2020-MSCA-ITN-2019 Project 860744 "BiD4BESt: Big Data applications for black hole Evolution STudies" and from project PID2022-141105NB-I00 "Tracking active galactic nuclei feedback from parsec to kiloparsec scales", funded by MICINN-AEI/10.13039/501100011033.

Data Availability Statement: Not applicable.

Acknowledgments: We thank the participants of "AGN on the Beach" and "The Importance of Jet-induced Feedback on Galaxy Scales", meetings for inspiring the contents of this article. We thank Ilaria Ruffa for the organisation of the first meeting and both the organising committee and the Lorentz Center staff for the latter. We thank Houda Haidar and Tiago Costa for providing insightful comments and suggestions and Gabriel Pérez Díaz (IAC MultiMedia Service, SMM) for producing the digital version of Figures 1–3. We thank the referees for their reports which helped us improve the clarity of this article.

Conflicts of Interest: The authors declare no conflicts of interest. The funders had no role in study or in the decision to publish the results.

Notes

1 We ignore the possibility of three regimes, with the addition of a distinct extremely highly accretion rate state ('super Eddington'), which may be particularly relevant for early black hole growth, including the origin and accretion rates of massive black holes in the early Universe that have been observed with *JWST* [76–78].

2 The variety of multi-wavelength methods used to identify radiatively efficient AGN results in a menagerie of AGN classifications and terminology used in the literature [89]. Here, we consider two broad classes: quasars and Seyferts. Quasars have high bolometric luminosities ($L_{\rm AGN} \gtrsim 10^{45}\,{\rm erg\,s^{-1}}$) and lower-luminosity AGN are considered to be Seyferts. Both classes can be type-1, where direct accretion emission and the broad line region (BLR) are observed, or type-2 where no accretion emission or BLR is detected.

3 We note that the inverse Compton effect could be important for cooling gas that is above the Compton temperature, including the effect of the AGN radiation field on the outflowing shocked gas, which could be crucial for establishing the potential impact of AGN-driven winds [53,98].

4 Although the values in [106] are generated from a population model, with values not calculated for individual sources, they do represent the general observed quasar population trends. We also note that, unlike in [106], we do not assume that radio emission associated with AGN (as opposed to star formation) is necessarily attributed to jets.

5 We note that "Low-ionisation nuclear emission-line regions" (LINERs) are another class of galaxies, where the observed emission is not dominated by photoionisation from AGN, nor from star-forming regions and a subset of these sources may be associated with radiatively inefficient AGN (e.g., [109–111]).

References

1. Ghez, A.M.; Salim, S.; Weinberg, N.N.; Lu, J.R.; Do, T.; Dunn, J.K.; Matthews, K.; Morris, M.R.; Yelda, S.; Becklin, E.E.; et al. Measuring Distance and Properties of the Milky Way's Central Supermassive Black Hole with Stellar Orbits. *Astrophys. J.* **2008**, *689*, 1044–1062. [CrossRef]
2. Kormendy, J.; Ho, L.C. Coevolution (or Not) of Supermassive Black Holes and Host Galaxies. *Annu. Rev. Astron. Astrophys.* **2013**, *51*, 511–653. [CrossRef]
3. Akiyama, K. et al. [Event Horizon Telescope Collaboration]. First M87 Event Horizon Telescope Results. I. The Shadow of the Supermassive Black Hole. *Astrophys. J. Lett.* **2019**, *875*, L1. [CrossRef]
4. Abuter, R. et al. [GRAVITY Collaboration]. Detection of the Schwarzschild precession in the orbit of the star S2 near the Galactic centre massive black hole. *Astron. Astrophys.* **2020**, *636*, L5. [CrossRef]
5. Nightingale, J.W.; Smith, R.J.; He, Q.; O'Riordan, C.M.; Kegerreis, J.A.; Amvrosiadis, A.; Edge, A.C.; Etherington, A.; Hayes, R.G.; Kelly, A.; et al. Abell 1201: Detection of an ultramassive black hole in a strong gravitational lens. *Mon. Not. R. Astron. Soc.* **2023**, *521*, 3298–3322. [CrossRef]
6. Soltan, A. Masses of quasars. *Mon. Not. R. Astron. Soc.* **1982**, *200*, 115–122. [CrossRef]
7. Marconi, A.; Risaliti, G.; Gilli, R.; Hunt, L.K.; Maiolino, R.; Salvati, M. Local supermassive black holes, relics of active galactic nuclei and the X-ray background. *Mon. Not. R. Astron. Soc.* **2004**, *351*, 169–185. [CrossRef]

8. Alexander, D.M.; Hickox, R.C. What drives the growth of black holes? *New Astron. Rev.* **2012**, *56*, 93–121. [CrossRef]
9. Harrison, C.M.; Costa, T.; Tadhunter, C.N.; Flütsch, A.; Kakkad, D.; Perna, M.; Vietri, G. AGN outflows and feedback twenty years on. *Nat. Astron.* **2018**, *2*, 198–205. [CrossRef]
10. Magorrian, J.; Tremaine, S.; Richstone, D.; Bender, R.; Bower, G.; Dressler, A.; Faber, S.M.; Gebhardt, K.; Green, R.; Grillmair, C.; et al. The Demography of Massive Dark Objects in Galaxy Centers. *Astron. J.* **1998**, *115*, 2285–2305. [CrossRef]
11. Gebhardt, K.; Bender, R.; Bower, G.; Dressler, A.; Faber, S.M.; Filippenko, A.V.; Green, R.; Grillmair, C.; Ho, L.C.; Kormendy, J.; et al. A Relationship between Nuclear Black Hole Mass and Galaxy Velocity Dispersion. *Astrophys. J. Lett.* **2000**, *539*, L13–L16. [CrossRef]
12. Ferrarese, L.; Merritt, D. A Fundamental Relation between Supermassive Black Holes and Their Host Galaxies. *Astrophys. J. Lett.* **2000**, *539*, L9–L12. [CrossRef]
13. Silk, J.; Rees, M.J. Quasars and galaxy formation. *Astron. Astrophys.* **1998**, *331*, L1–L4. [CrossRef]
14. King, A. Black Holes, Galaxy Formation, and the M_{BH}-σ Relation. *Astrophys. J. Lett.* **2003**, *596*, L27–L29. [CrossRef]
15. King, A. The AGN-Starburst Connection, Galactic Superwinds, and M_{BH}-σ. *Astrophys. J. Lett.* **2005**, *635*, L121–L123. [CrossRef]
16. Binney, J.; Tabor, G. Evolving cooling flows. *Mon. Not. R. Astron. Soc.* **1995**, *276*, 663–678. [CrossRef]
17. Ciotti, L.; Ostriker, J.P. Cooling Flows and Quasars: Different Aspects of the Same Phenomenon? I. Concepts. *Astrophys. J. Lett.* **1997**, *487*, L105–L108. [CrossRef]
18. Peterson, J.R.; Kahn, S.M.; Paerels, F.B.S.; Kaastra, J.S.; Tamura, T.; Bleeker, J.A.M.; Ferrigno, C.; Jernigan, J.G. High-Resolution X-Ray Spectroscopic Constraints on Cooling-Flow Models for Clusters of Galaxies. *Astrophys. J.* **2003**, *590*, 207–224. [CrossRef]
19. Benson, A.J.; Bower, R.G.; Frenk, C.S.; Lacey, C.G.; Baugh, C.M.; Cole, S. What Shapes the Luminosity Function of Galaxies? *Astrophys. J.* **2003**, *599*, 38–49. [CrossRef]
20. Springel, V.; Di Matteo, T.; Hernquist, L. Modelling feedback from stars and black holes in galaxy mergers. *Mon. Not. R. Astron. Soc.* **2005**, *361*, 776–794. [CrossRef]
21. Croton, D.J.; Springel, V.; White, S.D.M.; De Lucia, G.; Frenk, C.S.; Gao, L.; Jenkins, A.; Kauffmann, G.; Navarro, J.F.; Yoshida, N. The many lives of active galactic nuclei: Cooling flows, black holes and the luminosities and colours of galaxies. *Mon. Not. R. Astron. Soc.* **2006**, *365*, 11–28. [CrossRef]
22. Bower, R.G.; Benson, A.J.; Malbon, R.; Helly, J.C.; Frenk, C.S.; Baugh, C.M.; Cole, S.; Lacey, C.G. Breaking the hierarchy of galaxy formation. *Mon. Not. R. Astron. Soc.* **2006**, *370*, 645–655. [CrossRef]
23. Somerville, R.S.; Hopkins, P.F.; Cox, T.J.; Robertson, B.E.; Hernquist, L. A semi-analytic model for the co-evolution of galaxies, black holes and active galactic nuclei. *Mon. Not. R. Astron. Soc.* **2008**, *391*, 481–506. [CrossRef]
24. Schaye, J.; Crain, R.A.; Bower, R.G.; Furlong, M.; Schaller, M.; Theuns, T.; Dalla Vecchia, C.; Frenk, C.S.; McCarthy, I.G.; Helly, J.C.; et al. The EAGLE project: Simulating the evolution and assembly of galaxies and their environments. *Mon. Not. R. Astron. Soc.* **2015**, *446*, 521–554. [CrossRef]
25. Khandai, N.; Di Matteo, T.; Croft, R.; Wilkins, S.; Feng, Y.; Tucker, E.; DeGraf, C.; Liu, M.S. The MassiveBlack-II simulation: The evolution of haloes and galaxies to $z \sim 0$. *Mon. Not. R. Astron. Soc.* **2015**, *450*, 1349–1374. [CrossRef]
26. Dubois, Y.; Peirani, S.; Pichon, C.; Devriendt, J.; Gavazzi, R.; Welker, C.; Volonteri, M. The HORIZON-AGN simulation: Morphological diversity of galaxies promoted by AGN feedback. *Mon. Not. R. Astron. Soc.* **2016**, *463*, 3948–3964. [CrossRef]
27. Naab, T.; Ostriker, J.P. Theoretical Challenges in Galaxy Formation. *Annu. Rev. Astron. Astrophys.* **2017**, *55*, 59–109. [CrossRef]
28. McCarthy, I.G.; Schaye, J.; Bird, S.; Le Brun, A.M.C. The BAHAMAS project: Calibrated hydrodynamical simulations for large-scale structure cosmology. *Mon. Not. R. Astron. Soc.* **2017**, *465*, 2936–2965. [CrossRef]
29. Nelson, D.; Pillepich, A.; Springel, V.; Weinberger, R.; Hernquist, L.; Pakmor, R.; Genel, S.; Torrey, P.; Vogelsberger, M.; Kauffmann, G.; et al. First results from the IllustrisTNG simulations: The galaxy colour bimodality. *Mon. Not. R. Astron. Soc.* **2018**, *475*, 624–647. [CrossRef]
30. Davé, R.; Anglés-Alcázar, D.; Narayanan, D.; Li, Q.; Rafieferantsoa, M.H.; Appleby, S. SIMBA: Cosmological simulations with black hole growth and feedback. *Mon. Not. R. Astron. Soc.* **2019**, *486*, 2827–2849. [CrossRef]
31. Choi, E.; Somerville, R.S.; Ostriker, J.P.; Naab, T.; Hirschmann, M. The Role of Black Hole Feedback on Size and Structural Evolution in Massive Galaxies. *Astrophys. J.* **2018**, *866*, 91. [CrossRef]
32. Davies, J.J.; Crain, R.A.; McCarthy, I.G.; Oppenheimer, B.D.; Schaye, J.; Schaller, M.; McAlpine, S. The gas fractions of dark matter haloes hosting simulated $\sim L^{\star}$ galaxies are governed by the feedback history of their black holes. *Mon. Not. R. Astron. Soc.* **2019**, *485*, 3783–3793. [CrossRef]
33. Peirani, S.; Sonnenfeld, A.; Gavazzi, R.; Oguri, M.; Dubois, Y.; Silk, J.; Pichon, C.; Devriendt, J.; Kaviraj, S. Total density profile of massive early-type galaxies in HORIZON-AGN simulation: Impact of AGN feedback and comparison with observations. *Mon. Not. R. Astron. Soc.* **2019**, *483*, 4615–4627. [CrossRef]
34. van der Vlugt, D.; Costa, T. How AGN feedback drives the size growth of the first quasars. *Mon. Not. R. Astron. Soc.* **2019**, *490*, 4918–4934. [CrossRef]
35. Wright, R.J.; Lagos, C.d.P.; Power, C.; Mitchell, P.D. The impact of stellar and AGN feedback on halo-scale baryonic and dark matter accretion in the EAGLE simulations. *Mon. Not. R. Astron. Soc.* **2020**, *498*, 1668–1692. [CrossRef]
36. Cochrane, R.K.; Anglés-Alcázar, D.; Mercedes-Feliz, J.; Hayward, C.C.; Faucher-Giguère, C.A.; Wellons, S.; Terrazas, B.A.; Wetzel, A.; Hopkins, P.F.; Moreno, J.; et al. The impact of AGN-driven winds on physical and observable galaxy sizes. *Mon. Not. R. Astron. Soc.* **2023**, *523*, 2409–2421. [CrossRef]

37. Obreja, A.; Arrigoni Battaia, F.; Macciò, A.V.; Buck, T. AGN radiation imprints on the circumgalactic medium of massive galaxies. *Mon. Not. R. Astron. Soc.* **2024**, *527*, 8078–8102. [CrossRef]
38. Zakamska, N.L.; Lampayan, K.; Petric, A.; Dicken, D.; Greene, J.E.; Heckman, T.M.; Hickox, R.C.; Ho, L.C.; Krolik, J.H.; Nesvadba, N.P.H.; et al. Star formation in quasar hosts and the origin of radio emission in radio-quiet quasars. *Mon. Not. R. Astron. Soc.* **2016**, *455*, 4191–4211. [CrossRef]
39. Ishibashi, W.; Fabian, A.C.; Maiolino, R. The energetics of AGN radiation pressure-driven outflows. *Mon. Not. R. Astron. Soc.* **2018**, *476*, 512–519. [CrossRef]
40. Rakshit, S.; Woo, J.H. A Census of Ionized Gas Outflows in Type 1 AGNs: Gas Outflows in AGNs. V. *Astrophys. J.* **2018**, *865*, 5. [CrossRef]
41. Jarvis, M.E.; Harrison, C.M.; Thomson, A.P.; Circosta, C.; Mainieri, V.; Alexander, D.M.; Edge, A.C.; Lansbury, G.B.; Molyneux, S.J.; Mullaney, J.R. Prevalence of radio jets associated with galactic outflows and feedback from quasars. *Mon. Not. R. Astron. Soc.* **2019**, *485*, 2710–2730. [CrossRef]
42. Costa, T.; Pakmor, R.; Springel, V. Powering galactic superwinds with small-scale AGN winds. *Mon. Not. R. Astron. Soc.* **2020**, *497*, 5229–5255. [CrossRef]
43. Brüggen, M.; Scannapieco, E. The Launching of Cold Clouds by Galaxy Outflows. IV. Cosmic-Ray-driven Acceleration. *Astrophys. J.* **2020**, *905*, 19. [CrossRef]
44. Venturi, G.; Cresci, G.; Marconi, A.; Mingozzi, M.; Nardini, E.; Carniani, S.; Mannucci, F.; Marasco, A.; Maiolino, R.; Perna, M.; et al. MAGNUM survey: Compact jets causing large turmoil in galaxies. Enhanced line widths perpendicular to radio jets as tracers of jet-ISM interaction. *Astron. Astrophys.* **2021**, *648*, A17. [CrossRef]
45. Mandal, A.; Mukherjee, D.; Federrath, C.; Nesvadba, N.P.H.; Bicknell, G.V.; Wagner, A.Y.; Meenakshi, M. Impact of relativistic jets on the star formation rate: A turbulence-regulated framework. *Mon. Not. R. Astron. Soc.* **2021**, *508*, 4738–4757. [CrossRef]
46. Meena, B.; Crenshaw, D.M.; Schmitt, H.R.; Revalski, M.; Chapman, Z.; Fischer, T.C.; Kraemer, S.B.; Robinson, J.H.; Falcone, J.; Polack, G.E. Investigating the Narrow-line Region Dynamics in Nearby Active Galaxies. *Astrophys. J.* **2023**, *943*, 98. [CrossRef]
47. Calistro Rivera, G.; Alexander, D.M.; Harrison, C.M.; Fawcett, V.A.; Best, P.N.; Williams, W.L.; Hardcastle, M.J.; Rosario, D.J.; Smith, D.J.B.; Arnaudova, M.I.; et al. Ubiquitous radio emission in quasars: Predominant AGN origin and a connection to jets, dust and winds. *arXiv* **2023**, arXiv:2312.10177. [CrossRef]
48. Fabian, A.C. Observational Evidence of Active Galactic Nuclei Feedback. *Annu. Rev. Astron. Astrophys.* **2012**, *50*, 455–489. [CrossRef]
49. McNamara, B.R.; Nulsen, P.E.J. Mechanical feedback from active galactic nuclei in galaxies, groups and clusters. *New J. Phys.* **2012**, *14*, 055023. [CrossRef]
50. Heckman, T.M.; Best, P.N. The Coevolution of Galaxies and Supermassive Black Holes: Insights from Surveys of the Contemporary Universe. *Annu. Rev. Astron. Astrophys.* **2014**, *52*, 589–660. [CrossRef]
51. Harrison, C.M. Impact of supermassive black hole growth on star formation. *Nat. Astron.* **2017**, *1*, 0165. [CrossRef]
52. Morganti, R. The many routes to AGN feedback. *Front. Astron. Space Sci.* **2017**, *4*, 42. [CrossRef]
53. King, A.; Pounds, K. Powerful Outflows and Feedback from Active Galactic Nuclei. *Annu. Rev. Astron. Astrophys.* **2015**, *53*, 115–154. [CrossRef]
54. Hardcastle, M.J.; Croston, J.H. Radio galaxies and feedback from AGN jets. *New Astron. Rev.* **2020**, *88*, 101539. [CrossRef]
55. Veilleux, S.; Maiolino, R.; Bolatto, A.D.; Aalto, S. Cool outflows in galaxies and their implications. *Astron. Astrophys. Rev.* **2020**, *28*, 2. [CrossRef]
56. Bourne, M.A.; Yang, H.Y.K. Recent Progress in Modeling the Macro- and Micro-Physics of Radio Jet Feedback in Galaxy Clusters. *Galaxies* **2023**, *11*, 73. [CrossRef]
57. Krause, M.G.H. Jet Feedback in Star-Forming Galaxies. *Galaxies* **2023**, *11*, 29. [CrossRef]
58. Ramos Almeida, C.; Ricci, C. Nuclear obscuration in active galactic nuclei. *Nat. Astron.* **2017**, *1*, 679–689. [CrossRef]
59. Tacconi, L.J.; Genzel, R.; Sternberg, A. The Evolution of the Star-Forming Interstellar Medium Across Cosmic Time. *Annu. Rev. Astron. Astrophys.* **2020**, *58*, 157–203. [CrossRef]
60. Tumlinson, J.; Peeples, M.S.; Werk, J.K. The Circumgalactic Medium. *Annu. Rev. Astron. Astrophys.* **2017**, *55*, 389–432. [CrossRef]
61. Davies, J.J.; Crain, R.A.; Oppenheimer, B.D.; Schaye, J. The quenching and morphological evolution of central galaxies is facilitated by the feedback-driven expulsion of circumgalactic gas. *Mon. Not. R. Astron. Soc.* **2020**, *491*, 4462–4480. [CrossRef]
62. Ganguly, S.; Li, Y.; Olivares, V.; Su, Y.; Combes, F.; Prakash, S.; Hamer, S.; Guillard, P.; Ha, T. The nature of the motions of multiphase filaments in the centers of galaxy clusters. *Front. Astron. Space Sci.* **2023**, *10*, 1138613. [CrossRef]
63. García-Burillo, S.; Alonso-Herrero, A.; Ramos Almeida, C.; González-Martín, O.; Combes, F.; Usero, A.; Hönig, S.; Querejeta, M.; Hicks, E.K.S.; Hunt, L.K.; et al. The Galaxy Activity, Torus, and Outflow Survey (GATOS). I. ALMA images of dusty molecular tori in Seyfert galaxies. *Astron. Astrophys.* **2021**, *652*, A98. [CrossRef]
64. Mukherjee, D.; Bicknell, G.V.; Wagner, A.Y.; Sutherland, R.S.; Silk, J. Relativistic jet feedback—III. Feedback on gas discs. *Mon. Not. R. Astron. Soc.* **2018**, *479*, 5544–5566. [CrossRef]
65. Richings, A.J.; Faucher-Giguère, C.A. The origin of fast molecular outflows in quasars: Molecule formation in AGN-driven galactic winds. *Mon. Not. R. Astron. Soc.* **2018**, *474*, 3673–3699. [CrossRef]
66. Peterson, B.M. Variability of Active Galactic Nuclei. *In Proceedings of the Advanced Lectures on the Starburst-AGN*; Aretxaga, I., Kunth, D., Mújica, R., Eds.; World Scientific: Singapore, 2001; p. 3. [CrossRef]

67. Lintott, C.J.; Schawinski, K.; Keel, W.; van Arkel, H.; Bennert, N.; Edmondson, E.; Thomas, D.; Smith, D.J.B.; Herbert, P.D.; Jarvis, M.J.; et al. Galaxy Zoo: 'Hanny's Voorwerp', a quasar light echo? *Mon. Not. R. Astron. Soc.* **2009**, *399*, 129–140. [CrossRef]
68. Mostert, R.I.J.; Morganti, R.; Brienza, M.; Duncan, K.J.; Oei, M.S.S.L.; Röttgering, H.J.A.; Alegre, L.; Hardcastle, M.J.; Jurlin, N. Finding AGN remnant candidates based on radio morphology with machine learning. *Astron. Astrophys.* **2023**, *674*, A208. [CrossRef]
69. King, A.; Nixon, C. AGN flickering and chaotic accretion. *Mon. Not. R. Astron. Soc.* **2015**, *453*, L46–L47. [CrossRef]
70. Schawinski, K.; Koss, M.; Berney, S.; Sartori, L.F. Active galactic nuclei flicker: An observational estimate of the duration of black hole growth phases of $\sim 10^5$ yr. *Mon. Not. R. Astron. Soc.* **2015**, *451*, 2517–2523. [CrossRef]
71. Sabater, J.; Best, P.N.; Hardcastle, M.J.; Shimwell, T.W.; Tasse, C.; Williams, W.L.; Brüggen, M.; Cochrane, R.K.; Croston, J.H.; de Gasperin, F.; et al. The LoTSS view of radio AGN in the local Universe. The most massive galaxies are always switched on. *Astron. Astrophys.* **2019**, *622*, A17. [CrossRef]
72. Hickox, R.C.; Mullaney, J.R.; Alexander, D.M.; Chen, C.T.J.; Civano, F.M.; Goulding, A.D.; Hainline, K.N. Black Hole Variability and the Star Formation-Active Galactic Nucleus Connection: Do All Star-forming Galaxies Host an Active Galactic Nucleus? *Astrophys. J.* **2014**, *782*, 9. [CrossRef]
73. Novak, G.S.; Ostriker, J.P.; Ciotti, L. Feedback from Central Black Holes in Elliptical Galaxies: Two-dimensional Models Compared to One-dimensional Models. *Astrophys. J.* **2011**, *737*, 26. [CrossRef]
74. Hickox, R.C.; Alexander, D.M. Obscured Active Galactic Nuclei. *Annu. Rev. Astron. Astrophys.* **2018**, *56*, 625–671. [CrossRef]
75. Aird, J.; Coil, A.L.; Georgakakis, A. X-rays across the galaxy population—II. The distribution of AGN accretion rates as a function of stellar mass and redshift. *Mon. Not. R. Astron. Soc.* **2018**, *474*, 1225–1249. [CrossRef]
76. Rennehan, D.; Babul, A.; Moa, B.; Davé, R. Three regimes of black hole feedback. *arXiv* **2023**, arXiv:2309.15898. [CrossRef]
77. Lupi, A.; Quadri, G.; Volonteri, M.; Colpi, M.; Regan, J.A. Sustained super-Eddington accretion in high-redshift quasars. *arXiv* **2023**, arXiv:2312.08422. [CrossRef]
78. Maiolino, R.; Scholtz, J.; Witstok, J.; Carniani, S.; D'Eugenio, F.; de Graaff, A.; Uebler, H.; Tacchella, S.; Curtis-Lake, E.; Arribas, S.; et al. A small and vigorous black hole in the early Universe. *arXiv* **2023**, arXiv:2305.12492. [CrossRef]
79. Shakura, N.I.; Sunyaev, R.A. Black holes in binary systems. Observational appearance. *Astron. Astrophys.* **1973**, *24*, 337–355.
80. Esin, A.A.; McClintock, J.E.; Narayan, R. Advection-Dominated Accretion and the Spectral States of Black Hole X-Ray Binaries: Application to Nova Muscae 1991. *Astrophys. J.* **1997**, *489*, 865–889. [CrossRef]
81. Narayan, R.; Yi, I. Advection-dominated Accretion: A Self-similar Solution. *Astrophys. J. Lett.* **1994**, *428*, L13. [CrossRef]
82. Blandford, R.D.; Znajek, R.L. Electromagnetic extraction of energy from Kerr black holes. *Mon. Not. R. Astron. Soc.* **1977**, *179*, 433–456. [CrossRef]
83. Sikora, M.; Stawarz, Ł.; Lasota, J.P. Radio Loudness of Active Galactic Nuclei: Observational Facts and Theoretical Implications. *Astrophys. J.* **2007**, *658*, 815–828. [CrossRef]
84. Blandford, R.D.; Payne, D.G. Hydromagnetic flows from accretion disks and the production of radio jets. *Mon. Not. R. Astron. Soc.* **1982**, *199*, 883–903. [CrossRef]
85. Begelman, M.C.; McKee, C.F.; Shields, G.A. Compton heated winds and coronae above accretion disks. I. Dynamics. *Astrophys. J.* **1983**, *271*, 70–88. [CrossRef]
86. Yuan, F.; Narayan, R. Hot Accretion Flows around Black Holes. *Annu. Rev. Astron. Astrophys.* **2014**, *52*, 529–588. [CrossRef]
87. Gan, Z.; Yuan, F.; Ostriker, J.P.; Ciotti, L.; Novak, G.S. Active Galactic Nucleus Feedback in an Isolated Elliptical Galaxy: The Effect of Strong Radiative Feedback in the Kinetic Mode. *Astrophys. J.* **2014**, *789*, 150. [CrossRef]
88. Almeida, I.; Nemmen, R.; Riffel, R.A. Quenching star formation with low-luminosity AGN winds. *Mon. Not. R. Astron. Soc.* **2023**, *526*, 217–223. [CrossRef]
89. Padovani, P.; Alexander, D.M.; Assef, R.J.; De Marco, B.; Giommi, P.; Hickox, R.C.; Richards, G.T.; Smolčić, V.; Hatziminaoglou, E.; Mainieri, V.; et al. Active galactic nuclei: What is in a name? *Astron. Astrophys. Rev.* **2017**, *25*, 2. [CrossRef]
90. Arrigoni Battaia, F.; Hennawi, J.F.; Prochaska, J.X.; Oñorbe, J.; Farina, E.P.; Cantalupo, S.; Lusso, E. QSO MUSEUM I: A sample of 61 extended Ly α-emission nebulae surrounding $z \sim 3$ quasars. *Mon. Not. R. Astron. Soc.* **2019**, *482*, 3162–3205. [CrossRef]
91. Costa, T.; Arrigoni Battaia, F.; Farina, E.P.; Keating, L.C.; Rosdahl, J.; Kimm, T. AGN-driven outflows and the formation of Lyα nebulae around high-z quasars. *Mon. Not. R. Astron. Soc.* **2022**, *517*, 1767–1790. [CrossRef]
92. Costa, T.; Rosdahl, J.; Sijacki, D.; Haehnelt, M.G. Quenching star formation with quasar outflows launched by trapped IR radiation. *Mon. Not. R. Astron. Soc.* **2018**, *479*, 2079–2111. [CrossRef]
93. Murray, N.; Chiang, J.; Grossman, S.A.; Voit, G.M. Accretion Disk Winds from Active Galactic Nuclei. *Astrophys. J.* **1995**, *451*, 498. [CrossRef]
94. Proga, D.; Stone, J.M.; Kallman, T.R. Dynamics of Line-driven Disk Winds in Active Galactic Nuclei. *Astrophys. J.* **2000**, *543*, 686–696. [CrossRef]
95. Mizumoto, M.; Nomura, M.; Done, C.; Ohsuga, K.; Odaka, H. UV line-driven disc wind as the origin of UltraFast Outflows in AGN. *Mon. Not. R. Astron. Soc.* **2021**, *503*, 1442–1458. [CrossRef]
96. Tanner, R.; Weaver, K.A. Simulations of AGN-driven Galactic Outflow Morphology and Content. *Astron. J.* **2022**, *163*, 134. [CrossRef]
97. Sazonov, S.Y.; Ostriker, J.P.; Ciotti, L.; Sunyaev, R.A. Radiative feedback from quasars and the growth of massive black holes in stellar spheroids. *Mon. Not. R. Astron. Soc.* **2005**, *358*, 168–180. [CrossRef]

98. Bourne, M.A.; Nayakshin, S. Inverse Compton X-ray signature of AGN feedback. *Mon. Not. R. Astron. Soc.* **2013**, *436*, 2346–2351. [CrossRef]
99. Kellermann, K.I.; Sramek, R.; Schmidt, M.; Shaffer, D.B.; Green, R. VLA Observations of Objects in the Palomar Bright Quasar Survey. *Astron. J.* **1989**, *98*, 1195. [CrossRef]
100. Xu, C.; Livio, M.; Baum, S. Radio-loud and Radio-quiet Active Galactic Nuclei. *Astron. J.* **1999**, *118*, 1169–1176. [CrossRef]
101. Klindt, L.; Alexander, D.M.; Rosario, D.J.; Lusso, E.; Fotopoulou, S. Fundamental differences in the radio properties of red and blue quasars: Evolution strongly favoured over orientation. *Mon. Not. R. Astron. Soc.* **2019**, *488*, 3109–3128. [CrossRef]
102. Helou, G.; Soifer, B.T.; Rowan-Robinson, M. Thermal infrared and nonthermal radio: Remarkable correlation in disks of galaxies. *Astrophys. J. Lett.* **1985**, *298*, L7–L11. [CrossRef]
103. Ivison, R.J.; Magnelli, B.; Ibar, E.; Andreani, P.; Elbaz, D.; Altieri, B.; Amblard, A.; Arumugam, V.; Auld, R.; Aussel, H.; et al. The far-infrared/radio correlation as probed by Herschel. *Astron. Astrophys.* **2010**, *518*, L31. [CrossRef]
104. Best, P.N.; Heckman, T.M. On the fundamental dichotomy in the local radio-AGN population: Accretion, evolution and host galaxy properties. *Mon. Not. R. Astron. Soc.* **2012**, *421*, 1569–1582. [CrossRef]
105. Del Moro, A.; Alexander, D.M.; Mullaney, J.R.; Daddi, E.; Pannella, M.; Bauer, F.E.; Pope, A.; Dickinson, M.; Elbaz, D.; Barthel, P.D.; et al. GOODS-Herschel: Radio-excess signature of hidden AGN activity in distant star-forming galaxies. *Astron. Astrophys.* **2013**, *549*, A59. [CrossRef]
106. Macfarlane, C.; Best, P.N.; Sabater, J.; Gürkan, G.; Jarvis, M.J.; Röttgering, H.J.A.; Baldi, R.D.; Calistro Rivera, G.; Duncan, K.J.; Morabito, L.K.; et al. The radio loudness of SDSS quasars from the LOFAR Two-metre Sky Survey: Ubiquitous jet activity and constraints on star formation. *Mon. Not. R. Astron. Soc.* **2021**, *506*, 5888–5907. [CrossRef]
107. Zakamska, N.L.; Strauss, M.A.; Heckman, T.M.; Ivezić, Ž.; Krolik, J.H. Candidate Type II Quasars from the Sloan Digital Sky Survey. II. From Radio to X-Rays. *Astron. J.* **2004**, *128*, 1002–1016. [CrossRef]
108. Kellermann, K.I.; Condon, J.J.; Kimball, A.E.; Perley, R.A.; Ivezić, Ž. Radio-loud and Radio-quiet QSOs. *Astrophys. J.* **2016**, *831*, 168. [CrossRef]
109. Heckman, T.M. An Optical and Radio Survey of the Nuclei of Bright Galaxies—Activity in the Normal Galactic Nuclei. *Astron. Astrophys.* **1980**, *87*, 152.
110. Flohic, H.M.L.G.; Eracleous, M.; Chartas, G.; Shields, J.C.; Moran, E.C. The Central Engines of 19 LINERs as Viewed by Chandra. *Astrophys. J.* **2006**, *647*, 140–160. [CrossRef]
111. Ho, L.C. Radiatively Inefficient Accretion in Nearby Galaxies. *Astrophys. J.* **2009**, *699*, 626–637. [CrossRef]
112. Nims, J.; Quataert, E.; Faucher-Giguère, C.A. Observational signatures of galactic winds powered by active galactic nuclei. *Mon. Not. R. Astron. Soc.* **2015**, *447*, 3612–3622. [CrossRef]
113. Panessa, F.; Baldi, R.D.; Laor, A.; Padovani, P.; Behar, E.; McHardy, I. The origin of radio emission from radio-quiet active galactic nuclei. *Nat. Astron.* **2019**, *3*, 387–396. [CrossRef]
114. Morabito, L.K.; Sweijen, F.; Radcliffe, J.F.; Best, P.N.; Kondapally, R.; Bondi, M.; Bonato, M.; Duncan, K.J.; Prandoni, I.; Shimwell, T.W.; et al. Identifying active galactic nuclei via brightness temperature with sub-arcsecond international LOFAR telescope observations. *Mon. Not. R. Astron. Soc.* **2022**, *515*, 5758–5774. [CrossRef]
115. Jarvis, M.E.; Harrison, C.M.; Mainieri, V.; Alexander, D.M.; Arrigoni Battaia, F.; Calistro Rivera, G.; Circosta, C.; Costa, T.; De Breuck, C.; Edge, A.C.; et al. The quasar feedback survey: Discovering hidden Radio-AGN and their connection to the host galaxy ionized gas. *Mon. Not. R. Astron. Soc.* **2021**, *503*, 1780–1797. [CrossRef]
116. Fischer, T.C.; Johnson, M.C.; Secrest, N.J.; Crenshaw, D.M.; Kraemer, S.B. No Small-scale Radio Jets Here: Multiepoch Observations of Radio Continuum Structures in NGC 1068 with the VLBA. *Astrophys. J.* **2023**, *953*, 87. [CrossRef]
117. Bicknell, G.V.; Mukherjee, D.; Wagner, A.Y.; Sutherland, R.S.; Nesvadba, N.P.H. Relativistic jet feedback - II. Relationship to gigahertz peak spectrum and compact steep spectrum radio galaxies. *Mon. Not. R. Astron. Soc.* **2018**, *475*, 3493–3501. [CrossRef]
118. Meenakshi, M.; Mukherjee, D.; Wagner, A.Y.; Nesvadba, N.P.H.; Bicknell, G.V.; Morganti, R.; Janssen, R.M.J.; Sutherland, R.S.; Mandal, A. Modelling observable signatures of jet-ISM interaction: Thermal emission and gas kinematics. *Mon. Not. R. Astron. Soc.* **2022**, *516*, 766–786. [CrossRef]
119. Sijacki, D.; Springel, V.; Di Matteo, T.; Hernquist, L. A unified model for AGN feedback in cosmological simulations of structure formation. *Mon. Not. R. Astron. Soc.* **2007**, *380*, 877–900. [CrossRef]
120. Hopkins, P.F.; Hernquist, L.; Cox, T.J.; Di Matteo, T.; Robertson, B.; Springel, V. A Unified, Merger-driven Model of the Origin of Starbursts, Quasars, the Cosmic X-Ray Background, Supermassive Black Holes, and Galaxy Spheroids. *Astrophys. J. Suppl.* **2006**, *163*, 1–49. [CrossRef]
121. Dubois, Y.; Devriendt, J.; Slyz, A.; Teyssier, R. Jet-regulated cooling catastrophe. *Mon. Not. R. Astron. Soc.* **2010**, *409*, 985–1001. [CrossRef]
122. Bower, R.G.; Benson, A.J.; Crain, R.A. What shapes the galaxy mass function? Exploring the roles of supernova-driven winds and active galactic nuclei. *Mon. Not. R. Astron. Soc.* **2012**, *422*, 2816–2840. [CrossRef]
123. Oppenheimer, B.D.; Davies, J.J.; Crain, R.A.; Wijers, N.A.; Schaye, J.; Werk, J.K.; Burchett, J.N.; Trayford, J.W.; Horton, R. Feedback from supermassive black holes transforms centrals into passive galaxies by ejecting circumgalactic gas. *Mon. Not. R. Astron. Soc.* **2020**, *491*, 2939–2952. [CrossRef]
124. Cicone, C.; Mainieri, V.; Circosta, C.; Kakkad, D.; Vietri, G.; Perna, M.; Bischetti, M.; Carniani, S.; Cresci, G.; Harrison, C.; et al. SUPER. VI. A giant molecular halo around a z$\sim$2 quasar. *Astron. Astrophys.* **2021**, *654*, L8. [CrossRef]

125. Audibert, A.; Ramos Almeida, C.; García-Burillo, S.; Combes, F.; Bischetti, M.; Meenakshi, M.; Mukherjee, D.; Bicknell, G.; Wagner, A.Y. Jet-induced molecular gas excitation and turbulence in the Teacup. *Astron. Astrophys.* **2023**, *671*, L12. [CrossRef]
126. Longinotti, A.L.; Krongold, Y.; Kriss, G.A.; Ely, J.; Gallo, L.; Grupe, D.; Komossa, S.; Mathur, S.; Pradhan, A. The Rise of an Ionized Wind in the Narrow-line Seyfert 1 Galaxy Mrk 335 Observed by XMM-Newton and HST. *Astrophys. J.* **2013**, *766*, 104. [CrossRef]
127. Tombesi, F.; Cappi, M.; Reeves, J.N.; Nemmen, R.S.; Braito, V.; Gaspari, M.; Reynolds, C.S. Unification of X-ray winds in Seyfert galaxies: From ultra-fast outflows to warm absorbers. *Mon. Not. R. Astron. Soc.* **2013**, *430*, 1102–1117. [CrossRef]
128. Chartas, G.; Cappi, M.; Vignali, C.; Dadina, M.; James, V.; Lanzuisi, G.; Giustini, M.; Gaspari, M.; Strickland, S.; Bertola, E. Multiphase Powerful Outflows Detected in High-z Quasars. *Astrophys. J.* **2021**, *920*, 24. [CrossRef]
129. Murray, N.; Chiang, J. Active Galactic Nuclei Disk Winds, Absorption Lines, and Warm Absorbers. *Astrophys. J. Lett.* **1995**, *454*, L105. [CrossRef]
130. Sulentic, J.W.; Bachev, R.; Marziani, P.; Negrete, C.A.; Dultzin, D. C IV λ1549 as an Eigenvector 1 Parameter for Active Galactic Nuclei. *Astrophys. J.* **2007**, *666*, 757–777. [CrossRef]
131. Coil, A.L.; Weiner, B.J.; Holz, D.E.; Cooper, M.C.; Yan, R.; Aird, J. Outflowing Galactic Winds in Post-starburst and Active Galactic Nucleus Host Galaxies at $0.2 < z < 0.8$. *Astrophys. J.* **2011**, *743*, 46. [CrossRef]
132. Martin, C.L.; Dijkstra, M.; Henry, A.; Soto, K.T.; Danforth, C.W.; Wong, J. The Lyα Line Profiles of Ultraluminous Infrared Galaxies: Fast Winds and Lyman Continuum Leakage. *Astrophys. J.* **2015**, *803*, 6. [CrossRef]
133. Harrison, C.M.; Alexander, D.M.; Mullaney, J.R.; Swinbank, A.M. Kiloparsec-scale outflows are prevalent among luminous AGN: Outflows and feedback in the context of the overall AGN population. *Mon. Not. R. Astron. Soc.* **2014**, *441*, 3306–3347. [CrossRef]
134. Villar-Martín, M.; Arribas, S.; Emonts, B.; Humphrey, A.; Tadhunter, C.; Bessiere, P.; Cabrera Lavers, A.; Ramos Almeida, C. Ionized outflows in luminous type 2 AGNs at $z < 0.6$: No evidence for significant impact on the host galaxies. *Mon. Not. R. Astron. Soc.* **2016**, *460*, 130–162. [CrossRef]
135. Förster Schreiber, N.M.; Genzel, R.; Newman, S.F.; Kurk, J.D.; Lutz, D.; Tacconi, L.J.; Wuyts, S.; Bandara, K.; Burkert, A.; Buschkamp, P.; et al. The Sins/zC-Sinf Survey of z ~2 Galaxy Kinematics: Evidence for Powerful Active Galactic Nucleus-Driven Nuclear Outflows in Massive Star-Forming Galaxies. *Astrophys. J.* **2014**, *787*, 38. [CrossRef]
136. Förster Schreiber, N.M.; Übler, H.; Davies, R.L.; Genzel, R.; Wisnioski, E.; Belli, S.; Shimizu, T.; Lutz, D.; Fossati, M.; Herrera-Camus, R.; et al. The KMOS^{3D} Survey: Demographics and Properties of Galactic Outflows at z = 0.6–2.7. *Astrophys. J.* **2019**, *875*, 21. [CrossRef]
137. Rose, M.; Tadhunter, C.; Ramos Almeida, C.; Rodríguez Zaurín, J.; Santoro, F.; Spence, R. Quantifying the AGN-driven outflows in ULIRGs (QUADROS)–I: VLT/Xshooter observations of nine nearby objects. *Mon. Not. R. Astron. Soc.* **2018**, *474*, 128–156. [CrossRef]
138. Scholtz, J.; Alexander, D.M.; Harrison, C.M.; Rosario, D.J.; McAlpine, S.; Mullaney, J.R.; Stanley, F.; Simpson, J.; Theuns, T.; Bower, R.G.; et al. Identifying the subtle signatures of feedback from distant AGN using ALMA observations and the EAGLE hydrodynamical simulations. *Mon. Not. R. Astron. Soc.* **2018**, *475*, 1288–1305. [CrossRef]
139. Scholtz, J.; Harrison, C.M.; Rosario, D.J.; Alexander, D.M.; Chen, C.C.; Kakkad, D.; Mainieri, V.; Tiley, A.L.; Turner, O.; Cirasuolo, M.; et al. KASHz: No evidence for ionised outflows instantaneously suppressing star formation in moderate luminosity AGN at $z\sim1.4$–2.6. *Mon. Not. R. Astron. Soc.* **2020**, *492*, 3194–3216. [CrossRef]
140. Rupke, D.S.N.; Veilleux, S. Breaking the Obscuring Screen: A Resolved Molecular Outflow in a Buried QSO. *Astrophys. J. Lett.* **2013**, *775*, L15. [CrossRef]
141. Ramos Almeida, C.; Acosta-Pulido, J.A.; Tadhunter, C.N.; González-Fernández, C.; Cicone, C.; Fernández-Torreiro, M. A near-infrared study of the multiphase outflow in the type-2 quasar J1509+0434. *Mon. Not. R. Astron. Soc.* **2019**, *487*, L18–L23. [CrossRef]
142. Speranza, G.; Ramos Almeida, C.; Acosta-Pulido, J.A.; Riffel, R.A.; Tadhunter, C.; Pierce, J.C.S.; Rodríguez-Ardila, A.; Coloma Puga, M.; Brusa, M.; Musiimenta, B.; et al. Warm molecular and ionized gas kinematics in the type-2 quasar J0945+1737. *Astron. Astrophys.* **2022**, *665*, A55. [CrossRef]
143. Riffel, R.A.; Storchi-Bergmann, T.; Riffel, R.; Bianchin, M.; Zakamska, N.L.; Ruschel-Dutra, D.; Bentz, M.C.; Burtscher, L.; Crenshaw, D.M.; Dahmer-Hahn, L.G.; et al. The AGNIFS survey: Spatially resolved observations of hot molecular and ionized outflows in nearby active galaxies. *Mon. Not. R. Astron. Soc.* **2023**, *521*, 1832–1848. [CrossRef]
144. Morganti, R.; Tadhunter, C.N.; Oosterloo, T.A. Fast neutral outflows in powerful radio galaxies: A major source of feedback in massive galaxies. *Astron. Astrophys.* **2005**, *444*, L9–L13. [CrossRef]
145. Morganti, R.; Veilleux, S.; Oosterloo, T.; Teng, S.H.; Rupke, D. Another piece of the puzzle: The fast H I outflow in Mrk 231. *Astron. Astrophys.* **2016**, *593*, A30. [CrossRef]
146. Rupke, D.S.; Veilleux, S.; Sanders, D.B. Outflows in Active Galactic Nucleus/Starburst-Composite Ultraluminous Infrared Galaxies1,. *Astrophys. J.* **2005**, *632*, 751–780. [CrossRef]
147. Cazzoli, S.; Arribas, S.; Maiolino, R.; Colina, L. Neutral gas outflows in nearby [U]LIRGs via optical NaD feature. *Astron. Astrophys.* **2016**, *590*, A125. [CrossRef]
148. Concas, A.; Popesso, P.; Brusa, M.; Mainieri, V.; Thomas, D. Two-face(s): Ionized and neutral gas winds in the local Universe. *Astron. Astrophys.* **2019**, *622*, A188. [CrossRef]

149. Maiolino, R.; Gallerani, S.; Neri, R.; Cicone, C.; Ferrara, A.; Genzel, R.; Lutz, D.; Sturm, E.; Tacconi, L.J.; Walter, F.; et al. Evidence of strong quasar feedback in the early Universe. *Mon. Not. R. Astron. Soc.* **2012**, *425*, L66–L70. [CrossRef]
150. Bischetti, M.; Maiolino, R.; Carniani, S.; Fiore, F.; Piconcelli, E.; Fluetsch, A. Widespread QSO-driven outflows in the early Universe. *Astron. Astrophys.* **2019**, *630*, A59. [CrossRef]
151. Dasyra, K.M.; Combes, F. Turbulent and fast motions of H_2 gas in active galactic nuclei. *Astron. Astrophys.* **2011**, *533*, L10. [CrossRef]
152. Riffel, R.A.; Zakamska, N.L.; Riffel, R. Active galactic nuclei winds as the origin of the H_2 emission excess in nearby galaxies. *Mon. Not. R. Astron. Soc.* **2020**, *491*, 1518–1529. [CrossRef]
153. Sturm, E.; González-Alfonso, E.; Veilleux, S.; Fischer, J.; Graciá-Carpio, J.; Hailey-Dunsheath, S.; Contursi, A.; Poglitsch, A.; Sternberg, A.; Davies, R.; et al. Massive Molecular Outflows and Negative Feedback in ULIRGs Observed by Herschel-PACS. *Astrophys. J. Lett.* **2011**, *733*, L16. [CrossRef]
154. Veilleux, S.; Meléndez, M.; Sturm, E.; Gracia-Carpio, J.; Fischer, J.; González-Alfonso, E.; Contursi, A.; Lutz, D.; Poglitsch, A.; Davies, R.; et al. Fast Molecular Outflows in Luminous Galaxy Mergers: Evidence for Quasar Feedback from Herschel. *Astrophys. J.* **2013**, *776*, 27. [CrossRef]
155. Spoon, H.W.W.; Farrah, D.; Lebouteiller, V.; González-Alfonso, E.; Bernard-Salas, J.; Urrutia, T.; Rigopoulou, D.; Westmoquette, M.S.; Smith, H.A.; Afonso, J.; et al. Diagnostics of AGN-Driven Molecular Outflows in ULIRGs from Herschel-PACS Observations of OH at 119 μm. *Astrophys. J.* **2013**, *775*, 127. [CrossRef]
156. González-Alfonso, E.; Fischer, J.; Spoon, H.W.W.; Stewart, K.P.; Ashby, M.L.N.; Veilleux, S.; Smith, H.A.; Sturm, E.; Farrah, D.; Falstad, N.; et al. Molecular Outflows in Local ULIRGs: Energetics from Multitransition OH Analysis. *Astrophys. J.* **2017**, *836*, 11. [CrossRef]
157. Feruglio, C.; Maiolino, R.; Piconcelli, E.; Menci, N.; Aussel, H.; Lamastra, A.; Fiore, F. Quasar feedback revealed by giant molecular outflows. *Astron. Astrophys.* **2010**, *518*, L155. [CrossRef]
158. Cicone, C.; Maiolino, R.; Sturm, E.; Graciá-Carpio, J.; Feruglio, C.; Neri, R.; Aalto, S.; Davies, R.; Fiore, F.; Fischer, J.; et al. Massive molecular outflows and evidence for AGN feedback from CO observations. *Astron. Astrophys.* **2014**, *562*, A21. [CrossRef]
159. Pereira-Santaella, M.; Colina, L.; García-Burillo, S.; Combes, F.; Emonts, B.; Aalto, S.; Alonso-Herrero, A.; Arribas, S.; Henkel, C.; Labiano, A.; et al. Spatially resolved cold molecular outflows in ULIRGs. *Astron. Astrophys.* **2018**, *616*, A171. [CrossRef]
160. Fluetsch, A.; Maiolino, R.; Carniani, S.; Marconi, A.; Cicone, C.; Bourne, M.A.; Costa, T.; Fabian, A.C.; Ishibashi, W.; Venturi, G. Cold molecular outflows in the local Universe and their feedback effect on galaxies. *Mon. Not. R. Astron. Soc.* **2019**, *483*, 4586–4614. [CrossRef]
161. Lamperti, I.; Pereira-Santaella, M.; Perna, M.; Colina, L.; Arribas, S.; García-Burillo, S.; González-Alfonso, E.; Aalto, S.; Alonso-Herrero, A.; Combes, F.; et al. Physics of ULIRGs with MUSE and ALMA: The PUMA project. IV. No tight relation between cold molecular outflow rates and AGN luminosities. *Astron. Astrophys.* **2022**, *668*, A45. [CrossRef]
162. Ramos Almeida, C.; Bischetti, M.; García-Burillo, S.; Alonso-Herrero, A.; Audibert, A.; Cicone, C.; Feruglio, C.; Tadhunter, C.N.; Pierce, J.C.S.; Pereira-Santaella, M.; et al. The diverse cold molecular gas contents, morphologies, and kinematics of type-2 quasars as seen by ALMA. *Astron. Astrophys.* **2022**, *658*, A155. [CrossRef]
163. Scholtz, J.; Maiolino, R.; Jones, G.C.; Carniani, S. Evidence of extended cold molecular gas and dust haloes around z 2.3 extremely red quasars with ALMA. *Mon. Not. R. Astron. Soc.* **2023**, *519*, 5246–5262. [CrossRef]
164. Cicone, C.; Brusa, M.; Ramos Almeida, C.; Cresci, G.; Husemann, B.; Mainieri, V. The largely unconstrained multiphase nature of outflows in AGN host galaxies. *Nat. Astron.* **2018**, *2*, 176–178. [CrossRef]
165. Rupke, D.S.N.; Gültekin, K.; Veilleux, S. Quasar-mode Feedback in Nearby Type 1 Quasars: Ubiquitous Kiloparsec-scale Outflows and Correlations with Black Hole Properties. *Astrophys. J.* **2017**, *850*, 40. [CrossRef]
166. Fluetsch, A.; Maiolino, R.; Carniani, S.; Arribas, S.; Belfiore, F.; Bellocchi, E.; Cazzoli, S.; Cicone, C.; Cresci, G.; Fabian, A.C.; et al. Properties of the multiphase outflows in local (ultra)luminous infrared galaxies. *Mon. Not. R. Astron. Soc.* **2021**, *505*, 5753–5783. [CrossRef]
167. Speranza, G.; Ramos Almeida, C.; Acosta-Pulido, J.A.; Audibert, A.; Holden, L.R.; Tadhunter, C.N.; Lapi, A.; González-Martín, O.; Brusa, M.; López, I.E.; et al. Multiphase characterization of AGN winds in five local type-2 quasars. *Astron. Astrophys.* **2024**, *681*, A63. [CrossRef]
168. Nesvadba, N.P.H.; Lehnert, M.D.; De Breuck, C.; Gilbert, A.M.; van Breugel, W. Evidence for powerful AGN winds at high redshift: Dynamics of galactic outflows in radio galaxies during the "Quasar Era". *Astron. Astrophys.* **2008**, *491*, 407–424. :200810346 [CrossRef]
169. Carniani, S.; Marconi, A.; Maiolino, R.; Balmaverde, B.; Brusa, M.; Cano-Díaz, M.; Cicone, C.; Comastri, A.; Cresci, G.; Fiore, F.; et al. Ionised outflows in z $\sim$2.4 quasar host galaxies. *Astron. Astrophys.* **2015**, *580*, A102. [CrossRef]
170. Kakkad, D.; Mainieri, V.; Vietri, G.; Carniani, S.; Harrison, C.M.; Perna, M.; Scholtz, J.; Circosta, C.; Cresci, G.; Husemann, B.; et al. SUPER. II. Spatially resolved ionised gas kinematics and scaling relations in z$\sim$2 AGN host galaxies. *Astron. Astrophys.* **2020**, *642*, A147. [CrossRef]
171. Vayner, A.; Zakamska, N.L.; Riffel, R.A.; Alexandroff, R.; Cosens, M.; Hamann, F.; Perrotta, S.; Rupke, D.S.N.; Bergmann, T.S.; Veilleux, S.; et al. Powerful winds in high-redshift obscured and red quasars. *Mon. Not. R. Astron. Soc.* **2021**, *504*, 4445–4459. [CrossRef]

172. Concas, A.; Maiolino, R.; Curti, M.; Hayden-Pawson, C.; Cirasuolo, M.; Jones, G.C.; Mercurio, A.; Belfiore, F.; Cresci, G.; Cullen, F.; et al. Being KLEVER at cosmic noon: Ionized gas outflows are inconspicuous in low-mass star-forming galaxies but prominent in massive AGN hosts. *Mon. Not. R. Astron. Soc.* **2022**, *513*, 2535–2562. [CrossRef]
173. Wylezalek, D.; Vayner, A.; Rupke, D.S.N.; Zakamska, N.L.; Veilleux, S.; Ishikawa, Y.; Bertemes, C.; Liu, W.; Barrera-Ballesteros, J.K.; Chen, H.W.; et al. First Results from the JWST Early Release Science Program Q3D: Turbulent Times in the Life of a z 3 Extremely Red Quasar Revealed by NIRSpec IFU. *Astrophys. J. Lett.* **2022**, *940*, L7. [CrossRef]
174. Vayner, A.; Zakamska, N.L.; Ishikawa, Y.; Sankar, S.; Wylezalek, D.; Rupke, D.S.N.; Veilleux, S.; Bertemes, C.; Barrera-Ballesteros, J.K.; Chen, H.W.; et al. First Results from the JWST Early Release Science Program Q3D: Powerful Quasar-driven Galactic Scale Outflow at z = 3. *Astrophys. J.* **2024**, *960*, 126. [CrossRef]
175. Cicone, C.; Maiolino, R.; Gallerani, S.; Neri, R.; Ferrara, A.; Sturm, E.; Fiore, F.; Piconcelli, E.; Feruglio, C. Very extended cold gas, star formation and outflows in the halo of a bright quasar at z > 6. *Astron. Astrophys.* **2015**, *574*, A14. [CrossRef]
176. Stacey, H.R.; Costa, T.; McKean, J.P.; Sharon, C.E.; Calistro Rivera, G.; Glikman, E.; van der Werf, P.P. Red quasars blow out molecular gas from galaxies during the peak of cosmic star formation. *Mon. Not. R. Astron. Soc.* **2022**, *517*, 3377–3391. [CrossRef]
177. Decarli, R.; Walter, F.; Venemans, B.P.; Bañados, E.; Bertoldi, F.; Carilli, C.; Fan, X.; Farina, E.P.; Mazzucchelli, C.; Riechers, D.; et al. An ALMA [C II] Survey of 27 Quasars at z > 5.94. *Astrophys. J.* **2018**, *854*, 97. [CrossRef]
178. Novak, M.; Venemans, B.P.; Walter, F.; Neeleman, M.; Kaasinen, M.; Liang, L.; Feldmann, R.; Bañados, E.; Carilli, C.; Decarli, R.; et al. No Evidence for [C II] Halos or High-velocity Outflows in z $\gtrsim$ 6 Quasar Host Galaxies. *Astrophys. J.* **2020**, *904*, 131. [CrossRef]
179. Herrera-Camus, R.; Tacconi, L.; Genzel, R.; Förster Schreiber, N.; Lutz, D.; Bolatto, A.; Wuyts, S.; Renzini, A.; Lilly, S.; Belli, S.; et al. Molecular and Ionized Gas Phases of an AGN-driven Outflow in a Typical Massive Galaxy at z $\approx$ 2. *Astrophys. J.* **2019**, *871*, 37. [CrossRef]
180. D'Eugenio, F.; Perez-Gonzalez, P.; Maiolino, R.; Scholtz, J.; Perna, M.; Circosta, C.; Uebler, H.; Arribas, S.; Boeker, T.; Bunker, A.; et al. A fast-rotator post-starburst galaxy quenched by supermassive black-hole feedback at z = 3. *arXiv* **2023**, arXiv:2308.06317. [CrossRef]
181. Vayner, A.; Zakamska, N.; Wright, S.A.; Armus, L.; Murray, N.; Walth, G. Multiphase Outflows in High-redshift Quasar Host Galaxies. *Astrophys. J.* **2021**, *923*, 59. [CrossRef]
182. Veilleux, S.; Teng, S.H.; Rupke, D.S.N.; Maiolino, R.; Sturm, E. Half-megasecond Chandra Spectral Imaging of the Hot Circumgalactic Nebula around Quasar Mrk 231. *Astrophys. J.* **2014**, *790*, 116. [CrossRef]
183. Greene, J.E.; Pooley, D.; Zakamska, N.L.; Comerford, J.M.; Sun, A.L. Extended X-Ray Emission from a Quasar-driven Superbubble. *Astrophys. J.* **2014**, *788*, 54. [CrossRef]
184. Lansbury, G.B.; Jarvis, M.E.; Harrison, C.M.; Alexander, D.M.; Del Moro, A.; Edge, A.C.; Mullaney, J.R.; Thomson, A.P. Storm in a Teacup: X-Ray View of an Obscured Quasar and Superbubble. *Astrophys. J. Lett.* **2018**, *856*, L1. [CrossRef]
185. Morganti, R.; Fogasy, J.; Paragi, Z.; Oosterloo, T.; Orienti, M. Radio Jets Clearing the Way Through a Galaxy: Watching Feedback in Action. *Science* **2013**, *341*, 1082–1085. [CrossRef]
186. Russell, H.R.; Lopez, L.A.; Allen, S.W.; Chartas, G.; Choudhury, P.P.; Dupke, R.A.; Fabian, A.C.; Flores, A.M.; Garofali, K.; Hodges-Kluck, E.; et al. The evolution of galaxies and clusters at high spatial resolution with AXIS. *arXiv* **2023**, arXiv:2311.07661. [CrossRef]
187. Simionescu, A.; ZuHone, J.; Zhuravleva, I.; Churazov, E.; Gaspari, M.; Nagai, D.; Werner, N.; Roediger, E.; Canning, R.; Eckert, D.; et al. Constraining Gas Motions in the Intra-Cluster Medium. *Space Sci. Rev.* **2019**, *215*, 24. [CrossRef]
188. Brownson, S.; Maiolino, R.; Tazzari, M.; Carniani, S.; Henden, N. Detecting the halo heating from AGN feedback with ALMA. *Mon. Not. R. Astron. Soc.* **2019**, *490*, 5134–5146. [CrossRef]
189. Richter, P. Hot Gas in Galaxy Halos Traced by Coronal Broad Lyα Absorbers. *Astrophys. J.* **2020**, *892*, 33. [CrossRef]
190. Richings, A.J.; Faucher-Giguère, C.A.; Stern, J. Unravelling the physics of multiphase AGN winds through emission line tracers. *Mon. Not. R. Astron. Soc.* **2021**, *503*, 1568–1585. [CrossRef]
191. Fiore, F.; Feruglio, C.; Shankar, F.; Bischetti, M.; Bongiorno, A.; Brusa, M.; Carniani, S.; Cicone, C.; Duras, F.; Lamastra, A.; et al. AGN wind scaling relations and the co-evolution of black holes and galaxies. *Astron. Astrophys.* **2017**, *601*, A143. [CrossRef]
192. Feruglio, C.; Fiore, F.; Carniani, S.; Piconcelli, E.; Zappacosta, L.; Bongiorno, A.; Cicone, C.; Maiolino, R.; Marconi, A.; Menci, N.; et al. The multi-phase winds of Markarian 231: From the hot, nuclear, ultra-fast wind to the galaxy-scale, molecular outflow. *Astron. Astrophys.* **2015**, *583*, A99. [CrossRef]
193. García-Burillo, S.; Combes, F.; Usero, A.; Aalto, S.; Krips, M.; Viti, S.; Alonso-Herrero, A.; Hunt, L.K.; Schinnerer, E.; Baker, A.J.; et al. Molecular line emission in NGC 1068 imaged with ALMA. I. An AGN-driven outflow in the dense molecular gas. *Astron. Astrophys.* **2014**, *567*, A125. [CrossRef]
194. Saito, T.; Takano, S.; Harada, N.; Nakajima, T.; Schinnerer, E.; Liu, D.; Taniguchi, A.; Izumi, T.; Watanabe, Y.; Bamba, K.; et al. The Kiloparsec-scale Neutral Atomic Carbon Outflow in the Nearby Type 2 Seyfert Galaxy NGC 1068: Evidence for Negative AGN Feedback. *Astrophys. J. Lett.* **2022**, *927*, L32. [CrossRef]
195. Tadhunter, C.; Morganti, R.; Rose, M.; Oonk, J.B.R.; Oosterloo, T. Jet acceleration of the fast molecular outflows in the Seyfert galaxy IC 5063. *Nature* **2014**, *511*, 440–443. [CrossRef]
196. Morganti, R.; Oosterloo, T.; Oonk, J.B.R.; Frieswijk, W.; Tadhunter, C. The fast molecular outflow in the Seyfert galaxy IC 5063 as seen by ALMA. *Astron. Astrophys.* **2015**, *580*, A1. [CrossRef]

197. Talbot, R.Y.; Bourne, M.A.; Sijacki, D. Blandford-Znajek jets in galaxy formation simulations: Method and implementation. *Mon. Not. R. Astron. Soc.* **2021**, *504*, 3619–3650. [CrossRef]
198. Talbot, R.Y.; Sijacki, D.; Bourne, M.A. Blandford-Znajek jets in galaxy formation simulations: Exploring the diversity of outflows produced by spin-driven AGN jets in Seyfert galaxies. *Mon. Not. R. Astron. Soc.* **2022**, *514*, 4535–4559. [CrossRef]
199. Couto, G.S.; Storchi-Bergmann, T.; Schnorr-Müller, A. Gas rotation, shocks and outflow within the inner 3 kpc of the radio galaxy 3C 33. *Mon. Not. R. Astron. Soc.* **2017**, *469*, 1573–1586. [CrossRef]
200. Balmaverde, B.; Capetti, A.; Marconi, A.; Venturi, G.; Chiaberge, M.; Baldi, R.D.; Baum, S.; Gilli, R.; Grandi, P.; Meyer, E.; et al. The MURALES survey. II. Presentation of MUSE observations of 20 3C low-z radio galaxies and first results. *Astron. Astrophys.* **2019**, *632*, A124. [CrossRef]
201. Balmaverde, B.; Capetti, A.; Baldi, R.D.; Baum, S.; Chiaberge, M.; Gilli, R.; Jimenez-Gallardo, A.; Marconi, A.; Massaro, F.; Meyer, E.; et al. The MURALES survey. VI. Properties and origin of the extended line emission structures in radio galaxies. *Astron. Astrophys.* **2022**, *662*, A23. [CrossRef]
202. Girdhar, A.; Harrison, C.M.; Mainieri, V.; Bittner, A.; Costa, T.; Kharb, P.; Mukherjee, D.; Arrigoni Battaia, F.; Alexander, D.M.; Calistro Rivera, G.; et al. Quasar feedback survey: Multiphase outflows, turbulence, and evidence for feedback caused by low power radio jets inclined into the galaxy disc. *Mon. Not. R. Astron. Soc.* **2022**, *512*, 1608–1628. [CrossRef]
203. Venturi, G.; Treister, E.; Finlez, C.; D'Ago, G.; Bauer, F.; Harrison, C.M.; Ramos Almeida, C.; Revalski, M.; Ricci, F.; Sartori, L.F.; et al. Complex AGN feedback in the Teacup galaxy. A powerful ionised galactic outflow, jet-ISM interaction, and evidence for AGN-triggered star formation in a giant bubble. *Astron. Astrophys.* **2023**, *678*, A127. [CrossRef]
204. Nyland, K.; Harwood, J.J.; Mukherjee, D.; Jagannathan, P.; Rujopakarn, W.; Emonts, B.; Alatalo, K.; Bicknell, G.V.; Davis, T.A.; Greene, J.E.; et al. Revolutionizing Our Understanding of AGN Feedback and its Importance to Galaxy Evolution in the Era of the Next Generation Very Large Array. *Astrophys. J.* **2018**, *859*, 23. [CrossRef]
205. Harrison, C.M.; Thomson, A.P.; Alexander, D.M.; Bauer, F.E.; Edge, A.C.; Hogan, M.T.; Mullaney, J.R.; Swinbank, A.M. Storm in a "Teacup": A Radio-quiet Quasar with $\approx$10 kpc Radio-emitting Bubbles and Extreme Gas Kinematics. *Astrophys. J.* **2015**, *800*, 45. [CrossRef]
206. Girdhar, A.; Harrison, C.M.; Mainieri, V.; Fernández Aranda, R.; Alexander, D.M.; Arrigoni Battaia, F.; Bianchin, M.; Calistro Rivera, G.; Circosta, C.; Costa, T.; et al. Quasar feedback survey: Molecular gas affected by central outflows and by 10 kpc radio lobes reveal dual feedback effects in 'radio quiet' quasars. *Mon. Not. R. Astron. Soc.* **2023**, *527*, 9322–9342. [CrossRef]
207. Cresci, G.; Tozzi, G.; Perna, M.; Brusa, M.; Marconcini, C.; Marconi, A.; Carniani, S.; Brienza, M.; Giroletti, M.; Belfiore, F.; et al. Bubbles and outflows: The novel JWST/NIRSpec view of the z = 1.59 obscured quasar XID2028. *Astron. Astrophys.* **2023**, *672*, A128. [CrossRef]
208. Veilleux, S.; Liu, W.; Vayner, A.; Wylezalek, D.; Rupke, D.S.N.; Zakamska, N.L.; Ishikawa, Y.; Bertemes, C.; Barrera-Ballesteros, J.K.; Chen, H.W.; et al. First Results from the JWST Early Release Science Program Q3D: The Warm Ionized Gas Outflow in z 1.6 Quasar XID 2028 and Its Impact on the Host Galaxy. *Astrophys. J.* **2023**, *953*, 56. [CrossRef]
209. Wagner, A.Y.; Bicknell, G.V.; Umemura, M.; Sutherland, R.S.; Silk, J. Galaxy-scale AGN feedback—Theory. *Astron. Nachrichten* **2016**, *337*, 167. [CrossRef]
210. Venanzi, M.; Hönig, S.; Williamson, D. The Role of Infrared Radiation Pressure in Shaping Dusty Winds in AGNs. *Astrophys. J.* **2020**, *900*, 174. [CrossRef]
211. Glikman, E.; Urrutia, T.; Lacy, M.; Djorgovski, S.G.; Mahabal, A.; Myers, A.D.; Ross, N.P.; Petitjean, P.; Ge, J.; Schneider, D.P.; et al. FIRST-2MASS Red Quasars: Transitional Objects Emerging from the Dust. *Astrophys. J.* **2012**, *757*, 51. [CrossRef]
212. Banerji, M.; Alaghband-Zadeh, S.; Hewett, P.C.; McMahon, R.G. Heavily reddened type 1 quasars at z > 2–I. Evidence for significant obscured black hole growth at the highest quasar luminosities. *Mon. Not. R. Astron. Soc.* **2015**, *447*, 3368–3389. [CrossRef]
213. Mercedes-Feliz, J.; Anglés-Alcázar, D.; Kiat Oh, B.; Hayward, C.C.; Cochrane, R.K.; Richings, A.J.; Faucher-Giguère, C.A.; Wellons, S.; Terrazas, B.A.; Moreno, J.; et al. Dense stellar clump formation driven by strong quasar winds in the FIRE cosmological hydrodynamic simulations. *arXiv* **2023**, arXiv:2310.19863. [CrossRef]
214. Rosario, D.J.; Togi, A.; Burtscher, L.; Davies, R.I.; Shimizu, T.T.; Lutz, D. An Accreting Supermassive Black Hole Irradiating Molecular Gas in NGC 2110. *Astrophys. J. Lett.* **2019**, *875*, L8. [CrossRef]
215. Feruglio, C.; Fabbiano, G.; Bischetti, M.; Elvis, M.; Travascio, A.; Fiore, F. Multiphase Gas Flows in the Nearby Seyfert Galaxy ESO428-G014. Paper I. *Astrophys. J.* **2020**, *890*, 29. [CrossRef]
216. Bessiere, P.S.; Ramos Almeida, C. Spatially resolved evidence of the impact of quasar-driven outflows on recent star formation: The case of Mrk 34. *Mon. Not. R. Astron. Soc. Lett.* **2022**, *512*, L54–L59. [CrossRef]
217. Hervella Seoane, K.; Ramos Almeida, C.; Acosta-Pulido, J.A.; Speranza, G.; Tadhunter, C.N.; Bessiere, P.S. Investigating the impact of quasar-driven outflows on galaxies at z$\sim$0.3–0.4. *Astron. Astrophys.* **2023**, *680*, A71. [CrossRef]
218. Whittle, M. The narrow line region of active galaxies–I. Nuclear [O III] profiles. *Mon. Not. R. Astron. Soc.* **1985**, *213*, 1–31. [CrossRef]
219. Speranza, G.; Balmaverde, B.; Capetti, A.; Massaro, F.; Tremblay, G.; Marconi, A.; Venturi, G.; Chiaberge, M.; Baldi, R.; Baum, S.; et al. The MURALES survey-IV. Searching for nuclear outflows in 3C radio galaxies at z < 0.3 with MUSE observations. *Astron. Astrophys.* **2021**, *653*, A150.

220. Audibert, A.; Combes, F.; García-Burillo, S.; Hunt, L.; Eckart, A.; Aalto, S.; Casasola, V.; Boone, F.; Krips, M.; Viti, S.; et al. ALMA captures feeding and feedback from the active galactic nucleus in NGC 613. *Astron. Astrophys.* **2019**, *632*, A33. [CrossRef]
221. Domínguez-Fernández, A.J.; Alonso-Herrero, A.; García-Burillo, S.; Davies, R.I.; Usero, A.; Labiano, A.; Levenson, N.A.; Pereira-Santaella, M.; Imanishi, M.; Ramos Almeida, C.; et al. Searching for molecular gas inflows and outflows in the nuclear regions of five Seyfert galaxies. *Astron. Astrophys.* **2020**, *643*, A127. [CrossRef]
222. Husemann, B.; Scharwächter, J.; Bennert, V.N.; Mainieri, V.; Woo, J.H.; Kakkad, D. Large-scale outflows in luminous QSOs revisited. The impact of beam smearing on AGN feedback efficiencies. *Astron. Astrophys.* **2016**, *594*, A44. [CrossRef]
223. Lutz, D.; Sturm, E.; Janssen, A.; Veilleux, S.; Aalto, S.; Cicone, C.; Contursi, A.; Davies, R.I.; Feruglio, C.; Fischer, J.; et al. Molecular outflows in local galaxies: Method comparison and a role of intermittent AGN driving. *Astron. Astrophys.* **2020**, *633*, A134. [CrossRef]
224. Bolatto, A.D.; Wolfire, M.; Leroy, A.K. The CO-to-H_2 Conversion Factor. *Annu. Rev. Astron. Astrophys.* **2013**, *51*, 207–268. [CrossRef]
225. Netzer, H. Bolometric correction factors for active galactic nuclei. *Mon. Not. R. Astron. Soc.* **2019**, *488*, 5185–5191. [CrossRef]
226. Merloni, A.; Heinz, S. Measuring the kinetic power of active galactic nuclei in the radio mode. *Mon. Not. R. Astron. Soc.* **2007**, *381*, 589–601. [CrossRef]
227. Bîrzan, L.; McNamara, B.R.; Nulsen, P.E.J.; Carilli, C.L.; Wise, M.W. Radiative Efficiency and Content of Extragalactic Radio Sources: Toward a Universal Scaling Relation between Jet Power and Radio Power. *Astrophys. J.* **2008**, *686*, 859–880. [CrossRef]
228. Cavagnolo, K.W.; McNamara, B.R.; Nulsen, P.E.J.; Carilli, C.L.; Jones, C.; Bîrzan, L. A Relationship Between AGN Jet Power and Radio Power. *Astrophys. J.* **2010**, *720*, 1066–1072. [CrossRef]
229. Kokotanekov, G.; Wise, M.; Heald, G.H.; McKean, J.P.; Bîrzan, L.; Rafferty, D.A.; Godfrey, L.E.H.; de Vries, M.; Intema, H.T.; Broderick, J.W.; et al. LOFAR MSSS: The scaling relation between AGN cavity power and radio luminosity at low radio frequencies. *Astron. Astrophys.* **2017**, *605*, A48. [CrossRef]
230. Croston, J.H.; Ineson, J.; Hardcastle, M.J. Particle content, radio-galaxy morphology, and jet power: All radio-loud AGN are not equal. *Mon. Not. R. Astron. Soc.* **2018**, *476*, 1614–1623. [CrossRef]
231. Di Matteo, T.; Springel, V.; Hernquist, L. Energy input from quasars regulates the growth and activity of black holes and their host galaxies. *Nature* **2005**, *433*, 604–607. [CrossRef]
232. Crain, R.A.; Schaye, J.; Bower, R.G.; Furlong, M.; Schaller, M.; Theuns, T.; Dalla Vecchia, C.; Frenk, C.S.; McCarthy, I.G.; Helly, J.C.; et al. The EAGLE simulations of galaxy formation: Calibration of subgrid physics and model variations. *Mon. Not. R. Astron. Soc.* **2015**, *450*, 1937–1961. [CrossRef]
233. Silk, J.; Begelman, M.C.; Norman, C.; Nusser, A.; Wyse, R.F.G. Which Came First: Supermassive Black Holes or Galaxies? Insights from JWST. *Astrophys. J. Lett.* **2024**, *961*, L39. [CrossRef]
234. Bruzual, G.; Charlot, S. Stellar population synthesis at the resolution of 2003. *Mon. Not. R. Astron. Soc.* **2003**, *344*, 1000–1028. [CrossRef]
235. Peeters, E.; Spoon, H.W.W.; Tielens, A.G.G.M. Polycyclic Aromatic Hydrocarbons as a Tracer of Star Formation? *Astrophys. J.* **2004**, *613*, 986–1003. [CrossRef]
236. Stanley, F.; Harrison, C.M.; Alexander, D.M.; Simpson, J.; Knudsen, K.K.; Mullaney, J.R.; Rosario, D.J.; Scholtz, J. Deep ALMA photometry of distant X-ray AGN: Improvements in star formation rate constraints, and AGN identification. *Mon. Not. R. Astron. Soc.* **2018**, *478*, 3721–3739. [CrossRef]
237. Scholtz, J.; Harrison, C.M.; Rosario, D.J.; Alexander, D.M.; Knudsen, K.K.; Stanley, F.; Chen, C.C.; Kakkad, D.; Mainieri, V.; Mullaney, J. The impact of ionized outflows from z 2.5 quasars is not through instantaneous in situ quenching: The evidence from ALMA and VLT/SINFONI. *Mon. Not. R. Astron. Soc.* **2021**, *505*, 5469–5487. [CrossRef]
238. Lamperti, I.; Harrison, C.M.; Mainieri, V.; Kakkad, D.; Perna, M.; Circosta, C.; Scholtz, J.; Carniani, S.; Cicone, C.; Alexander, D.M.; et al. SUPER. V. ALMA continuum observations of $z \sim 2$ AGN and the elusive evidence of outflows influencing star formation. *Astron. Astrophys.* **2021**, *654*, A90. [CrossRef]
239. Howell, J.H.; Mazzarella, J.M.; Chan, B.H.P.; Lord, S.; Surace, J.A.; Frayer, D.T.; Appleton, P.N.; Armus, L.; Evans, A.S.; Bothun, G.; et al. Tracing Polycyclic Aromatic Hydrocarbons and Warm Dust Emission in the Seyfert Galaxy NGC 1068. *Astron. J.* **2007**, *134*, 2086–2097. [CrossRef]
240. Jensen, J.J.; Hönig, S.F.; Rakshit, S.; Alonso-Herrero, A.; Asmus, D.; Gandhi, P.; Kishimoto, M.; Smette, A.; Tristram, K.R.W. PAH features within few hundred parsecs of active galactic nuclei. *Mon. Not. R. Astron. Soc.* **2017**, *470*, 3071–3094. [CrossRef]
241. Bessiere, P.S.; Tadhunter, C.N.; Ramos Almeida, C.; Villar Martín, M.; Cabrera-Lavers, A. Young stellar populations in type II quasars: Timing the onset of star formation and nuclear activity. *Mon. Not. R. Astron. Soc.* **2017**, *466*, 3887–3917. [CrossRef]
242. Dickson, R.; Tadhunter, C.; Shaw, M.; Clark, N.; Morganti, R. The nebular contribution to the extended UV continua of powerful radio galaxies. *Mon. Not. R. Astron. Soc.* **1995**, *273*, L29–L33. [CrossRef]
243. Zakamska, N.L.; Gómez, L.; Strauss, M.A.; Krolik, J.H. Mid-Infrared Spectra of Optically-Selected Type 2 Quasars. *Astron. J.* **2008**, *136*, 1607–1622. [CrossRef]
244. Deo, R.P.; Richards, G.T.; Crenshaw, D.M.; Kraemer, S.B. The Mid-Infrared Continua of Seyfert Galaxies. *Astrophys. J.* **2009**, *705*, 14–31. [CrossRef]
245. Diamond-Stanic, A.M.; Rieke, G.H. The Relationship between Black Hole Growth and Star Formation in Seyfert Galaxies. *Astrophys. J.* **2012**, *746*, 168. [CrossRef]

246. Sales, D.A.; Pastoriza, M.G.; Riffel, R.; Winge, C. Polycyclic aromatic hydrocarbon in the central region of the Seyfert 2 galaxy NGC 1808. *Mon. Not. R. Astron. Soc.* **2013**, *429*, 2634–2642. [CrossRef]
247. Alonso-Herrero, A.; Ramos Almeida, C.; Esquej, P.; Roche, P.F.; Hernán-Caballero, A.; Hönig, S.F.; González-Martín, O.; Aretxaga, I.; Mason, R.E.; Packham, C.; et al. Nuclear 11.3 μm PAH emission in local active galactic nuclei. *Mon. Not. R. Astron. Soc.* **2014**, *443*, 2766–2782. [CrossRef]
248. Esquej, P.; Alonso-Herrero, A.; González-Martín, O.; Hönig, S.F.; Hernán-Caballero, A.; Roche, P.; Ramos Almeida, C.; Mason, R.E.; Díaz-Santos, T.; Levenson, N.A.; et al. Nuclear Star Formation Activity and Black Hole Accretion in Nearby Seyfert Galaxies. *Astrophys. J.* **2014**, *780*, 86. [CrossRef]
249. Ramos Almeida, C.; Alonso-Herrero, A.; Esquej, P.; González-Martín, O.; Riffel, R.A.; García-Bernete, I.; Rodríguez Espinosa, J.M.; Packham, C.; Levenson, N.A.; Roche, P.; et al. A mid-infrared view of the inner parsecs of the Seyfert galaxy Mrk 1066 using CanariCam/GTC. *Mon. Not. R. Astron. Soc.* **2014**, *445*, 1130–1143. [CrossRef]
250. Esparza-Arredondo, D.; González-Martín, O.; Dultzin, D.; Alonso-Herrero, A.; Ramos Almeida, C.; Díaz-Santos, T.; García-Bernete, I.; Martinez-Paredes, M.; Rodríguez-Espinosa, J.M. Circumnuclear Star Formation and AGN Activity: Clues from Surface Brightness Radial Profile of PAHs and [S IV]. *Astrophys. J.* **2018**, *859*, 124. [CrossRef]
251. Smith, J.D.T.; Draine, B.T.; Dale, D.A.; Moustakas, J.; Kennicutt, R.C.J.; Helou, G.; Armus, L.; Roussel, H.; Sheth, K.; Bendo, G.J.; et al. The Mid-Infrared Spectrum of Star-forming Galaxies: Global Properties of Polycyclic Aromatic Hydrocarbon Emission. *Astrophys. J.* **2007**, *656*, 770–791. [CrossRef]
252. Diamond-Stanic, A.M.; Rieke, G.H. The Effect of Active Galactic Nuclei on the Mid-infrared Aromatic Features. *Astrophys. J.* **2010**, *724*, 140–153. [CrossRef]
253. García-Bernete, I.; Rigopoulou, D.; Alonso-Herrero, A.; Pereira-Santaella, M.; Roche, P.F.; Kerkeni, B. Polycyclic aromatic hydrocarbons in Seyfert and star-forming galaxies. *Mon. Not. R. Astron. Soc.* **2022**, *509*, 4256–4275. [CrossRef]
254. Martínez-Paredes, M.; Aretxaga, I.; González-Martín, O.; Alonso-Herrero, A.; Levenson, N.A.; Ramos Almeida, C.; López-Rodríguez, E. Quantifying Star Formation Activity in the Inner 1 kpc of Local MIR Bright QSOs. *Astrophys. J.* **2019**, *871*, 190. [CrossRef]
255. Xie, Y.; Ho, L.C. The Ionization and Destruction of Polycyclic Aromatic Hydrocarbons in Powerful Quasars. *Astrophys. J.* **2022**, *925*, 218. [CrossRef]
256. Ramos Almeida, C.; Esparza-Arredondo, D.; González-Martín, O.; García-Bernete, I.; Pereira-Santaella, M.; Alonso-Herrero, A.; Acosta-Pulido, J.A.; Bessiere, P.S.; Levenson, N.A.; Tadhunter, C.N.; et al. Absence of nuclear polycyclic aromatic hydrocarbon emission from a compact starburst: The case of the type-2 quasar Mrk 477. *Astron. Astrophys.* **2023**, *669*, L5. [CrossRef]
257. García-Bernete, I.; Rigopoulou, D.; Alonso-Herrero, A.; Donnan, F.R.; Roche, P.F.; Pereira-Santaella, M.; Labiano, A.; Peralta de Arriba, L.; Izumi, T.; Ramos Almeida, C.; et al. A high angular resolution view of the PAH emission in Seyfert galaxies using JWST/MRS data. *Astron. Astrophys.* **2022**, *666*, L5. [CrossRef]
258. Lai, T.S.Y.; Armus, L.; U, V.; Díaz-Santos, T.; Larson, K.L.; Evans, A.; Malkan, M.A.; Appleton, P.; Rich, J.; Müller-Sánchez, F.; et al. GOALS-JWST: Tracing AGN Feedback on the Star-forming Interstellar Medium in NGC 7469. *Astrophys. J. Lett.* **2022**, *941*, L36. [CrossRef]
259. Lai, T.S.Y.; Armus, L.; Bianchin, M.; Díaz-Santos, T.; Linden, S.T.; Privon, G.C.; Inami, H.; U, V.; Bohn, T.; Evans, A.S.; et al. GOALS-JWST: Small Neutral Grains and Enhanced 3.3 μm PAH Emission in the Seyfert Galaxy NGC 7469. *Astrophys. J. Lett.* **2023**, *957*, L26. [CrossRef]
260. Silk, J. Unleashing Positive Feedback: Linking the Rates of Star Formation, Supermassive Black Hole Accretion, and Outflows in Distant Galaxies. *Astrophys. J.* **2013**, *772*, 112. [CrossRef]
261. Zubovas, K.; Bourne, M.A. Do AGN outflows quench or enhance star formation? *Mon. Not. R. Astron. Soc.* **2017**, *468*, 4956–4967. [CrossRef]
262. Raouf, M.; Viti, S.; García-Burillo, S.; Richings, A.J.; Schaye, J.; Bemis, A.; Nobels, F.S.J.; Guainazzi, M.; Huang, K.Y.; Schaller, M.; et al. Hydrodynamic simulations of the disc of gas around supermassive black holes (HDGAS) - I. Molecular gas dynamics. *Mon. Not. R. Astron. Soc.* **2023**, *524*, 786–800. [CrossRef]
263. Cresci, G.; Marconi, A.; Zibetti, S.; Risaliti, G.; Carniani, S.; Mannucci, F.; Gallazzi, A.; Maiolino, R.; Balmaverde, B.; Brusa, M.; et al. The MAGNUM survey: Positive feedback in the nuclear region of NGC 5643 suggested by MUSE. *Astron. Astrophys.* **2015**, *582*, A63. [CrossRef]
264. Salomé, Q.; Salomé, P.; Miville-Deschênes, M.A.; Combes, F.; Hamer, S. Inefficient jet-induced star formation in Centaurus A. High resolution ALMA observations of the northern filaments. *Astron. Astrophys.* **2017**, *608*, A98. [CrossRef]
265. Carniani, S.; Marconi, A.; Maiolino, R.; Balmaverde, B.; Brusa, M.; Cano-Díaz, M.; Cicone, C.; Comastri, A.; Cresci, G.; Fiore, F.; et al. Fast outflows and star formation quenching in quasar host galaxies. *Astron. Astrophys.* **2016**, *591*, A28. [CrossRef]
266. Shin, J.; Woo, J.H.; Chung, A.; Baek, J.; Cho, K.; Kang, D.; Bae, H.J. Positive and Negative Feedback of AGN Outflows in NGC 5728. *Astrophys. J.* **2019**, *881*, 147. [CrossRef]
267. Torrey, P.; Hopkins, P.F.; Faucher-Giguère, C.A.; Anglés-Alcázar, D.; Quataert, E.; Ma, X.; Feldmann, R.; Keres, D.; Murray, N. The impact of AGN wind feedback in simulations of isolated galaxies with a multiphase ISM. *Mon. Not. R. Astron. Soc.* **2020**, *497*, 5292–5308. [CrossRef]

268. Sánchez, S.F.; Avila-Reese, V.; Hernandez-Toledo, H.; Cortes-Suárez, E.; Rodríguez-Puebla, A.; Ibarra-Medel, H.; Cano-Díaz, M.; Barrera-Ballesteros, J.K.; Negrete, C.A.; Calette, A.R.; et al. SDSS IV MaNGA - Properties of AGN Host Galaxies. *Rev. Mex. Astron. Astrofis.* **2018**, *54*, 217–260. [CrossRef]
269. Ellison, S.L.; Wong, T.; Sánchez, S.F.; Colombo, D.; Bolatto, A.; Barrera-Ballesteros, J.; García-Benito, R.; Kalinova, V.; Luo, Y.; Rubio, M.; et al. The EDGE-CALIFA survey: Central molecular gas depletion in AGN host galaxies—A smoking gun for quenching? *Mon. Not. R. Astron. Soc.* **2021**, *505*, L46–L51. [CrossRef]
270. Lammers, C.; Iyer, K.G.; Ibarra-Medel, H.; Pacifici, C.; Sánchez, S.F.; Tacchella, S.; Woo, J. Active Galactic Nuclei Feedback in SDSS-IV MaNGA: AGNs Have Suppressed Central Star Formation Rates. *Astrophys. J.* **2023**, *953*, 26. [CrossRef]
271. García-Burillo, S.; Combes, F.; Ramos Almeida, C.; Usero, A.; Alonso-Herrero, A.; Hunt, L.K.; Rouan, D.; Aalto, S.; Querejeta, M.; Viti, S.; et al. ALMA images the many faces of the NGC 1068 torus and its surroundings. *Astron. Astrophys.* **2019**, *632*, A61. [CrossRef]
272. Ruffa, I.; Prandoni, I.; Davis, T.A.; Laing, R.A.; Paladino, R.; Casasola, V.; Parma, P.; Bureau, M. The AGN fuelling/feedback cycle in nearby radio galaxies—IV. Molecular gas conditions and jet-ISM interaction in NGC 3100. *Mon. Not. R. Astron. Soc.* **2022**, *510*, 4485–4503. [CrossRef]
273. García-Bernete, I.; Alonso-Herrero, A.; García-Burillo, S.; Pereira-Santaella, M.; García-Lorenzo, B.; Carrera, F.J.; Rigopoulou, D.; Ramos Almeida, C.; Villar Martín, M.; González-Martín, O.; et al. Multiphase feedback processes in the Sy2 galaxy NGC 5643. *Astron. Astrophys.* **2021**, *645*, A21. [CrossRef]
274. Tripodi, R.; D'Eugenio, F.; Maiolino, R.; Curti, M.; Scholtz, J.; Tacchella, S.; Bunker, A.J.; Trussler, J.A.A.; Cameron, A.J.; Arribas, S.; et al. Spatially resolved emission lines in galaxies at $4 \leq z < 10$ from the JADES survey: Evidence for enhanced central star formation. *arXiv* **2024**, arXiv:2403.08431. [CrossRef]
275. Übler, H.; D'Eugenio, F.; Perna, M.; Arribas, S.; Jones, G.C.; Bunker, A.J.; Carniani, S.; Charlot, S.; Maiolino, R.; Rodríguez del Pino, B.; et al. GA-NIFS: NIRSpec reveals evidence for non-circular motions and AGN feedback in GN20. *arXiv* **2024**, arXiv:2403.03192. [CrossRef]
276. Boehringer, H.; Voges, W.; Fabian, A.C.; Edge, A.C.; Neumann, D.M. A ROSAT HRI study of the interaction of the X-ray emitting gas and radio lobes of NGC 1275. *Mon. Not. R. Astron. Soc.* **1993**, *264*, L25–L28. [CrossRef]
277. Hlavacek-Larrondo, J.; Li, Y.; Churazov, E. AGN Feedback in Groups and Clusters of Galaxies. In *Handbook of X-ray and Gamma-ray Astrophysics*; Springer: Berlin/Heidelberg, Germany, 2022; p. 5. [CrossRef]
278. Best, P.N.; Kaiser, C.R.; Heckman, T.M.; Kauffmann, G. AGN-controlled cooling in elliptical galaxies. *Mon. Not. R. Astron. Soc.* **2006**, *368*, L67–L71. [CrossRef]
279. Danielson, A.L.R.; Lehmer, B.D.; Alexander, D.M.; Brandt, W.N.; Luo, B.; Miller, N.; Xue, Y.Q.; Stott, J.P. The cosmic history of hot gas cooling and radio active galactic nucleus activity in massive early-type galaxies. *Mon. Not. R. Astron. Soc.* **2012**, *422*, 494–509. [CrossRef]
280. Hlavacek-Larrondo, J.; McDonald, M.; Benson, B.A.; Forman, W.R.; Allen, S.W.; Bleem, L.E.; Ashby, M.L.N.; Bocquet, S.; Brodwin, M.; Dietrich, J.P.; et al. X-ray Cavities in a Sample of 83 SPT-selected Clusters of Galaxies: Tracing the Evolution of AGN Feedback in Clusters of Galaxies out to z = 1.2. *Astrophys. J.* **2015**, *805*, 35. [CrossRef]
281. Smolčić, V.; Novak, M.; Delvecchio, I.; Ceraj, L.; Bondi, M.; Delhaize, J.; Marchesi, S.; Murphy, E.; Schinnerer, E.; Vardoulaki, E.; et al. The VLA-COSMOS 3 GHz Large Project: Cosmic evolution of radio AGN and implications for radio-mode feedback since z 5. *Astron. Astrophys.* **2017**, *602*, A6. [CrossRef]
282. McDonald, M.; Gaspari, M.; McNamara, B.R.; Tremblay, G.R. Revisiting the Cooling Flow Problem in Galaxies, Groups, and Clusters of Galaxies. *Astrophys. J.* **2018**, *858*, 45. [CrossRef]
283. Hardcastle, M.J.; Williams, W.L.; Best, P.N.; Croston, J.H.; Duncan, K.J.; Röttgering, H.J.A.; Sabater, J.; Shimwell, T.W.; Tasse, C.; Callingham, J.R.; et al. Radio-loud AGN in the first LoTSS data release. The lifetimes and environmental impact of jet-driven sources. *Astron. Astrophys.* **2019**, *622*, A12. [CrossRef]
284. Bîrzan, L.; Rafferty, D.A.; Brüggen, M.; Intema, H.T. A study of high-redshift AGN feedback in SZ cluster samples. *Mon. Not. R. Astron. Soc.* **2017**, *471*, 1766–1787. [CrossRef]
285. Werner, N.; McNamara, B.R.; Churazov, E.; Scannapieco, E. Hot Atmospheres, Cold Gas, AGN Feedback and the Evolution of Early Type Galaxies: A Topical Perspective. *Space Sci. Rev.* **2019**, *215*, 5. [CrossRef]
286. McCarthy, I.G.; Schaye, J.; Bower, R.G.; Ponman, T.J.; Booth, C.M.; Dalla Vecchia, C.; Springel, V. Gas expulsion by quasar-driven winds as a solution to the overcooling problem in galaxy groups and clusters. *Mon. Not. R. Astron. Soc.* **2011**, *412*, 1965–1984. [CrossRef]
287. Zinger, E.; Pillepich, A.; Nelson, D.; Weinberger, R.; Pakmor, R.; Springel, V.; Hernquist, L.; Marinacci, F.; Vogelsberger, M. Ejective and preventative: The IllustrisTNG black hole feedback and its effects on the thermodynamics of the gas within and around galaxies. *Mon. Not. R. Astron. Soc.* **2020**, *499*, 768–792. [CrossRef]
288. Ward, S.R.; Harrison, C.M.; Costa, T.; Mainieri, V. Cosmological simulations predict that AGN preferentially live in gas-rich, star-forming galaxies despite effective feedback. *Mon. Not. R. Astron. Soc.* **2022**, *514*, 2936–2957. [CrossRef]
289. Kirkpatrick, A.; Sharon, C.; Keller, E.; Pope, A. CO Emission in Infrared-selected Active Galactic Nuclei. *Astrophys. J.* **2019**, *879*, 41. [CrossRef]

290. Circosta, C.; Mainieri, V.; Lamperti, I.; Padovani, P.; Bischetti, M.; Harrison, C.M.; Kakkad, D.; Zanella, A.; Vietri, G.; Lanzuisi, G.; et al. SUPER. IV. CO(J = 3-2) properties of active galactic nucleus hosts at cosmic noon revealed by ALMA. *Astron. Astrophys.* **2021**, *646*, A96. [CrossRef]
291. Bischetti, M.; Feruglio, C.; Piconcelli, E.; Duras, F.; Pérez-Torres, M.; Herrero, R.; Venturi, G.; Carniani, S.; Bruni, G.; Gavignaud, I.; et al. The WISSH quasars project—IX. Cold gas content and environment of luminous QSOs at z 4–4.7. *Astron. Astrophys.* **2021**, *645*, A33. [CrossRef]
292. Molyneux, S.J.; Calistro Rivera, G.; De Breuck, C.; Harrison, C.M.; Mainieri, V.; Lundgren, A.; Kakkad, D.; Circosta, C.; Girdhar, A.; Costa, T.; et al. The Quasar Feedback Survey: Characterizing CO excitation in quasar host galaxies. *Mon. Not. R. Astron. Soc.* **2024**, *527*, 4420–4439. [CrossRef]
293. Rosario, D.J.; Santini, P.; Lutz, D.; Netzer, H.; Bauer, F.E.; Berta, S.; Magnelli, B.; Popesso, P.; Alexander, D.M.; Brandt, W.N.; et al. Nuclear activity is more prevalent in star-forming galaxies. *Astrophys. J.* **2013**, *771*, 63. [CrossRef]
294. Stanley, F.; Harrison, C.M.; Alexander, D.M.; Swinbank, A.M.; Aird, J.A.; Moro, A.D.; Hickox, R.C.; Mullaney, J.R. A remarkably flat relationship between the average star formation rate and AGN luminosity for distant X-ray AGN. *Mon. Not. R. Astron. Soc.* **2015**, *453*, 591–604. [CrossRef]
295. Balmaverde, B.; Marconi, A.; Brusa, M.; Carniani, S.; Cresci, G.; Lusso, E.; Maiolino, R.; Mannucci, F.; Nagao, T. Is there any evidence that ionized outflows quench star formation in type 1 quasars at z < 1? *Astron. Astrophys.* **2016**, *585*, A148. [CrossRef]
296. Woo, J.H.; Son, D.; Bae, H.J. Delayed or No Feedback? Gas Outflows in Type 2 AGNs. III. *Astrophys. J.* **2017**, *839*, 120. [CrossRef]
297. Stanley, F.; Alexander, D.M.; Harrison, C.M.; Rosario, D.J.; Wang, L.; Aird, J.A.; Bourne, N.; Dunne, L.; Dye, S.; Eales, S.; et al. The mean star formation rates of unobscured QSOs: Searching for evidence of suppressed or enhanced star formation. *Mon. Not. R. Astron. Soc.* **2017**, *472*, 2221–2240. [CrossRef]
298. Wylezalek, D.; Morganti, R. Questions and challenges of what powers galactic outflows in active galactic nuclei. *Nat. Astron.* **2018**, *2*, 181–182. [CrossRef]
299. Shangguan, J.; Ho, L.C. Testing the Evolutionary Link between Type 1 and Type 2 Quasars with Measurements of the Interstellar Medium. *Astrophys. J.* **2019**, *873*, 90. [CrossRef]
300. Zhuang, M.Y.; Ho, L.C.; Shangguan, J. Black Hole Accretion Correlates with Star Formation Rate and Star Formation Efficiency in Nearby Luminous Type 1 Active Galaxies. *Astrophys. J.* **2021**, *906*, 38. [CrossRef]
301. Kim, C.; Woo, J.H.; Jadhav, Y.; Chung, A.; Baek, J.; Lee, J.A.; Shin, J.; Hwang, H.S.; Luo, R.; Son, D.; et al. Determining Star Formation Rates of Active Galactic Nucleus Host Galaxies Based on SED Fitting with Submillimeter Data. *Astrophys. J.* **2022**, *928*, 73. [CrossRef]
302. Perna, M.; Sargent, M.T.; Brusa, M.; Daddi, E.; Feruglio, C.; Cresci, G.; Lanzuisi, G.; Lusso, E.; Comastri, A.; Coogan, R.T.; et al. Molecular gas content in obscured AGN at z > 1. *Astron. Astrophys.* **2018**, *619*, A90. [CrossRef]
303. Aird, J.; Coil, A.L.; Georgakakis, A. X-rays across the galaxy population—III. The incidence of AGN as a function of star formation rate. *Mon. Not. R. Astron. Soc.* **2019**, *484*, 4360–4378. [CrossRef]
304. Grimmett, L.P.; Mullaney, J.R.; Bernhard, E.P.; Harrison, C.M.; Alexander, D.M.; Stanley, F.; Masoura, V.A.; Walters, K. A binning-free method reveals a continuous relationship between galaxies' AGN power and offset from main sequence. *Mon. Not. R. Astron. Soc.* **2020**, *495*, 1392–1402. [CrossRef]
305. Miraghaei, H.; Best, P.N. The nuclear properties and extended morphologies of powerful radio galaxies: The roles of host galaxy and environment. *Mon. Not. R. Astron. Soc.* **2017**, *466*, 4346–4363. [CrossRef]
306. Piotrowska, J.M.; Bluck, A.F.L.; Maiolino, R.; Peng, Y. On the quenching of star formation in observed and simulated central galaxies: Evidence for the role of integrated AGN feedback. *Mon. Not. R. Astron. Soc.* **2022**, *512*, 1052–1090. [CrossRef]
307. Bluck, A.F.L.; Piotrowska, J.M.; Maiolino, R. The Fundamental Signature of Star Formation Quenching from AGN Feedback: A Critical Dependence of Quiescence on Supermassive Black Hole Mass, Not Accretion Rate. *Astrophys. J.* **2023**, *944*, 108. [CrossRef]
308. Bluck, A.F.L.; Conselice, C.J.; Ormerod, K.; Piotrowska, J.M.; Adams, N.; Austin, D.; Caruana, J.; Duncan, K.J.; Ferreira, L.; Goubert, P.; et al. Galaxy quenching at the high redshift frontier: A fundamental test of cosmological models in the early universe with JWST-CEERS. *arXiv* **2023**, arXiv:2311.02526. [CrossRef]
309. Martín-Navarro, I.; Brodie, J.P.; Romanowsky, A.J.; Ruiz-Lara, T.; van de Ven, G. Black-hole-regulated star formation in massive galaxies. *Nature* **2018**, *553*, 307–309. [CrossRef]
310. Martín-Navarro, I.; Pillepich, A.; Nelson, D.; Rodriguez-Gomez, V.; Donnari, M.; Hernquist, L.; Springel, V. Anisotropic satellite galaxy quenching modulated by black hole activity. *Nature* **2021**, *594*, 187–190. [CrossRef]
311. Terrazas, B.A.; Bell, E.F.; Henriques, B.M.B.; White, S.D.M.; Cattaneo, A.; Woo, J. Quiescence Correlates Strongly with Directly Measured Black Hole Mass in Central Galaxies. *Astrophys. J. Lett.* **2016**, *830*, L12. [CrossRef]
312. Voit, G.M.; Oppenheimer, B.D.; Bell, E.F.; Terrazas, B.; Donahue, M. Black Hole Growth, Baryon Lifting, Star Formation, and IllustrisTNG. *Astrophys. J.* **2024**, *960*, 28. [CrossRef]
313. Chen, Z.; Faber, S.M.; Koo, D.C.; Somerville, R.S.; Primack, J.R.; Dekel, A.; Rodríguez-Puebla, A.; Guo, Y.; Barro, G.; Kocevski, D.D.; et al. Quenching as a Contest between Galaxy Halos and Their Central Black Holes. *Astrophys. J.* **2020**, *897*, 102. [CrossRef]
314. Habouzit, M.; Li, Y.; Somerville, R.S.; Genel, S.; Pillepich, A.; Volonteri, M.; Davé, R.; Rosas-Guevara, Y.; McAlpine, S.; Peirani, S.; et al. Supermassive black holes in cosmological simulations I: M_{BH} - $M_{\star}$ relation and black hole mass function. *Mon. Not. R. Astron. Soc.* **2021**, *503*, 1940–1975. [CrossRef]

315. Kondapally, R.; Best, P.N.; Raouf, M.; Thomas, N.L.; Davé, R.; Shabala, S.S.; Röttgering, H.J.A.; Hardcastle, M.J.; Bonato, M.; Cochrane, R.K.; et al. Cosmic evolution of radio-AGN feedback: Confronting models with data. *Mon. Not. R. Astron. Soc.* **2023**, *523*, 5292–5305. [CrossRef]
316. Bluck, A.F.L.; Maiolino, R.; Piotrowska, J.M.; Trussler, J.; Ellison, S.L.; Sánchez, S.F.; Thorp, M.D.; Teimoorinia, H.; Moreno, J.; Conselice, C.J. How do central and satellite galaxies quench?—Insights from spatially resolved spectroscopy in the MaNGA survey. *Mon. Not. R. Astron. Soc.* **2020**, *499*, 230–268. [CrossRef]
317. Khaire, V.; Hu, T.; Hennawi, J.F.; Burchett, J.N.; Walther, M.; Davies, F. Searching for the Imprints of AGN Feedback on the Lyman Alpha Forest Around Luminous Red Galaxies. *arXiv* **2023**, arXiv:2311.08470. [CrossRef]
318. Wright, R.J.; Somerville, R.S.; Lagos, C.d.P.; Schaller, M.; Davé, R.; Anglés-Alcázar, D.; Genel, S. The baryon cycle in modern cosmological hydrodynamical simulations. *arXiv* **2024**, arXiv:2402.08408. [CrossRef]
319. Davis, F.; Kaviraj, S.; Hardcastle, M.J.; Martin, G.; Jackson, R.A.; Kraljic, K.; Malek, K.; Peirani, S.; Smith, D.J.B.; Volonteri, M.; et al. Radio AGN in nearby dwarf galaxies: The important role of AGN in dwarf galaxy evolution. *Mon. Not. R. Astron. Soc.* **2022**, *511*, 4109–4122. [CrossRef]
320. Bohn, T.; Canalizo, G.; Veilleux, S.; Liu, W. Near-infrared Coronal Line Observations of Dwarf Galaxies Hosting AGN-driven Outflows. *Astrophys. J.* **2021**, *911*, 70. [CrossRef]
321. Koudmani, S.; Sijacki, D.; Smith, M.C. Two can play at that game: Constraining the role of supernova and AGN feedback in dwarf galaxies with cosmological zoom-in simulations. *Mon. Not. R. Astron. Soc.* **2022**, *516*, 2112–2141. [CrossRef]
322. Arjona-Galvez, E.; Di Cintio, A.; Grand, R.J.J. The role of AGN feedback on the evolution of dwarf galaxies from cosmological simulations: SMBHs suppress star formation in low-mass galaxies. *arXiv* **2024**, arXiv:2402.00929. [CrossRef]
323. Carniani, S.; Venturi, G.; Parlanti, E.; de Graaff, A.; Maiolino, R.; Arribas, S.; Bonaventura, N.; Boyett, K.; Bunker, A.J.; Cameron, A.J.; et al. JADES: The incidence rate and properties of galactic outflows in low-mass galaxies across $3 < z < 9$. *arXiv* **2023**, arXiv:2306.11801. [CrossRef]
324. Dutta, R.; Sharma, P.; Sarkar, K.C.; Stone, J.M. Dissipation of AGN jets in a clumpy interstellar medium. *arXiv* **2023**, arXiv:2401.00446. [CrossRef]

galaxies

MDPI

Review

AGN Feedback Signatures in UV Emission

K. Rubinur

Institute of Theoretical Astrophysics, University of Oslo, P.O. Box 1029, Blindern, 0315 Oslo, Norway; rubinur.khatun@astro.uio.no

Abstract: Supermassive black holes (SMBH) are believed to influence galaxy evolution through AGN (active galactic nuclei) feedback. Galaxy mergers are key processes of galaxy formation that lead to AGN activity and star formation. The relative contribution of AGN feedback and mergers to star formation is not yet well understood. In radio-loud objects, AGN outflows are dominated by large jets. However, in radio-quiet objects, outflows are more complex and involve jet, wind, and radiation. In this review, we discuss the signatures of AGN feedback through the alignment of radio and UV emissions. Current research on AGN feedback is discussed, along with a few examples of studies such as the galaxy merger system MRK 212, the radio-quiet AGN NGC 2639, and the radio-loud system Centaurus A. Multi-frequency observations of MRK 212 indicate the presence of dual AGN, as well as feedback-induced star-forming UV clumps. The fourth episode of AGN activity was detected in radio observations of the Seyfert galaxy NGC 2639, which also showed a central cavity of 6 kpc radius in CO and UV maps. This indicates that multi-epoch jets of radio-quiet AGN can blow out cold molecular gas, which can further reduce star formation in the center of the galaxies. Recent UV observations of Cen A have revealed two sets of stellar population in the northern star-forming region, which may have two different origins. Recent studies have shown that there is evidence that both positive and negative feedback can be present in galaxies at different scales and times. High-resolution, multi-band observations of large samples of different types of AGN and their host galaxies are important for understanding the two types of AGN feedback and their effect on the host galaxies. Future instruments like INSIST and UVEX will be able to help achieve some of these goals.

Keywords: galaxies; active; AGN feedback; Seyfert; galaxy merger; UV continuum; radio continuum emission

1. Introduction

From an observational perspective, two important findings in extra-galactic astronomy are that (i) the star-formation density and AGN activity peak around redshift 2–3, which is known as the cosmic noon [1]; (ii) the mass of the central super-massive black holes correlates with galaxy properties like the SMBH mass-bulge velocity dispersion (M-$\sigma_\star$) relation [2–4]. We still do not fully understand the reason for these correlations. AGN are believed to expel energy through jets, wind, or radiation [5]. These processes can blow out star-forming material and hence decrease the star-formation rate (SFR) in the galaxy, known as negative feedback. On the other hand, these outflows can also compress the star-forming material, making this an ideal scenario for star-formation causing positive feedback. Hence, to understand galaxy evolution, estimation of the effect of AGN feedback is important [6–9].

Most galaxy formation models require negative feedback to produce the present day's quenched massive galaxies. This ejects interstellar medium (ISM) gas and prevents the circumgalactic material gas from cooling and becoming accreted in the galaxies. This process helps to regulate the size and density of galaxies and produces the observed bimodality in galaxy colors. In the past two decades, theoretical models and simulations have included AGN feedback to improve galaxy formation and evolution [10–15]. The constraints provided by observations are important to improve theoretical models. The observations of negative AGN feedback come from X-ray and radio studies of galaxy clusters. X-ray cavities and hot shock gas fronts are detected in these clusters, which coincide with the radio lobes. These studies suggest that a radio jet powers the surrounding medium and hence suppresses the gas cooling [16,17]. These studies suggest that even radio jets in weak AGN can be mechanically powerful. Hence, the effect of such radio jets can also be important in isolated galaxies [18,19]. However, in radio-quiet AGN (which is the majority of the galaxy population), the origin of radio emission is still not settled. Hence, the AGN outflows (jet, wind, radiation) are complex and the relative contribution from each component is still not understood [20–23].

Along with negative feedback, in the past decade, there have been detections of positive feedback in galaxies with evidence of jet-induced star formation [24,25]. Earlier studies of AGN feedback signatures were mostly based on ionized gas, which is mostly observed as long-slit spectroscopy (emission or absorption) [26]. There has been growing evidence of AGN feedback signatures (both positive and negative) in galaxies [27–33].

A few well-studied AGN outflows in radio-quiet AGN include MRK 231 [34], IC 5063 [35], NGC 1068 [32], etc. MRK 231 is the nearest quasar and shows wind-driven outflows. It has nuclear ultra-fast outflows (UFO) detected in X-ray, sub-kpc HII outflows, as well as kpc-scale molecular outflows [27,36–38]. These outflows show a wide angle and the radio jet in MRK 231 does not seem to drive the outflows [9]. Disturbed molecular gas and high excitation compared to the disk were detected along the radio jet in IC 5063 [39]. The spatial coincidence of the radio jet and outflow supports the radio jet being the dominant mechanism [40,41]. The galaxy merger, IRAS F23128-5919 hosts an obscured AGN. A multi-wavelength diagnostic showed star formation within the outflowing material [42]. Gallagher et al. [43] compiled a sample from MaNGA data and identified outflows in a sub-sample of 37 galaxies where 1/3rd of those showed signatures of star formation within the outflowing gas. The SFR ranged from 0.1 to 1 $M_{\odot}$ yr $^{-1}$, which could contribute to 5–30% of the total SFR in the galaxy. Recent observations, as well as simulations, suggest that both negative and positive feedback can work together and can be observed in some galaxies [44,45].

The alignment of non-thermal radio emission and UV emission also suggests jet-induced star formation [46,47]. UV emission arises from massive O-, B-, and A-type stars and hence traces star-formation within the age range of 10^6 to 10^8 yr [48,49]. Recently, Tamhane et al. [45] explored jet-ISM interaction in the central elliptical galaxy of the Abell 1975 cluster using multi-frequency observations. They concluded that radio jets that heat the gas on a large scale and over a large timescale can also trigger star formation through positive feedback over a shorter time scale. Duggal et al. [50] studied a sample of seven powerful compact steep spectrum (CSS) sources that have sub-galactic scale jets with a projected size of 1–20 kpc. UV continuum emission was detected in six sources. The UV emission showed alignment with the radio jets in five of those sources.

In this review, we discuss a few recent studies of possible AGN feedback signatures found in UV bands in different types of galaxy systems. These studies are mostly from UVIT [51] and GALEX [52]. UVIT is one of the five payloads onboard India's first astronomical satellite AstroSat [51]. The expected spatial resolution of UVIT is $\sim$1.2$''$ to $\sim$1.5$''$, while GALEX has 5$''$. Both telescopes have FUV and NUV filters.

2. Galaxy Merger and Feedback

It is believed that blue star-forming galaxies move to red quiescent galaxies in the galaxy main sequence [53–56]. There are several processes through which the star-formation quenching happens, like AGN feedback [5,10,57–59], galaxy mergers [60–62], and environment effects such as ram-pressure stripping [63]. Our current understanding of galaxy formation is that galaxies grow through mergers [61,64]. The non-axisymmetric gravitational forces change the potential of the galaxies [65]. Simulations have found that mergers lead to gas inflow towards the nuclei as well as on the disk, hence increasing AGN activity and star formation [66]. Several studies have tried to understand the effect of mergers on star formation and AGN activity, as well as on galaxy evolution [67–69]. However, this is a complex process to understand. (i) Merger-induced star-burst and AGN activity consume gas, (ii) they also blow out the gas through negative feedback, and (iii) AGN contributes to star-formation (positive feedback) during the merger in some galaxies like ultra-luminous infrared galaxies (ULIRGs) (e.g., [70]). So, galaxy mergers and AGN may work together [71,72]. However, the contribution of each mechanism is not well understood.

A sample of dual-nuclei galaxies was investigated with UVIT [73,74]. The motivation for this work was to understand star formation, as well as to look for any signatures of feedback. This pilot study had a sample of 10 galaxies where the nuclei are separate $\leq$11 kpc. It included four galaxies with confirmed AGN and the remaining six galaxies had star-forming nuclei. However, the nature of the star-forming nuclei came mostly from optical spectroscopy and a Baldwin, Phillips, and Terlevich (BPT) diagram [75]. It was recently found that many of the star-forming galaxies from the BPT diagram show AGN nuclei in X-ray or IR analysis [76,77]. Here, an investigation of one of the sample galaxies, MRK 212 with multi-wavelength data revealed interesting results on the nature of nuclei and star formation [73].

Galaxy Merger System: MRK 212

MRK 212 (Figure 1A) is a pinwheel galaxy (UGC 07593 NED02: S1) in an ongoing merger system with its companion (UGC 07593 NED01: S2) at redshift 0.022893. The SDSS image shows that the nuclei projected separation is 11.8 arcsec ($\sim$5.6 kpc). Both nuclei show radio emission in the 1.4 GHz Faint Images of the Radio Sky at Twenty cm image (http://www.cv.nrao.edu/first/), making it a good candidate for being DAGN (dual AGN). DAGN are close pairs of AGN, which is also an expected output of galaxy mergers where both the parent galaxies host central SMBH [78–83]. The coalescence of the pair of SMBH will generate gravitational waves that are of great interest for future astronomy [84]. On the other hand, the pair of AGN at different separations can help us to understand the evolution of SMBH, as well as the galaxy [83]. This requires very high-resolution multi-waveband observations to detect DAGN and there are still only a handful of detections. However, the number has started to increase in the past few years [80–82,85–88]. For MRK 212, the SDSS has only one spectrum assigned to S2 that is star-forming in nature.

The 8.5 GHz image (Figure 1B; [73]) revealed two radio structures at a separation of $\sim$6 kpc, where S1 shows an unresolved structure with some extension in the north–south direction and S2 shows a double radio structure (A, B) of size $\sim$2 Kpc, which is $1''$ offset from the optical nuclei (C). The Ku-band 15 GHz data showed a resolved extended structure associated with S1, while a compact core was associated with S2 (C) (Figure 1B). The 1.4 GHz upgraded Giant Metrewave Radio Telescope (uGMRT) (http://www.gmrt.ncra.tifr.res.in/) image shows two unresolved structures associated with S1 and S2. The optical spectra of both nuclei were obtained from the Himalayan Chandra Telescope (HCT; [89]). A deep 15 ksec UVIT image detected several UV-bright clumps (Figure 1C; [73]).

The double-lobe structure of S1 of size 750 pc at 15 GHz and a spectral index value of -0.81 ± 0.10 resembles a compact symmetric object [90]. However, the 1.4 GHz radio power ($\sim10^{21}$ W Hz^{-1}) is lower than the compact spectrum object (CSO: [90]) and overlaps with the low-power AGN. The total radio power was compared with the average power of an individual supernova (2×10^{20} W Hz^{-1}) [91]. While a number of SNe could produce the observed radio luminosity, it is unrealistic to show a double structure around the nuclei. Rubinur et al. [73] concluded that the radio source associated with S1 is a CSO-like weak AGN. The compact core detected at the centre of S2 has a radio luminosity of $\sim10^{21}$ W Hz^{-1}. This core is not detected at lower frequency images, and the detection at the highest observed frequency suggests that an inverted spectrum supports the synchrotron self-absorbed emission from the base of an unresolved jet. The HCT spectra put S1 and S2 in the composite region of the BPT diagram. From the fitted SDSS spectrum of S2, Hernández-Ibarra et al. [92] also classified it as an AGN. Based on the detection of the compact radio core, as well as optical spectra, Rubinur et al. [73] concluded that S2 is a weak AGN, hence confirming the dual AGN in MRK 212.

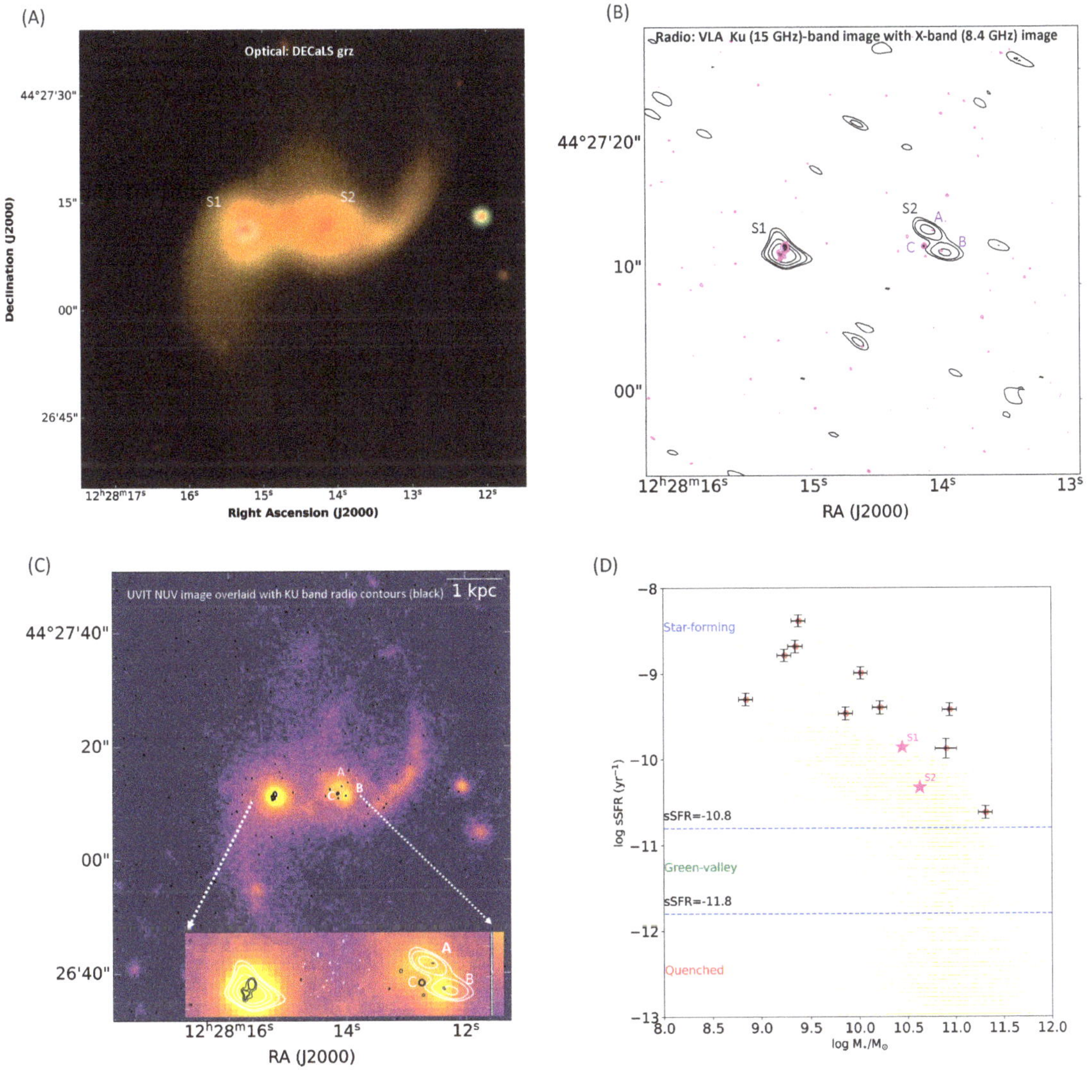

Figure 1. Multi-band images of MRK 212. Note that images are not at the same scale. (**A**) Optical color composite DECalS image. The parent galaxies are defined as S1 and S2. The separation is ~5.6 kpc. (**B**) VLA X band contours overlaid with Ku-band (15 GHz) image and contours. The X-band contour levels are 20, 40, 60, and 80% of the peak flux density. Both sources S1 and S2 are detected. The extended structure is detected at 8.5 GHz (**A**,**B**), which is situated 1$''$ (~500 pc) away from the optical center of S2. (**C**) The UVIT NUV image is overlaid with Ku-band 15 GHz radio contours (black). The inset shows a zoomed version of the radio structures. The NUV image revealed two clumps near the center of S2, which coincide with A and B in the radio 8.5 GHz image (white contours). The compact core (**C**) detected in the 15 GHz image associated with S2 coincides with the optical center. (**D**) The sSFR-M$_\star$ plot. Black points are all 10 dual-nuclei UVIT sample galaxies where MRK 212 (S1, S2) are shown with magenta star symbols. The data were adapted from [73,74] and images were reproduced.

An extended radio structure is detected in the 8.5 GHz VLA image (components A, B in Figure 1B,C). The spectral indices for components A and B are $\alpha = -0.53 \pm 0.10$

and -0.34 ± 0.10, respectively [73]. The relatively flat spectral indices are consistent with the emission from SF regions where free–free emission could be mixed with synchrotron emission from supernovae. The UVIT NUV image shows UV bright knots (Figure 1C). These two clumps are situated $1''$ away from S2 and coincide with the radio-extended structure (A, B) at 8.5 GHz. The calculated SFR is higher compared to other selected regions on the disk or spiral arms. The 1.4 GHz uGMRT flux densities of A and B are 0.53 mJy and 0.37 mJy, respectively. As these clumps (A, B) are situated near the AGN nuclei S2 and on both sides, the star-formation in these clumps could have been triggered by the jet of S2 (if present) through AGN feedback. At the same time, these could be merger-induced star formation. A more sensitive high-resolution radio study could reveal a connection, if present, between the structures detected at 15 GHz and 8.4 GHz, and confirm or rule out the feedback scenario. The overall extinction corrected specific SFR (sSFR) and stellar mass ($M_\star$) show that both S1 and S2 are actively star-forming systems in the galaxy main-sequence plot (purple stars: Figure 1D). This overlaps with the results found for the rest of the nine dual-nuclei galaxies of the sample (black points: Figure 1D) [74].

3. AGN Feedback in Radio-Quiet AGN

The majority of galaxies host radio-quiet AGN [93–97]. There are several questions related to radio-quiet sources that are still debatable, like the dominant process that produces radio emission, the mechanism through which they transfer energy, and their impact on the host galaxies [98,99].

Seyfert galaxies host radio-quiet AGN. Several Seyfert galaxies show kilo-parsec radio structures, which are sometimes distorted from the parsec scale structure [97]. A few Seyfert galaxies also show episodic AGN activity. Mrk6 exhibits a highly complex morphology: two pairs of self-similar bipolar bubble-like radio structures that are oriented nearly perpendicular to each other [100]. Another such example is NGC2992 [101]. However, most of these studies were performed with the GHz band observation, which can miss the total extent of radio emission as well as some emissions from the old epoch. Recently, a sample of Seyfert was observed with GMRT at 325 and 735 MHz to study the same (Rubinur et al. in preparation). One of the sample galaxies, NGC 2639, was studied with radio, UV, and CO data by [102].

3.1. *Episodic AGN activity and feedback: NGC 2639*

This is a Seyfert 2 type galaxy with a LINER (low-ionization nuclear emission-line region) like spectrum at a luminosity distance of 48.2 Mpc (redshift 0.01113) [103]. This galaxy hosts a SMBH of mass $M_{BH} = 1.48 \times 10^8\ M_\odot$ [104] with an Eddington ratio 4×10^{-4} [102].

The uGMRT 735 MHz observation of NGC 2639 revealed a jet of size ∼9 kpc (white contours in Figure 2A) in the NE–SW direction [102], while the 325 MHz image showed an unresolved structure in the same direction (Rubinur et al., in preparation). Rao et al. [102] proposed that the uGMRT detected jet (∼9 kpc) was the fourth episode of AGN activity, as NGC 2639 is known to have hosted three episodes of AGN activity before [105]. These include a (i) 1.5 Kpc jet in the north–south direction (Black contours in Figure 2A); (ii) 360 pc jet in the east–west direction detected at 5 GHz VLA map (Figure 2B); and (iii) the 3pc jet detected with VLBA in the NE–SW direction (Figure 2C). These four jets do not show any connecting emissions and have clear hotspots or edges that distinguish them from each other.

The spectral age derived from BRATS (https://www.askanastronomer.co.uk/brats/) (Broadband Radio Astronomy Tools) found the age of the jet episodes with sizes of 9 kpc, 1.5 kpc, and 360 pc to be 34^{+4}_{-6} Myr, $11.8^{+1.7}_{-1.4}$ Myr, and $2.8^{+0.7}_{-0.5}$ Myr. These spectral ages proved the AGN jet duty cycle to be $\sim 60\%$. Sanders [106] proposed that Seyfert galaxies can have 100s of episodic AGN activity with a timescale 10^4–10^5 yr during their statistical lifetime of 3–7×10^8 yr. However, as mentioned earlier, episodic AGN activities are detected only in a few Seyfert galaxies, and the fourth episode in NGC 2639 is the highest number of detected epochs. This high-sensitivity uGMRT MHz observation shows that the detection of episodic activity might be a matter of detecting low surface brightness emission from Seyfert/LINERs. While looking for the origin of these multiple jet episodes using the minor merger rate from the stellar mass and an assumed mass ratio, Rao et al. [102] found that the timescale favors three minor mergers that may have resulted in a new accretion disk and hence outflows.

The CARMA CO(1-0) (https://mmwave.astro.illinois.edu/edgedata/), GALEX FUV, and NUV images show a deficiency of molecular gas and UV emission in the central ~6 kpc of the galaxy, which is surrounded by a brighter ring (Figure 3). Using the molecular gas data from the CALIFA (Calar Alto Legacy Integral Field Area) (https://califa.caha.es/) survey, Ellison et al. [107] found that the central region of NGC 2639 has a lower gas fraction than the star-forming regions. This suggests that AGN has blown out the central molecular gas which leads to low star formation, as seen by GALEX data. Hence, evidence for a star-formation quenching phase is found in the central region of NGC 2639. Rao et al. [102] calculated the amount of work (pressure time volume PV) required to push the CO gas ring and the time-averaged power from 5 GHz flux densities. It turned out that only 0.5% of east–west jet power or a small fraction from the north–south jet would have been sufficient to push back the CO gas and create the hole.

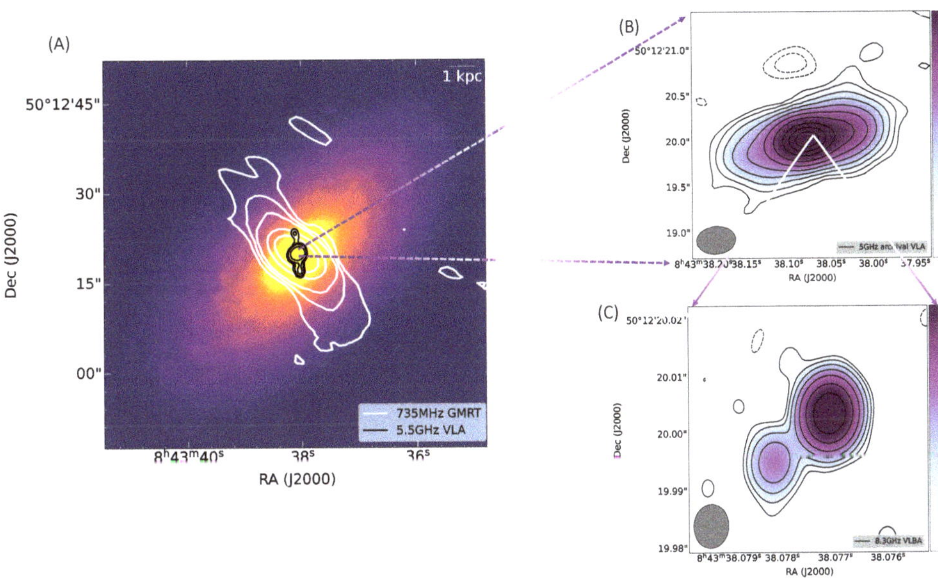

Figure 2. The four radio jet episodes of NGC2639. (**A**) The radio multi-band contours overlaid with the optical image. The color image is the SDSS r-band image. The white contours show the 735 MHz upgraded Giant Meterwave Telescope (uGMRT), which showed a jet of size ~9 kpc in the north–east–south–west direction. The beam of this image is 5.48 arcsec × 3.0 arcsec at PA = 54.6 degrees. Contour levels: $(1, 2, 4, 32) \times 0.6$ mJy beam^{-1}. The black contours are from the 5.5 GHz VLA total intensity image with the ~1.5 kpc jet in the north–south direction. The contour levels are $(2, 4, 8, 64) \times 0.03$ mJy beam^{-1} and the beam size is 1.02 arcsec × 0.89 arcsec at PA = −5.8 degrees. (**B**) The 5 GHz VLA radio image with beam size 0.43 arcsec × 0.30 arcsec at PA = −85.4 degrees. This image shows ~ 360 parsec lobe in the east-west direction. Contour levels: $(-2, -1, 1, 2, 4, 8, 16, 32, 64, 128) \times 0.164$ mJy beam^{-1}. (**C**) The 8.3 GHz VLBA image with a beam size of 7.7 mas × 6.2 mas at PA = −4.9 degrees. It shows ~3 parsec jet at PA = 130 degrees. Contour levels: $(-2, -1, 1, 2, 4, 8, 16, 32, 64) \times 0.239$ mJy beam^{-1}. Data are adapted and figures are reproduced from Rao et al. [102], Sebastian et al. [105].

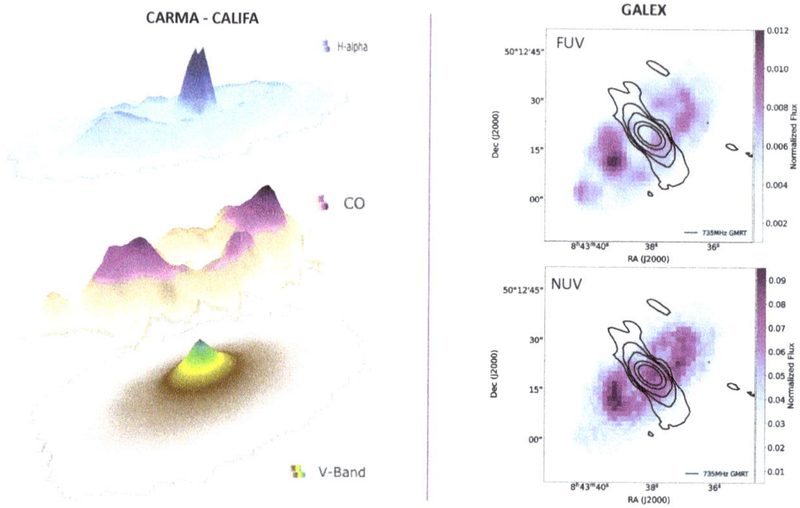

Figure 3. Multi-band images of NGC 2639 (**left**). This is a set of three-band images taken from the CARMA EDGE page for visualization: (**top**) the CALIFA Hα image; (**middle**) the CO moment-0 map (**bottom**) V-band image; the CO(1-0) moment 0 map shows the deficiency at the central region of radius ∼6 kpc. A hint of the same is seen in the Hα image. (**right**) The GALEX FUV and NUV images overlaid with the uGMRT 735 MHz radio contours in black. The contour labels are the same as in Figure 2A. The central ∼6 kpc shows a relative deficiency of FUV and NUV similar to the CO(1-0) map [102].

In NGC 7252 and the jellyfish galaxy JO201, AGN activities were found to suppress the star formation in the central region [108,109]. Both positive and negative feedback were observed in NGC 5728 [110]. The SFR surface density map from the GALEX NUV band showed a signature of suppressed SFR in the central region of NGC 3982 [111]. The Seyfert galaxy NGC 4051 shows the signature of multiple jet episodes [112,113], as well as a hole with a diameter of ∼10 parsecs in its molecular gas distribution [114]. A recent MaNGA study [115] suggested that AGN feedback can significantly affect central star formation, but may be inefficient in driving galaxy-wide quenching in low-redshift galaxies. These studies showed that a detailed study of low-power AGN outflows, molecular gas, and star formation in radio-quiet galaxies with high resolution is important to understand feedback processes and their time scales.

3.2. Radio Loud Galaxies

Radio-loud brightest cluster galaxies (BCGs) have been found to show strong signatures of negative feedback from radio and X-ray data. Here, we look for signatures of jet-induced star formation in some of these objects from the literature. Very few such radio-loud objects show positive feedback: the examples in the nearby Universe are 3C 34 [116], 3C 285, and Minkowski's object (MO) [117]. 3C 34 is a radio galaxy with FR-II type AGN at redshift 0.69 [118]. It is situated in the center of a compact galaxy cluster, and te [OIII] line emission of the host galaxy is extended, which overlaps with the radio lobes [119]. Hubble Space Telescope (HST) images reveal a long blue optical emission in between the western lobe. Using fitted SED with a galaxy stellar population and spectral aging estimation of the radio jets, Best et al. [120] concluded that the supersonic shocks associated with the jet propagation induced the star formation on its path. 3C 285 is another double-lobed FR-II radio galaxy with a complex structure with filaments. The eastern lobe shows unresolved knots. An H_α region known as 3C 285/09.6 [121] has been detected at a projected distance of 70 kpc on the eastern and shows jet-induced star formation. NGC 541, a galaxy in the

Abell 194 cluster, hosts double-lobed FR-I radio AGN. A star-forming region, known as Minkowski's Object (MO) is situated along the radio jet at a distance of 20 kpc [122] of NGC 541. The calculated SFR density, gas density, and depletion time support that star formation is triggered by the radio jet and hence supports positive feedback [117,123].

Centaurus A (Cen A: Figure 4) is a large radio source ($4^{\circ} \times 3^{\circ}$) hosted by the giant galaxy NGC 5128 [124,125]. This system has been studied through multi-band observations that revealed many interesting features, such as a warped disk and optical shells [126,127]. Recent simulations show that Cen A went through a merger approximately 2 Myr ago [128]. The radio structure shows radio lobes at different scales [129–132] (Figure 4B). The lobe at a distance of 15 kpc from the host galaxy is called the northern middle lobe (NML), which looks like an extension of the northern inner lobe (NIL) at a distance of about 8 kpc [133]. Different explanations exist for the origin of NML [134–136]. Cen A shows star formation along the jet, which is believed to be due to positive feedback [137,138] where the radio jet might interact with the gas around the NGC 5128 from the past merger activity [139]. However, it should be noted that the star-formation efficiency is quite low [140]. From the HI kinematics, Oosterloo and Morganti [141] found the signature of jet interaction with a large HI reservoir in the NE direction. UV bright filaments have been detected in the northern direction where young star clusters are also present. These filaments are known as inner and outer filaments, depending on the distance [136].

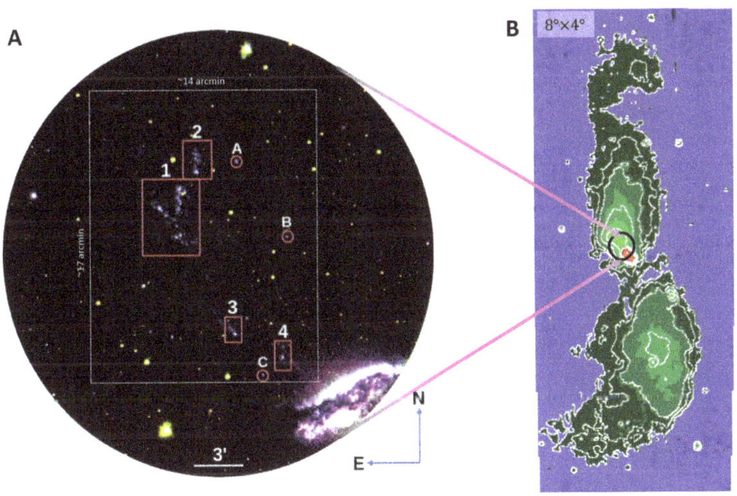

Figure 4. Image of Cen A: (**A**) color composite image with UVIT filters N245M in red, N219M in green, and F148W in blue colors. The circular image shows the 0.5-degree UVIT field of view. The scale is at the bottom. The Northern Star-forming Region (NSR) is shown in the dotted white box. The red boxes and the circular regions show a few interesting sources which were revealed by UVIT. Boxes 1 and 2 are part of the outer filament; Box 3 is part of the inner filament, and Box 4 shows new candidates for possible jet-induced star-forming sources. The circular regions A and C are UV sources, and a pair of stars are present in B. A detailed zoomed version of each thumbnail can be found in [142]. (**B**) The radio 6.3 cm total intensity image shows the full jet emission. These images are reproduced from [129,142].

Joseph et al. [142] studied the star-formation properties in the Northern Star-forming Region (NSR) of size 17×14 arcmin with the UVIT FUV and NUV multiple filters (F148W, N219M, and N245M) (white dotted box: Figure 4A). These UVIT observations revealed a complex structure of star-formation regions and some new candidates for star-forming regions. These images show that most of the UV bright objects are situated near the outer filament in a double structure, which is consistent with a two-segment structure: a V-shaped segment and an elongated segment. Fewer UV sources are observed near the inner filament compared to the outer filament, and they show an elongated structure. There is also UV emission present between the inner filament and NGC 5128. After excluding the Galactic sources and background galaxies, 83 star-forming regions were detected, of which 61 are situated in outer regions and 16 are in inner regions. They compared these star-forming structures with radio images and used the *starburst99* [143] model and FUV-NUV color to estimate the age of the star-forming sources. The main findings were (i) the star-forming structure is along the jet vector, (ii) two distinct populations of stars are present in inner and outer regions, which may have two different origins, and (iii) a catalog including new UV sources likely connected to Cen A was provided [142].

4C 41.17 is one of the best studied radio galaxies at z = 3.79 [144] and shows jet-induced star-formation at high redshift. Recently, regarding the gas mass density and SFR surface density of 4C 41.17, Nesvadba et al. [145] found that the jets are not currently forming stars and they argued that radio jets help in star formation, but without enhancing it, especially after the jet has broken out of the interstellar medium.

Recently, LOFAR 144 MHz images revealed a large-scale radio plume in the central galaxy GIN 049 in galaxy cluster Abell A160 that passes through the spiral galaxy J036 [146]. Further investigation of this system with uGMRT and MUSE data showed that the radio plume of GI 049 interacted with JO36 in the past 200–500 Myr, which resulted in a burst of star formation with SFR 14 $M_{\odot}$ yr^{-1}. This is another example of positive feedback [146].

4. Summary and Future Scope with UV Telescopes

In this article, we discussed the signatures of AGN feedback, mainly focusing on the alignment between observed radio and UV emission. However, it should be noted that it was not possible to include all AGN feedback-related discoveries from all the existing UV telescopes. The presented works are mainly from UVIT [73,111] and GALEX [102]. A few studies that used HST UV data are also mentioned [45,50]. The high-resolution UV data have detected star-forming clumps near the nuclei of MRK 212 showing a signature of possible positive AGN feedback. The UV deficiency in the center of NGC 2639 suggests AGN negative feedback. These studies show that detailed high-resolution multi-frequency observations of a large sample of galaxy mergers and radio-quiet galaxies are important for understanding the feedback effect throughout the lifetime of the galaxies. The detection of a fourth episodic jet of NGC 2639 from uGMRT also shows the importance of high-resolution, high-sensitive low-frequency radio observations. The detailed study of Cen A star-forming regions revealed two distinct stellar populations and new star-forming regions that might be triggered by the jet [111].

Next, we briefly discuss upcoming UV missions that are expected to provide better images and spectra in this field. The HST has been providing high-resolution images (in both UV and optical) and spectra since the 1990s. However, the field of view is very small (0.002 deg^2) and the HST is approaching the end of its lifetime. The GALEX mission, which worked between 2003 and 2013, had a resolution resolution of $5''$ and a FOV of 1.4 deg^2. The UVIT has been providing images with $\sim 1.5''$ resolution since the launch of Astrosat in 2015. Initially, it was planned to have a 5-year lifespan. However, some of the filters are still working and providing good results. A future UV mission, INdian Spectroscopic and Imaging Space Telescope (INSIST; PI: Annapurni Subramaniam: [147]) has been approved by the Indian Space Research Organisation (ISRO) (https://www.isro.gov.in/) in 2019. This is an observatory-class mission with an expected resolution of $\sim 0.2''$, FOV of $\sim 5'$–$15'$ and a detection limit of $\sim$26 mag in UV for an exposure time of 1 kilosec. More details about the instruments of the mission can be found in [147]. Ultraviolet Explorer (UVEX) is NASA's upcoming mission, targeted in 2030. This is a medium-class explorer mission (https://www.ipac.caltech.edu/project/uvex). UVEX will also have imaging and spectroscopy in both NUV and FUV bands. This is a dedicated mission for an all-sky survey in UV, with an expected PSF size of $2.5''$ and a FOV of 12 deg^2. This is expected to make 50/100-times deeper images than GALEX, as well as to provide spectroscopy with a resolution of ∼1000.

Both INSIST[147] and UVEX [148] are motivated by strong scientific goals, which include understanding galaxy formation and evolution, the nuclei of galaxies, stellar populations, transients, large sky survey, etc. AGN feedback is an important part of galaxy evolution and a slowly emerging picture is that both positive and negative feedback may be present in galaxies coupled with galaxy mergers or secular and environmental processes. In addition, AGN feedback may work differently in different objects (like radio-loud and radio-quiet) and in different time scales. Here, the future instruments INSIST and UVEX in the UV band will play a crucial role, along with JWST (https://webb.nasa.gov/), LSST (https://www.lsst.org/), and EUCLID (https://www.euclid-ec.org/) in other bands.

Funding: This research received no external funding

Institutional Review Board Statement: Not applicable

Informed Consent Statement: Not applicable.

Data Availability Statement: The data underlying this article will be shared on reasonable request to the corresponding author.

Acknowledgments: We would like to thank the "AGN on the beach" Tropea, Italy, organizers. We thank the anonymous reviewers for their suggestions that have significantly improved this paper. We want to acknowledge the help received from the Institute of Theoretical Astrophysics (ITA) at the University of Oslo. RK wants to thank N. Junkes for sharing the total intensity radio maps from Parkes radio telescope observations of Cen A. This publication uses data from the AstroSat mission of the Indian Space Research Organisation(ISRO), archived at the Indian Space Science DataCentre (ISSDC), which is a result of collaboration between IIA, Bengaluru, IUCAA, Pune, TIFR, Mumbai, several centers of ISRO, and CSA. GMRT is run by the National Centre for Radio Astrophysics of the Tata Institute of Fundamental Research. RK wants to thank Prasanta Kumar Nayak, Prajwel Joseph, Mousumi Das, Avijeet Prasad, and Akhil Lasrado for productive discussions.

Conflicts of Interest: The author declares no conflict of interest.

References

1. Madau, P.; Dickinson, M. Cosmic Star-Formation History. *Annu. Rev. Astron. Astrophys.* **2014**, *52*, 415–486. . [CrossRef]
2. Ferrarese, L.; Merritt, D. A Fundamental Relation between Supermassive Black Holes and Their Host Galaxies. *Astrophys. J.* **2000**, *539*, L9–L12. [CrossRef]
3. Gültekin, K.; Richstone, D.O.; Gebhardt, K.; Lauer, T.R.; Tremaine, S.; Aller, M.C.; Bender, R.; Dressler, A.; Faber, S.M.; Filippenko, A.V.; et al. The M-σ and M-L Relations in Galactic Bulges, and Determinations of Their Intrinsic Scatter. *Astrophys. J.* **2009**, *698*, 198–221. [CrossRef]
4. Kormendy, J.; Ho, L.C. Coevolution (Or Not) of Supermassive Black Holes and Host Galaxies. *Annu. Rev. Astron. Astrophys.* **2013**, *51*, 511–653. [CrossRef]
5. Fabian, A.C. Observational Evidence of Active Galactic Nuclei Feedback. *Annu. Rev. Astron. Astrophys.* **2012**, *50*, 455–489. [CrossRef]
6. Cattaneo, A.; Faber, S.M.; Binney, J.; Dekel, A.; Kormendy, J.; Mushotzky, R.; Babul, A.; Best, P.N.; Brüggen, M.; Fabian, A.C.; et al. The role of black holes in galaxy formation and evolution. *Nature* **2009**, *460*, 213–219. [CrossRef]
7. Alexander, D.M.; Hickox, R.C. What drives the growth of black holes? *New Astron. Rev.* **2012**, *56*, 93–121. [CrossRef]
8. King, A.; Pounds, K. Powerful Outflows and Feedback from Active Galactic Nuclei. *Annu. Rev. Astron. Astrophys.* **2015**, *53*, 115–154. [CrossRef]
9. Morganti, R. The many routes to AGN feedback. *Front. Astron. Space Sci.* **2017**, *4*, 42. [CrossRef]
10. Di Matteo, T.; Springel, V.; Hernquist, L. Energy input from quasars regulates the growth and activity of black holes and their host galaxies. *Nature* **2005**, *433*, 604–607. [CrossRef] [PubMed]
11. Baldry, I.K.; Balogh, M.L.; Bower, R.G.; Glazebrook, K.; Nichol, R.C.; Bamford, S.P.; Budavari, T. Galaxy bimodality versus stellar mass and environment. *Mon. Not. R. Astron. Soc.* **2006**, *373*, 469–483. [CrossRef]
12. Dubois, Y.; Devriendt, J.; Slyz, A.; Teyssier, R. Self-regulated growth of supermassive black holes by a dual jet-heating active galactic nucleus feedback mechanism: Methods, tests and implications for cosmological simulations. *Mon. Not. R. Astron. Soc.* **2012**, *420*, 2662–2683. [CrossRef]
13. Silk, J.; Mamon, G.A. The current status of galaxy formation. *Res. Astron. Astrophys.* **2012**, *12*, 917–946. [CrossRef]

14. Choi, E.; Somerville, R.S.; Ostriker, J.P.; Naab, T.; Hirschmann, M. The Role of Black Hole Feedback on Size and Structural Evolution in Massive Galaxies. *Astrophys. J.* **2018**, *866*, 91. [CrossRef]
15. Davé, R.; Anglés-Alcázar, D.; Narayanan, D.; Li, Q.; Rafieferantsoa, M.H.; Appleby, S. SIMBA: Cosmological simulations with black hole growth and feedback. *Mon. Not. R. Astron. Soc.* **2019**, *486*, 2827–2849. [CrossRef]
16. McNamara, B.R.; Nulsen, P.E.J. Heating Hot Atmospheres with Active Galactic Nuclei. *Annu. Rev. Astron. Astrophys.* **2007**, *45*, 117–175. [CrossRef]
17. McNamara, B.R.; Nulsen, P.E.J. Mechanical feedback from active galactic nuclei in galaxies, groups and clusters. *New J. Phys.* **2012**, *14*, 055023. [CrossRef]
18. Nulsen, P.E.J.; Jones, C.; Forman, W.R.; David, L.P.; McNamara, B.R.; Rafferty, D.A.; Bîrzan, L.; Wise, M.W. AGN Heating Through Cavities and Shocks. In *Heating versus Cooling in Galaxies and Clusters of Galaxies*; Böhringer, H., Pratt, G.W., Finoguenov, A., Schuecker, P., Eds.; Eso Astrophysics Symposia; Springer: Berlin/Heidelberg, Germany, 2007; p. 210. [CrossRef]
19. Cavagnolo, K.W.; McNamara, B.R.; Nulsen, P.E.J.; Carilli, C.L.; Jones, C.; Bîrzan, L. A Relationship Between AGN Jet Power and Radio Power. *Astrophys. J.* **2010**, *720*, 1066–1072. [CrossRef]
20. Faucher-Giguère, C.A.; Quataert, E. The physics of galactic winds driven by active galactic nuclei. *Mon. Not. R. Astron. Soc.* **2012**, *425*, 605–622. [CrossRef]
21. Costa, T.; Sijacki, D.; Trenti, M.; Haehnelt, M.G. The environment of bright QSOs at z $\sim$ 6: Star-forming galaxies and X-ray emission. *Mon. Not. R. Astron. Soc.* **2014**, *439*, 2146–2174. [CrossRef]
22. Mahony, E.K.; Morganti, R.; Emonts, B.H.C.; Oosterloo, T.A.; Tadhunter, C. The location and impact of jet-driven outflows of cold gas: The case of 3C 293. *Mon. Not. R. Astron. Soc.* **2013**, *435*, L58–L62. [CrossRef]
23. Hardcastle, M.J.; Croston, J.H. Radio galaxies and feedback from AGN jets. *New Astron. Rev.* **2020**, *88*, 101539. [CrossRef]
24. Ishibashi, W.; Fabian, A.C. Active galactic nucleus feedback and triggering of star formation in galaxies. *Mon. Not. R. Astron. Soc.* **2012**, *427*, 2998–3005. [CrossRef]
25. Zubovas, K.; Nayakshin, S.; Sazonov, S.; Sunyaev, R. Outflows of stars due to quasar feedback. *Mon. Not. R. Astron. Soc.* **2013**, *431*, 793–798. [CrossRef]
26. Rupke, D.S.; Veilleux, S.; Sanders, D.B. Outflows in Active Galactic Nucleus/Starburst-Composite Ultraluminous Infrared Galaxies1. *Astrophys. J.* **2005**, *632*, 751–780. [CrossRef]
27. Rupke, D.S.N.; Veilleux, S. The Multiphase Structure and Power Sources of Galactic Winds in Major Mergers. *Astrophys. J.* **2013**, *768*, 75. [CrossRef]
28. Liu, G.; Zakamska, N.L.; Greene, J.E.; Nesvadba, N.P.H.; Liu, X. Observations of feedback from radio-quiet quasars—II. Kinematics of ionized gas nebulae. *Mon. Not. R. Astron. Soc.* **2013**, *436*, 2576–2597. [CrossRef]
29. Cicone, C.; Maiolino, R.; Sturm, E.; Graciá-Carpio, J.; Feruglio, C.; Neri, R.; Aalto, S.; Davies, R.; Fiore, F.; Fischer, J.; et al. Massive molecular outflows and evidence for AGN feedback from CO observations. *Astron. Astrophys.* **2014**, *562*, A21. [CrossRef]
30. Harrison, C.M.; Costa, T.; Tadhunter, C.N.; Flütsch, A.; Kakkad, D.; Perna, M.; Vietri, G. AGN outflows and feedback twenty years on. *Nat. Astron.* **2018**, *2*, 198–205. [CrossRef]
31. Roy, N.; Bundy, K.; Rubin, K.H.R.; Rowlands, K.; Westfall, K.; Riffel, R.; Bizyaev, D.; Stark, D.V.; Riffel, R.A.; Lacerna, I.; et al. Signatures of Inflowing Gas in Red Geyser Galaxies Hosting Radio Active Galactic Nuclei. *Astrophys. J.* **2021**, *919*, 145. [CrossRef]
32. Venturi, G.; Cresci, G.; Marconi, A.; Mingozzi, M.; Nardini, E.; Carniani, S.; Mannucci, F.; Marasco, A.; Maiolino, R.; Perna, M.; et al. MAGNUM survey: Compact jets causing large turmoil in galaxies. Enhanced line widths perpendicular to radio jets as tracers of jet-ISM interaction. *Astron. Astrophys.* **2021**, *648*, A17. [CrossRef]
33. Ramos Almeida, C.; Bischetti, M.; García-Burillo, S.; Alonso-Herrero, A.; Audibert, A.; Cicone, C.; Feruglio, C.; Tadhunter, C.N.; Pierce, J.C.S.; Pereira-Santaella, M.; et al. The diverse cold molecular gas contents, morphologies, and kinematics of type-2 quasars as seen by ALMA. *Astron. Astrophys.* **2022**, *658*, A155. [CrossRef]
34. Fischer, J.; Sturm, E.; González-Alfonso, E.; Graciá-Carpio, J.; Hailey-Dunsheath, S.; Poglitsch, A.; Contursi, A.; Lutz, D.; Genzel, R.; Sternberg, A.; et al. Herschel-PACS spectroscopic diagnostics of local ULIRGs: Conditions and kinematics in Markarian 231. *Astron. Astrophys.* **2010**, *518*, L41. [CrossRef]
35. Oosterloo, T.A.; Morganti, R.; Tzioumis, A.; Reynolds, J.; King, E.; McCulloch, P.; Tsvetanov, Z. A Strong Jet-Cloud Interaction in the Seyfert Galaxy IC 5063: VLBI Observations. *Astron. J.* **2000**, *119*, 2085–2091. [CrossRef]
36. Cicone, C.; Feruglio, C.; Maiolino, R.; Fiore, F.; Piconcelli, E.; Menci, N.; Aussel, H.; Sturm, E. The physics and the structure of the quasar-driven outflow in Mrk 231. *Astron. Astrophys.* **2012**, *543*, A99. [CrossRef]
37. Feruglio, C.; Fiore, F.; Carniani, S.; Piconcelli, E.; Zappacosta, L.; Bongiorno, A.; Cicone, C.; Maiolino, R.; Marconi, A.; Menci, N.; et al. The multi-phase winds of Markarian 231: From the hot, nuclear, ultra-fast wind to the galaxy-scale, molecular outflow. *Astron. Astrophys.* **2015**, *583*, A99. [CrossRef]
38. Morganti, R.; Veilleux, S.; Oosterloo, T.; Teng, S.H.; Rupke, D. Another piece of the puzzle: The fast H I outflow in Mrk 231. *Astron. Astrophys.* **2016**, *593*, A30. [CrossRef]
39. Dasyra, K.M.; Combes, F.; Oosterloo, T.; Oonk, J.B.R.; Morganti, R.; Salomé, P.; Vlahakis, N. ALMA reveals optically thin, highly excited CO gas in the jet-driven winds of the galaxy IC 5063. *Astron. Astrophys.* **2016**, *595*, L7. [CrossRef]
40. Tadhunter, C.; Morganti, R.; Rose, M.; Oonk, J.B.R.; Oosterloo, T. Jet acceleration of the fast molecular outflows in the Seyfert galaxy IC 5063. *Nature* **2014**, *511*, 440–443. [CrossRef]

41. Morganti, R.; Oosterloo, T.; Oonk, J.B.R.; Frieswijk, W.; Tadhunter, C. The fast molecular outflow in the Seyfert galaxy IC 5063 as seen by ALMA. *Astron. Astrophys.* **2015**, *580*, A1. [CrossRef]
42. Maiolino, R.; Russell, H.R.; Fabian, A.C.; Carniani, S.; Gallagher, R.; Cazzoli, S.; Arribas, S.; Belfiore, F.; Bellocchi, E.; Colina, L.; et al. Star formation inside a galactic outflow. *Nature* **2017**, *544*, 202–206. [CrossRef]
43. Gallagher, R.; Maiolino, R.; Belfiore, F.; Drory, N.; Riffel, R.; Riffel, R.A. Widespread star formation inside galactic outflows. *Mon. Not. R. Astron. Soc.* **2019**, *485*, 3409–3429. [CrossRef]
44. Mukherjee, D.; Bicknell, G.V.; Wagner, A.Y.; Sutherland, R.S.; Silk, J. Relativistic jet feedback—III. Feedback on gas discs. *Mon. Not. R. Astron. Soc.* **2018**, *479*, 5544–5566. [CrossRef]
45. Tamhane, P.D.; McNamara, B.R.; Russell, H.R.; Combes, F.; Qiu, Y.; Edge, A.C.; Maiolino, R.; Fabian, A.C.; Nulsen, P.E.J.; Johnstone, R.; et al. Radio jet-ISM interaction and positive radio-mechanical feedback in Abell 1795. *Mon. Not. R. Astron. Soc.* **2023**, *519*, 3338–3356. [CrossRef]
46. Zirm, A.W.; Overzier, R.A.; Miley, G.K.; Blakeslee, J.P.; Clampin, M.; De Breuck, C.; Demarco, R.; Ford, H.C.; Hartig, G.F.; Homeier, N.; et al. Feedback and Brightest Cluster Galaxy Formation: ACS Observations of the Radio Galaxy TN J1338-1942 at z = 4.1. *Astrophys. J.* **2005**, *630*, 68–81. [CrossRef]
47. Drouart, G.; Rocca-Volmerange, B.; De Breuck, C.; Fioc, M.; Lehnert, M.; Seymour, N.; Stern, D.; Vernet, J. Disentangling star formation and AGN activity in powerful infrared luminous radio galaxies at 1 < z < 4. *Astron. Astrophys.* **2016**, *593*, A109. [CrossRef]
48. Kennicutt, R.C. Star Formation in Galaxies Along the Hubble Sequence. *Annu. Rev. Astron. Astrophys.* **1998**, *36*, 189–232. [CrossRef]
49. Calzetti, D. Star Formation Rate Indicators. In *Secular Evolution of Galaxies*; Falcón-Barroso, J., Knapen, J.H., Eds.; Proceedings of the XXIII Canary Islands Winter School of Astrophysics: 'Secular Evolution of Galaxies'; Cambridge University Press: Cambridge, UK, 2013; p. 419.
50. Duggal, C.; O'Dea, C.P.; Baum, S.A.; Labiano, A.; Tadhunter, C.; Worrall, D.M.; Morganti, R.; Tremblay, G.R.; Dicken, D. Optical- & UV-Continuum Morphologies of Compact Radio Source Hosts. *arXiv* **2023**, arXiv:2309.00110. [CrossRef]
51. Kumar, A.; Ghosh, S.K.; Hutchings, J.; Kamath, P.U.; Kathiravan, S.; Mahesh, P.K.; Murthy, J.; Nagbhushana, S.; Pati, A.K.; Rao, M.N. Ultra Violet Imaging Telescope (UVIT) on ASTROSAT. In *Society of Photo-Optical Instrumentation Engineers (SPIE) Conference Series*; Space Telescopes and Instrumentation 2012: Ultraviolet to Gamma Ray: Bellingham, WA, USA, 2012; Volume 8443. [CrossRef]
52. Martin, D.C.; Fanson, J.; Schiminovich, D.; Morrissey, P.; Friedman, P.G.; Barlow, T.A.; Conrow, T.; Grange, R.; Jelinsky, P.N.; Milliard, B.; et al. The Galaxy Evolution Explorer: A Space Ultraviolet Survey Mission. *Astrophys. J.* **2005**, *619*, L1–L6. [CrossRef]
53. Daddi, E.; Dickinson, M.; Morrison, G.; Chary, R.; Cimatti, A.; Elbaz, D.; Frayer, D.; Renzini, A.; Pope, A.; Alexander, D.M.; et al. Multiwavelength Study of Massive Galaxies at z∼2. I. Star Formation and Galaxy Growth. *Astrophys. J.* **2007**, *670*, 156–172. [CrossRef]
54. Salim, S.; Rich, R.M.; Charlot, S.; Brinchmann, J.; Johnson, B.D.; Schiminovich, D.; Seibert, M.; Mallery, R.; Heckman, T.M.; Forster, K.; et al. UV Star Formation Rates in the Local Universe. *Astrophys. J. Suppl. Ser.* **2007**, *173*, 267–292. [CrossRef]
55. Tomczak, A.R.; Quadri, R.F.; Tran, K.V.H.; Labbé, I.; Straatman, C.M.S.; Papovich, C.; Glazebrook, K.; Allen, R.; Brammer, G.B.; Cowley, M.; et al. The SFR-M* Relation and Empirical Star-Formation Histories from ZFOURGE* at 0.5 < z < 4. *Astrophys. J.* **2016**, *817*, 118. [CrossRef]
56. Elbaz, D.; Leiton, R.; Nagar, N.; Okumura, K.; Franco, M.; Schreiber, C.; Pannella, M.; Wang, T.; Dickinson, M.; Díaz-Santos, T., et al. Starbursts in and out of the star-formation main sequence. *Astron. Astrophys.* **2018**, *616*, A110. [CrossRef]
57. Schawinski, K.; Khochfar, S.; Kaviraj, S.; Yi, S.K.; Boselli, A.; Barlow, T.; Conrow, T.; Forster, K.; Friedman, P.G.; Martin, D.C.; et al. Suppression of star formation in early-type galaxies by feedback from supermassive black holes. *Nature* **2006**, *442*, 888–891. [CrossRef] [PubMed]
58. Hickox, R.C.; Jones, C.; Forman, W.R.; Murray, S.S.; Kochanek, C.S.; Eisenstein, D.; Jannuzi, B.T.; Dey, A.; Brown, M.J.I.; Stern, D.; et al. Host Galaxies, Clustering, Eddington Ratios, and Evolution of Radio, X-Ray, and Infrared-Selected AGNs. *Astrophys. J.* **2009**, *696*, 891–919. [CrossRef]
59. Kaviraj, S.; Laigle, C.; Kimm, T.; Devriendt, J.E.G.; Dubois, Y.; Pichon, C.; Slyz, A.; Chisari, E.; Peirani, S. The Horizon-AGN simulation: Evolution of galaxy properties over cosmic time. *Mon. Not. R. Astron. Soc.* **2017**, *467*, 4739–4752. [CrossRef]
60. Volonteri, M.; Haardt, F.; Madau, P. The Assembly and Merging History of Supermassive Black Holes in Hierarchical Models of Galaxy Formation. *Astrophys. J.* **2003**, *582*, 559–573. [CrossRef]
61. Springel, V. The cosmological simulation code GADGET-2. *Mon. Not. R. Astron. Soc.* **2005**, *364*, 1105–1134. [CrossRef]
62. Faisst, A.L.; Carollo, C.M.; Capak, P.L.; Tacchella, S.; Renzini, A.; Ilbert, O.; McCracken, H.J.; Scoville, N.Z. Constraints on Quenching of $Z \lesssim 2$ Massive Galaxies from the Evolution of the Average Sizes of Star-forming and Quenched Populations in COSMOS. *Astrophys. J.* **2017**, *839*, 71. [CrossRef]
63. Gunn, J.E.; Gott, J. Richard, I. On the Infall of Matter Into Clusters of Galaxies and Some Effects on Their Evolution. *Astrophys. J.* **1972**, *176*, 1. [CrossRef]
64. Barnes, J.E.; Hernquist, L. Dynamics of interacting galaxies. *Annu. Rev. Astron. Astrophys.* **1992**, *30*, 705–742. [CrossRef]
65. Bournaud, F. Star Formation and Structure Formation in Galaxy Interactions and Mergers. In Proceedings of the Galaxy Wars: Stellar Populations and Star Formation in Interacting Galaxies ASP Conference Series, Johnson City, TN, USA, 19–22 July 2009; 2010; Volume 423, p. 177.

66. Hopkins, P.F.; Hernquist, L.; Cox, T.J.; Di Matteo, T.; Robertson, B.; Springel, V. A Unified, Merger-driven Model of the Origin of Starbursts, Quasars, the Cosmic X-Ray Background, Supermassive Black Holes, and Galaxy Spheroids. *Astrophys. J. Suppl. Ser.* **2006**, *163*, 1–49. [CrossRef]
67. Ellison, S.L.; Patton, D.R.; Mendel, J.T.; Scudder, J.M. Galaxy pairs in the Sloan Digital Sky Survey—IV. Interactions trigger active galactic nuclei. *Mon. Not. R. Astron. Soc.* **2011**, *418*, 2043–2053. [CrossRef]
68. Satyapal, S.; Ellison, S.L.; McAlpine, W.; Hickox, R.C.; Patton, D.R.; Mendel, J.T. Galaxy pairs in the Sloan Digital Sky Survey—IX. Merger-induced AGN activity as traced by the Wide-field Infrared Survey Explorer. *Mon. Not. R. Astron. Soc.* **2014**, *441*, 1297–1304. [CrossRef]
69. Kaviraj, S.; Martin, G.; Silk, J. AGN in dwarf galaxies: Frequency, triggering processes and the plausibility of AGN feedback. *Mon. Not. R. Astron. Soc.* **2019**, *489*, L12–L16. [CrossRef]
70. Cazzoli, S. Negative and positive outflow-feedback in nearby (U)LIRGs. *Front. Astron. Space Sci.* **2017**, *4*, 62. [CrossRef]
71. Springel, V.; Di Matteo, T.; Hernquist, L. Black Holes in Galaxy Mergers: The Formation of Red Elliptical Galaxies. *Astrophys. J.* **2005**, *620*, L79–L82. [CrossRef]
72. Sparre, M.; Springel, V. The unorthodox evolution of major merger remnants into star-forming spiral galaxies. *Mon. Not. R. Astron. Soc.* **2017**, *470*, 3946–3958. [CrossRef]
73. Rubinur, K.; Kharb, P.; Das, M.; Rahna, P.T.; Honey, M.; Paswan, A.; Vaddi, S.; Murthy, J. A multiwavelength study of the dual nuclei in Mrk 212. *Mon. Not. R. Astron. Soc.* **2021**, *500*, 3908–3919. [CrossRef]
74. Rubinur, K.; Das, M.; Kharb, P.; Yadav, J.; Mondal, C.; Rahna, P.T. Study of star formation in dual nuclei galaxies using UVIT observations. *Mon. Not. R. Astron. Soc.* **2024**, *528*, 4432–4450. [CrossRef]
75. Baldwin, J.A.; Phillips, M.M.; Terlevich, R. Classification parameters for the emission-line spectra of extragalactic objects. *Publ. Astron. Soc. Pac.* **1981**, *93*, 5–19. [CrossRef]
76. Satyapal, S.; Secrest, N.J.; Ricci, C.; Ellison, S.L.; Rothberg, B.; Blecha, L.; Constantin, A.; Gliozzi, M.; McNulty, P.; Ferguson, J. Buried AGNs in Advanced Mergers: Mid-infrared Color Selection as a Dual AGN Candidate Finder. *Astrophys. J.* **2017**, *848*, 126. [CrossRef]
77. Agostino, C.J.; Salim, S. Crossing the Line: Active Galactic Nuclei in the Star-forming Region of the BPT Diagram. *Astrophys. J.* **2019**, *876*, 12. [CrossRef]
78. Begelman, M.C.; Blandford, R.D.; Rees, M.J. Massive black hole binaries in active galactic nuclei. *Nature* **1980**, *287*, 307–309. [CrossRef]
79. Komossa, S.; Burwitz, V.; Hasinger, G.; Predehl, P.; Kaastra, J.S.; Ikebe, Y. Discovery of a Binary Active Galactic Nucleus in the Ultraluminous Infrared Galaxy NGC 6240 Using Chandra. *Astrophys. J.* **2003**, *582*, L15–L19. [CrossRef]
80. Comerford, J.M.; Gerke, B.F.; Stern, D.; Cooper, M.C.; Weiner, B.J.; Newman, J.A.; Madsen, K.; Barrows, R.S. Kiloparsec-scale Spatial Offsets in Double-peaked Narrow-line Active Galactic Nuclei. I. Markers for Selection of Compelling Dual Active Galactic Nucleus Candidates. *Astrophys. J.* **2012**, *753*, 42. [CrossRef]
81. Müller-Sánchez, F.; Comerford, J.M.; Nevin, R.; Barrows, R.S.; Cooper, M.C.; Greene, J.E. The Origin of Double-peaked Narrow Lines in Active Galactic Nuclei. I. Very Large Array Detections of Dual AGNs and AGN Outflows. *Astrophys. J.* **2015**, *813*, 103. [CrossRef]
82. Rubinur, K.; Das, M.; Kharb, P. Searching for dual AGN in galaxies with double-peaked emission line spectra using radio observations. *Mon. Not. R. Astron. Soc.* **2019**, *484*, 4933–4950. [CrossRef]
83. De Rosa, A.; Vignali, C.; Bogdanović, T.; Capelo, P.R.; Charisi, M.; Dotti, M.; Husemann, B.; Lusso, E.; Mayer, L.; Paragi, Z.; et al. The quest for dual and binary supermassive black holes: A multi-messenger view. *New Astron. Rev.* **2019**, *86*, 101525. [CrossRef]
84. Thorne, K.S.; Braginskii, V.B. Gravitational-wave bursts from the nuclei of distant galaxies and quasars: Proposal for detection using Doppler tracking of interplanetary spacecraft. *Astrophys. J.* **1976**, *204*, L1–L6. [CrossRef]
85. Wang, J.M.; Chen, Y.M.; Hu, C.; Mao, W.M.; Zhang, S.; Bian, W.H. Active Galactic Nuclei with Double-Peaked Narrow Lines: Are they Dual Active Galactic Nuclei? *Astrophys. J.* **2009**, *705*, L76–L80. [CrossRef]
86. Smith, K.L.; Shields, G.A.; Bonning, E.W.; McMullen, C.C.; Rosario, D.J.; Salviander, S. A Search for Binary Active Galactic Nuclei: Double-peaked [O III] AGNs in the Sloan Digital Sky Survey. *Astrophys. J.* **2010**, *716*, 866–877. [CrossRef]
87. Teng, S.H.; Schawinski, K.; Urry, C.M.; Darg, D.W.; Kaviraj, S.; Oh, K.; Bonning, E.W.; Cardamone, C.N.; Keel, W.C.; Lintott, C.J.; et al. Chandra Observations of Galaxy Zoo Mergers: Frequency of Binary Active Nuclei in Massive Mergers. *Astrophys. J.* **2012**, *753*, 165. [CrossRef]
88. Liu, X.; Shen, Y.; Strauss, M.A. Active Galactic Nucleus Pairs from the Sloan Digital Sky Survey. II. Evidence for Tidally Enhanced Star Formation and Black Hole Accretion. *Astrophys. J.* **2012**, *745*, 94. [CrossRef]
89. Prabhu, T.P. Indian Astronomical Observatory, Leh-Hanle. *Proc. Indian Natl. Sci. Acad. Part A* **2014**, *80*, 887–912. [CrossRef]
90. Perucho, M.; Martí, J.M. Physical Parameters in the Hot Spots and Jets of Compact Symmetric Objects. *Astrophys. J.* **2002**, *568*, 639–650. [CrossRef]
91. Bondi, M.; Pérez-Torres, M.A.; Dallacasa, D.; Muxlow, T.W.B. A supernova factory in Mrk 273? *Mon. Not. R. Astron. Soc.* **2005**, *361*, 748–752. [CrossRef]
92. Hernández-Ibarra, F.J.; Krongold, Y.; Dultzin, D.; del Olmo, A.; Perea, J.; González, J.; Mendoza-Castrejón, S.; Bitsakis, T. The incidence of nuclear activity in galaxy pairs with different morphologies (E + E), (E + S) and (S + S). *Mon. Not. R. Astron. Soc.* **2016**, *459*, 291–309. [CrossRef]
93. de Bruyn, A.G.; Wilson, A.S. A 1415 MHz Survey of Seyfert and Related Galaxies. *Astron. Astrophys.* **1976**, *53*, 93.

94. Ulvestad, J.S.; Wilson, A.S. Radio structures of Seyfert galaxies. VI. VLA observations of a nearby sample. *Astrophys. J.* **1984**, *285*, 439–452. [CrossRef]
95. Roy, A.L.; Norris, R.P.; Kesteven, M.J.; Troup, E.R.; Reynolds, J.E. Compact Radio Cores in Seyfert Galaxies. *Astrophys. J.* **1994**, *432*, 496. [CrossRef]
96. Thean, A.; Pedlar, A.; Kukula, M.J.; Baum, S.A.; O'Dea, C.P. High-resolution radio observations of Seyfert galaxies in the extended 12-μm sample—II. The properties of compact radio components. *Mon. Not. R. Astron. Soc.* **2001**, *325*, 737–760. [CrossRef]
97. Gallimore, J.F.; Axon, D.J.; O'Dea, C.P.; Baum, S.A.; Pedlar, A. A Survey of Kiloparsec-Scale Radio Outflows in Radio-Quiet Active Galactic Nuclei. *Astron. J.* **2006**, *132*, 546–569. [CrossRef]
98. Jarvis, M.E.; Harrison, C.M.; Thomson, A.P.; Circosta, C.; Mainieri, V.; Alexander, D.M.; Edge, A.C.; Lansbury, G.B.; Molyneux, S.J.; Mullaney, J.R. Prevalence of radio jets associated with galactic outflows and feedback from quasars. *Mon. Not. R. Astron. Soc.* **2019**, *485*, 2710–2730. [CrossRef]
99. Silpa, S.; Kharb, P.; Ho, L.C.; Ishwara-Chandra, C.H.; Jarvis, M.E.; Harrison, C. Probing the origin of low-frequency radio emission in PG quasars with the uGMRT—I. *Mon. Not. R. Astron. Soc.* **2020**, *499*, 5826–5839. [CrossRef]
100. Kharb, P.; O'Dea, C.P.; Baum, S.A.; Colbert, E.J.M.; Xu, C. A Radio Study of the Seyfert Galaxy Markarian 6: Implications for Seyfert Life Cycles. *Astrophys. J.* **2006**, *652*, 177–188. [CrossRef]
101. Irwin, J.A.; Schmidt, P.; Damas-Segovia, A.; Beck, R.; English, J.; Heald, G.; Henriksen, R.N.; Krause, M.; Li, J.T.; Rand, R.J.; et al. CHANG-ES—VIII. Uncovering hidden AGN activity in radio polarization. *Mon. Not. R. Astron. Soc.* **2017**, *464*, 1333–1346. [CrossRef]
102. Rao, V.V.; Kharb, P.; Rubinur, K.; Silpa, S.; Roy, N.; Sebastian, B.; Singh, V.; Baghel, J.; Manna, S.; Ishwara-Chandra, C.H. AGN feedback through multiple jet cycles in the Seyfert galaxy NGC 2639. *Mon. Not. R. Astron. Soc.* **2023**, *524*, 1615–1624. [CrossRef]
103. Lacerda, E.A.D.; Sánchez, S.F.; Cid Fernandes, R.; López-Cobá, C.; Espinosa-Ponce, C.; Galbany, L. Galaxies hosting an active galactic nucleus: A view from the CALIFA survey. *Mon. Not. R. Astron. Soc.* **2020**, *492*, 3073–3090. [CrossRef]
104. Berrier, J.C.; Davis, B.L.; Kennefick, D.; Kennefick, J.D.; Seigar, M.S.; Barrows, R.S.; Hartley, M.; Shields, D.; Bentz, M.C.; Lacy, C.H.S. Further Evidence for a Supermassive Black Hole Mass-Pitch Angle Relation. *Astrophys. J.* **2013**, *769*, 132. [CrossRef]
105. Sebastian, B.; Kharb, P.; O'Dea, C.P.; Gallimore, J.F.; Baum, S.A. The discovery of secondary lobes in the Seyfert galaxy NGC 2639. *Mon. Not. R. Astron. Soc.* **2019**, *490*, L26–L31. [CrossRef]
106. Sanders, R.H. Seyfert nuclei as short-lived stochastic accretion events. *Astron. Astrophys.* **1984**, *140*, 52–54.
107. Ellison, S.L.; Wong, T.; Sánchez, S.F.; Colombo, D.; Bolatto, A.; Barrera-Ballesteros, J.; García-Benito, R.; Kalinova, V.; Luo, Y.; Rubio, M.; et al. The EDGE-CALIFA survey: Central molecular gas depletion in AGN host galaxies—A smoking gun for quenching? *Mon. Not. R. Astron. Soc.* **2021**, *505*, L46–L51. [CrossRef]
108. George, K.; Joseph, P.; Mondal, C.; Devaraj, A.; Subramaniam, A.; Stalin, C.S.; Côté, P.; Ghosh, S.K.; Hutchings, J.B.; Mohan, R.; et al. UVIT observations of the star-forming ring in NGC 7252: Evidence of possible AGN feedback suppressing central star formation. *Astron. Astrophys.* **2018**, *613*, L9. [CrossRef]
109. George, K.; Poggianti, B.M.; Bellhouse, C.; Radovich, M.; Fritz, J.; Paladino, R.; Bettoni, D.; Jaffé, Y.; Moretti, A.; Gullieuszik, M.; et al. GASP XVIII: Star formation quenching due to AGN feedback in the central region of a jellyfish galaxy. *Mon. Not. R. Astron. Soc.* **2019**, *487*, 3102–3111. [CrossRef]
110. Shin, J.; Woo, J.H.; Chung, A.; Baek, J.; Cho, K.; Kang, D.; Bae, H.J. Positive and Negative Feedback of AGN Outflows in NGC 5728. *Astrophys. J.* **2019**, *881*, 147. [CrossRef]
111. Joseph, P.; George, K.; Paul, K.T. Active galactic nucleus feedback in NGC 3982. *Astron. Astrophys.* **2022**, *667*, A88. [CrossRef]
112. Giroletti, M.; Panessa, F. The Faintest Seyfert Radio Cores Revealed by VLBI. *Astrophys. J.* **2009**, *706*, L260–L264. [CrossRef]
113. Jones, S.; McHardy, I.; Moss, D.; Seymour, N.; Breedt, E.; Uttley, P.; Körding, E.; Tudose, V. Radio and X-ray variability in the Seyfert galaxy NGC 4051. *Mon. Not. R. Astron. Soc.* **2011**, *412*, 2641–2652. [CrossRef]
114. Riffel, R.A.; Storchi-Bergmann, T.; Winge, C.; McGregor, P.J.; Beck, T.; Schmitt, H. Mapping of molecular gas inflow towards the Seyfert nucleus of NGC4051 using Gemini NIFS. *Mon. Not. R. Astron. Soc.* **2008**, *385*, 1129–1142. [CrossRef]
115. Lammers, C.; Iyer, K.G.; Ibarra-Medel, H.; Pacifici, C.; Sánchez, S.F.; Tacchella, S.; Woo, J. Active Galactic Nuclei Feedback in SDSS-IV MaNGA: AGNs Have Suppressed Central Star Formation Rates. *Astrophys. J.* **2023**, *953*, 26. [CrossRef]
116. Laing, R.A.; Riley, J.M.; Longair, M.S. Bright radio sources at 178 MHz: Flux densities, optical identifications and the cosmological evolution of powerful radio galaxies. *Mon. Not. R. Astron. Soc.* **1983**, *204*, 151–187. [CrossRef]
117. Salomé, Q.; Salomé, P.; Combes, F. Jet-induced star formation in 3C 285 and Minkowski's Object. *Astron. Astrophys.* **2015**, *574*, A34. [CrossRef]
118. Spinrad, H.; Djorgovski, S.; Marr, J.; Aguilar, L. A third update of the status of the 3 CR sources: Further new redshifts and new identifications of distant galaxies. *Publ. Astron. Soc. Pac.* **1985**, *97*, 932–961. [CrossRef]
119. McCarthy, P.J.; Spinrad, H.; van Breugel, W. Emission-Line Imaging of 3CR Radio Galaxies. I. Imaging Data. *Astrophys. J. Suppl. Ser.* **1995**, *99*, 27. [CrossRef]
120. Best, P.N.; Longair, M.S.; Rottgering, H.J.A. A jet-cloud interaction in 3C 34 at redshift z = 0.69. *Mon. Not. R. Astron. Soc.* **1997**, *286*, 785–794. [CrossRef]
121. van Breugel, W.J.M.; Dey, A. Induced Star Formation in a Radio Lobe of 3C 285? *Astrophys. J.* **1993**, *414*, 563. [CrossRef]
122. Croft, S.; van Breugel, W.; de Vries, W.; Dopita, M.; Martin, C.; Morganti, R.; Neff, S.; Oosterloo, T.; Schiminovich, D.; Stanford, S.A.; et al. Minkowski's Object: A Starburst Triggered by a Radio Jet, Revisited. *Astrophys. J.* **2006**, *647*, 1040–1055. [CrossRef]

123. Lacy, M.; Croft, S.; Fragile, C.; Wood, S.; Nyland, K. ALMA Observations of the Interaction of a Radio Jet with Molecular Gas in Minkowski's Object. *Astrophys. J.* **2017**, *838*, 146. [CrossRef]
124. Bolton, J.G.; Stanley, G.J.; Slee, O.B. Positions of Three Discrete Sources of Galactic Radio-Frequency Radiation. *Nature* **1949**, *164*, 101–102. [CrossRef]
125. Israel, F.P. Centaurus A—NGC 5128. *A&A Rev.* **1998**, *8*, 237–278. [CrossRef]
126. Charmandaris, V.; Combes, F.; van der Hulst, J.M. First detection of molecular gas in the shells of CenA. *Astron. Astrophys.* **2000**, *356*, L1–L4. [CrossRef]
127. Harris, G.L.H.; Rejkuba, M.; Harris, W.E. *The Distance to NGC 5128 (Centaurus A)*; Cambridge University Press: Cambridge, UK, 2010; Volume 27, pp. 457–462. [CrossRef]
128. Wang, J.; Hammer, F.; Rejkuba, M.; Crnojević, D.; Yang, Y. A recent major merger tale for the closest giant elliptical galaxy Centaurus A. *Mon. Not. R. Astron. Soc.* **2020**, *498*, 2766–2777. [CrossRef]
129. Junkes, N.; Haynes, R.F.; Harnett, J.I.; Jauncey, D.L. Radio polarization surveys of Centaurus A (NGC 5128). I. The complete radio source at lambda 6.3 cm. *Astron. Astrophys.* **1993**, *269*, 29–38.
130. Wykes, S.; Intema, H.T.; Hardcastle, M.J.; Achterberg, A.; Jones, T.W.; Jerjen, H.; Orrú, E.; Lazarian, A.; Shimwell, T.W.; Wise, M.W.; et al. Filaments in the southern giant lobe of Centaurus A: constraints on nature and origin from modelling and GMRT observations. *Mon. Not. R. Astron. Soc.* **2014**, *442*, 2867–2882. [CrossRef]
131. McKinley, B.; Tingay, S.J.; Carretti, E.; Ellis, S.; Bland-Hawthorn, J.; Morganti, R.; Line, J.; McDonald, M.; Veilleux, S.; Wahl Olsen, R.; et al. The jet/wind outflow in Centaurus A: a local laboratory for AGN feedback. *Mon. Not. R. Astron. Soc.* **2018**, *474*, 4056–4072. [CrossRef]
132. McKinley, B.; Tingay, S.J.; Gaspari, M.; Kraft, R.P.; Matherne, C.; Offringa, A.R.; McDonald, M.; Calzadilla, M.S.; Veilleux, S.; Shabala, S.S.; et al. Multi-scale feedback and feeding in the closest radio galaxy Centaurus A. *Nat. Astron.* **2022**, *6*, 109–120. [CrossRef]
133. Neff, S.G.; Eilek, J.A.; Owen, F.N. The Complex North Transition Region of Centaurus A: Radio Structure. *Astrophys. J.* **2015**, *802*, 87. [CrossRef]
134. Morganti, R.; Killeen, N.E.B.; Ekers, R.D.; Oosterloo, T.A. Centaurus A: Multiple outbursts or bursting bubble? *Mon. Not. R. Astron. Soc.* **1999**, *307*, 750–760. [CrossRef]
135. Kraft, R.P.; Forman, W.R.; Hardcastle, M.J.; Birkinshaw, M.; Croston, J.H.; Jones, C.; Nulsen, P.E.J.; Worrall, D.M.; Murray, S.S. The Jet Heated X-Ray Filament in the Centaurus A Northern Middle Radio Lobe. *Astrophys. J.* **2009**, *698*, 2036–2047. [CrossRef]
136. Neff, S.G.; Eilek, J.A.; Owen, F.N. The Complex North Transition Region of Centaurus A: A Galactic Wind. *Astrophys. J.* **2015**, *802*, 88. [CrossRef]
137. Rejkuba, M.; Minniti, D.; Silva, D.R.; Bedding, T.R. Stellar populations in NGC 5128 with the VLT: Evidence for recent star formation. *Astron. Astrophys.* **2001**, *379*, 781–797. [CrossRef]
138. Rejkuba, M.; Minniti, D.; Courbin, F.; Silva, D.R. Radio-Optical Alignment and Recent Star Formation Associated with Ionized Filaments in the Halo of NGC 5128 (Centaurus A). *Astrophys. J.* **2002**, *564*, 688–695. [CrossRef]
139. Eilek, J.A. The dynamic age of Centaurus A. *New J. Phys.* **2014**, *16*, 045001. [CrossRef]
140. Salomé, Q.; Salomé, P.; Combes, F.; Hamer, S.; Heywood, I. Star formation efficiency along the radio jet in Centaurus A. *Astron. Astrophys.* **2016**, *586*, A45. [CrossRef]
141. Oosterloo, T.A.; Morganti, R. Anomalous HI kinematics in Centaurus A: Evidence for jet-induced star formation. *Astron. Astrophys.* **2005**, *429*, 469–475. [CrossRef]
142. Joseph, P.; Sreekumar, P.; Stalin, C.S.; Paul, K.T.; Mondal, C.; George, K.; Mathew, B. UVIT view of Centaurus A: A detailed study on positive AGN feedback. *Mon. Not. R. Astron. Soc.* **2022**, *516*, 2300–2313. [CrossRef]
143. Leitherer, C.; Schaerer, D.; Goldader, J.D.; Delgado, R.M.G.; Robert, C.; Kune, D.F.; de Mello, D.F.; Devost, D.; Heckman, T.M. Starburst99: Synthesis Models for Galaxies with Active Star Formation. *Astrophys. J. Suppl. Ser.* **1999**, *123*, 3–40. [CrossRef]
144. Chambers, K.C.; Miley, G.K.; van Breugel, W.J.M. 4C 41.17: A Radio Galaxy at a Redshift of 3.8. *Astrophys. J.* **1990**, *363*, 21. [CrossRef]
145. Nesvadba, N.P.H.; Bicknell, G.V.; Mukherjee, D.; Wagner, A.Y. Gas, dust, and star formation in the positive AGN feedback candidate 4C 41.17 at z = 3.8. *Astron. Astrophys.* **2020**, *639*, L13. [CrossRef]
146. Ignesti, A.; Brienza, M.; Vulcani, B.; Poggianti, B.M.; Marasco, A.; Smith, R.; Hardcastle, M.J.; Botteon, A.; Roberts, I.D.; Fritz, J.; et al. On the Encounter between the GASP Galaxy JO36 and the Radio Plume of GIN 049. *Astrophys. J.* **2023**, *956*, 122. [CrossRef]
147. Subramaniam, A. An overview of the proposed Indian spectroscopic and imaging space telescope. *J. Astrophys. Astron.* **2022**, *43*, 80. [CrossRef]
148. Kulkarni, S.R.; Harrison, F.A.; Grefenstette, B.W.; Earnshaw, H.P.; Andreoni, I.; Berg, D.A.; Bloom, J.S.; Cenko, S.B.; Chornock, R.; Christiansen, J.L.; et al. Science with the Ultraviolet Explorer (UVEX). *arXiv* **2021**, arXiv:2111.15608. [CrossRef]

galaxies

MDPI

Review

Investigating the Properties of the Relativistic Jet and Hot Corona in AGN with X-ray Polarimetry

Dawoon E. Kim [1,2,3,*], Laura Di Gesu [4], Frédéric Marin [5], Alan P. Marscher [6], Giorgio Matt [7], Paolo Soffitta [1], Francesco Tombesi [3,8,9], Enrico Costa [1] and Immacolata Donnarumma [4]

1 INAF Istituto di Astrofisica e Planetologia Spaziali, Via del Fosso del Cavaliere 100, 00133 Roma, Italy
2 Dipartimento di Fisica, Università degli Studi di Roma "La Sapienza", Piazzale Aldo Moro 5, 00185 Roma, Italy
3 Dipartimento di Fisica, Universitaà degli Studi di Roma "Tor Vergata", Via della Ricerca Scientifica 1, 00133 Roma, Italy
4 ASI—Agenzia Spaziale Italiana, Via del Politecnico snc, 00133 Roma, Italy
5 Observatoire Astronomique de Strasbourg, CNRS, Université de Strasbourg, UMR 7550, 67000 Strasbourg, France
6 Institute for Astrophysical Research, Boston University, 725 Commonwealth Avenue, Boston, MA 02215, USA
7 Dipartimento di Matematica e Fisica, Università degli Studi Roma Tre, Via della Vasca Navale 84, 00146 Roma, Italy
8 Istituto Nazionale di Fisica Nucleare, Sezione di Roma "Tor Vergata", Via della Ricerca Scientifica 1, 00133 Roma, Italy
9 Department of Astronomy, University of Maryland, College Park, MD 20742, USA
* Correspondence: dawoon.kim@inaf.it

Abstract: X-ray polarimetry has been suggested as a prominent tool for investigating the geometrical and physical properties of the emissions from active galactic nuclei (AGN). The successful launch of the Imaging X-ray Polarimetry Explorer (IXPE) on 9 December 2021 has expanded the previously restricted scope of polarimetry into the X-ray domain, enabling X-ray polarimetric studies of AGN. Over a span of two years, IXPE has observed various AGN populations, including blazars and radio-quiet AGN. In this paper, we summarize the remarkable discoveries achieved thanks to the opening of the new window of X-ray polarimetry of AGN through IXPE observations. We will delve into two primary areas of interest: first, the magnetic field geometry and particle acceleration mechanisms in the jets of radio-loud AGN, such as blazars, where the relativistic acceleration process dominates the spectral energy distribution; and second, the geometry of the hot corona in radio-quiet AGN. Thus far, the IXPE results from blazars favor the energy-stratified shock acceleration model, and they provide evidence of helical magnetic fields inside the jet. Concerning the corona geometry, the IXPE results are consistent with a disk-originated slab-like or wedge-like shape, as could result from Comptonization around the accretion disk.

Keywords: X-ray polarimetry; active galactic nuclei; relativistic jets

1. Introduction

Active galactic nuclei (AGN) stand out as fascinating and powerful sources (with bolometric luminosities $\sim 10^{42}$–10^{46} erg s^{-1}; e.g., [1,2]) that emit electromagnetic radiation across the entire spectrum, ranging from radio wavelengths to gamma rays. Specifically, in the X-ray regime, emissions originate from the vicinity of the central engine, in either the inner part of the jet or in an X-ray corona. The high penetration power of X-ray radiation facilitates the unobstructed study of processes near black holes. So far, X-ray probes have been limited to spectral and timing techniques [3]. Although these methods provide clues about the radiative processes producing X-rays in the inner AGN region, they are insensitive to some key characteristics, such as the geometry of the emitting region or the magnetic field, which so far have remained largely unconstrained.

In this context, polarimetric studies have been suggested as a prominent tool for investigating the geometrical and physical properties of AGN emissions [4]. The successful launch of the Imaging X-ray Polarimetry Explorer (IXPE) on 9 December 2021 marked a significant advancement in polarimetry by extending it into the X-ray range [5]. IXPE is a collaborative mission between NASA and the Italian Space Agency (Agenzia Spaziale Italiana, ASI). It comprises three identical X-ray telescope systems, each one composed of one mirror module assembly and one detector unit hosting a gas pixel detector [6] with a full calibration and filtering system. IXPE is designed to measure the linear polarization in the 2–8 keV band using polarization mapping techniques. It offers a field of view greater than $11'$ and an angular resolution of less than or equal to $30''$. Over the first two years of operation, IXPE has detected X-ray polarization from several AGN, helping to address the long-standing uncertainty about the jet magnetic field structure and acceleration process and the geometry of the X-ray corona.

Traditionally, AGN are divided into two main classes: radio-loud (RLAGN) and radio-quiet (RQAGN), based on their radio to optical flux ratio [7,8]. In some cases, they are classified as jetted or non-jetted based on the presence or absence of strong relativistic jets [9]. Among the population of RLAGN, blazars are ideal targets for studying relativistic jets because, in these sources, the jet flow is relatively closely aligned to the observer's line of sight, and their emission is dominated by jet emission across the entire spectral energy distribution (SED) due to relativistic Doppler boosting, e.g., [10]. The SED of blazars exhibits two broad non-thermal radiation components as shown in Figure 1. One component, located in the lower-frequency range, covers emissions from radio to optical/UV (and, in some cases, to X-ray) and is generally ascribed to synchrotron emission from relativistic electrons. The other component may be explained in two different scenarios: leptonic or hadronic models. In the leptonic process, the high-energy emission is attributed to Compton scattering of synchrotron photons (synchrotron self-Compton or SSC; e.g., [11,12]) or photons originating outside the jet (external Compton or EC; e.g., [13–15]). In the hadronic scenario, proton synchrotron emission, π_0 decay photons, synchrotron, and Compton emission from secondary decay products of charged pions have been suggested, e.g., [16–18]. These emission processes are predicted to produce different X-ray polarization properties depending on the acceleration mechanism in both leptonic and hadronic scenarios [19]. Furthermore, depending on the position of the peak frequency of the electron synchrotron emission hump (ν_{peak}), blazars are separated into low- ($\nu_{peak} < 10^{14}$ Hz or 3 μm; LSP), intermediate- ($\nu_{peak} \sim 10^{14}$ Hz–10^{15} Hz; ISP), and high-synchrotron-peak ($\nu_{peak} > 10^{15}$ Hz or 0.3 μm; HSP) subclasses [20]. Therefore, studying various subclasses of blazars with X-ray polarimetry allows us to explore different physical processes. First, with HSP targets, the particle acceleration process and geometrical features of the magnetic field inside the jets can be studied. Second, with ISP and LSP blazars, the superposition of synchrotron emission and leptonic or hadronic emission in spectra enables tests of the acceleration mechanisms behind the high-energy emission hump.

In the case of RQAGN, given that they are not typically dominated by jet emissions, we can observe other emission components surrounding the central engine (see Figure 1). The primary X-ray emission of AGN is believed to originate from a corona of hot (with a temperature of 10–100 keV) and optically thin (or mildly thick) plasma that Compton scatters the optical/UV photons from the accretion disk, e.g., [21]. Additionally, these X-ray photons are reflected (backscattered) either by the accretion disk or by the more distant torus, which results in additional X-ray spectral components as secondary emissions [22]. RQAGN are divided into Seyfert 1 and 2 galaxies based on whether or not both broad and narrow emission lines are observed in the optical band, which partially depends on the differences in the line of sight direction [23]. In the X-ray regime, Seyfert 1 galaxies, which have a relatively small inclination angle between our line of sight and the axis of the system, averaging 39° [24,25], show unobscured X-ray coronae and disk reflection, while the emission from Seyfert 2s is dominated by photons reprocessed through reflection from the strongly absorbing dusty torus and from the polar winds due to the larger inclination angle.

The spectrum of the corona typically follows a power law with a high-energy cutoff, e.g., [26]. The slope of the power law is determined by the electron temperature and the optical depth, while the cutoff energy depends mostly on the electron temperature. The size and geometry of the corona remain elusive, although these characteristics are crucial for differentiating between models for the formation of AGN coronae, e.g., [27]. Regarding the geometric characteristics of the hot corona, several possibilities have been suggested, including lamppost, spherical, conical, slab, and wedge. In particular, spherical or conical structures located on the rotation axis above and below the black hole would suggest a potential connection between the corona and the jet, e.g., [28,29]. In this context, it is predicted that the X-ray polarization of the corona emission depends on the geometry, e.g., [30], and demonstrates that X-ray polarization measurements can serve as a powerful tool for probing the disk/corona geometry, e.g., [31,32]. Hence, the polarization observations of unobscured RQAGN can provide valuable insights into the geometric features of the corona.

In this short review, we will briefly summarize the theoretical expectations and remarkable discoveries obtained from the first two years of IXPE observations of RLAGN and RQAGN. These concern the magnetic field geometry and particle acceleration mechanisms in the jets, as well as the geometry of the hot corona. We will then discuss future prospects.

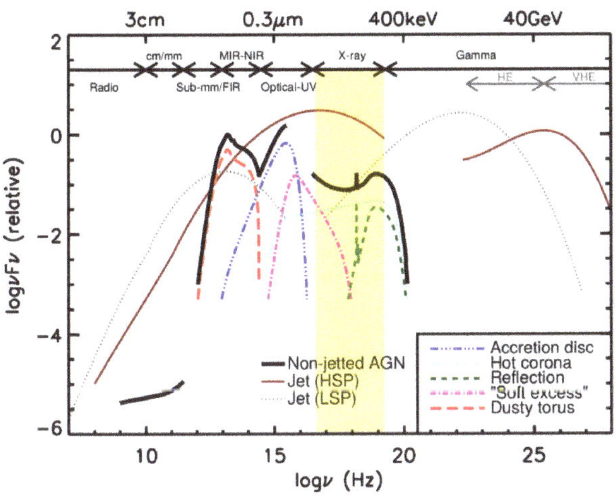

Figure 1. A schematic representation of SED for RLAGN (HSP: red solid line, and LSP: gray dotted line) and RQAGN (black solid line with labeled colored lines for different emission components). The yellow shaded area indicates the X-ray regime. The figure is reproduced from Figure 1 in [8].

2. X-ray Polarization from Relativistic Jets

2.1. Theoretical Expectations

The observed emission coming from relativistic jets is often well-described by a power law spectrum within a given frequency band. Theoretically, this can be explained as synchrotron radiation produced by a power law energy spectrum of electrons with Lorentz factor γ, $N_e(\gamma) \propto \gamma^{-p}$, which produces a power law synchrotron radiation spectrum $F_\nu \propto \nu^{-\alpha}$, with $\alpha = (p-1)/2$. This results in a polarization degree (Π) of [4]

$$\Pi_{Sync} = \frac{p+1}{p+7/3} = \frac{\alpha+1}{\alpha+5/3}. \tag{1}$$

Considering the typical power law index obtained from blazars $\sim p$ = 2–3, a polarization degree of the electron synchrotron emission in a homogeneous magnetic field is expected to be $\Pi \sim 70\%$. The polarization angle (ψ) is perpendicular to the averaged magnetic field direction projected onto the sky plane. However, in reality, polarization can be reduced by the geometrical and physical properties within the jet, such as temporal variation effects occurring over timescales shorter than the integration time of the polarization measurement [33–35]. Thereby, in order to explain the polarization measurements and variability, various geometrical and physical models have been suggested. For instance, for the geometrical model, there have been several suggestions: a bending jet [36], a spiral motion model in which a blob propagates in a spiral trajectory within a helical magnetic field [37], and the ballistic motion of an emission blob in a jet with a helical magnetic field [38]. In the case of the physical conditions, there have been several suggestions as well: many turbulent cells, the TEMZ model, [39–41], shock and kink instabilities that can efficiently dissipate jet bulk energy to accelerate particles [42–45], and magnetic reconnection [46–49]. Refs. [50,51] have presented the predicted X-ray and simultaneous multiwavelength polarization characteristics among various theoretical models as follows: (1) in the shock acceleration model, particle acceleration is characterized by a strong self-generated magnetic field. Hence, a large degree of polarization, $\Pi_X \sim 40\%$, and relatively stable Π_X and ψ_X over time (variability in a few days to a week) are predicted [52,53]. Moreover, a chromatic feature in X-ray and optical polarization, where $\Pi_X/\Pi_O \gtrsim 2$, is anticipated because particles emitting at longer wavelengths, due to their longer synchrotron cooling time, travel away from the shock front to regions with a more disordered magnetic field; (2) in the magnetic reconnection model, the reconnection episodes are triggered by instability induced, for example, by current sheets [46,54]. $\Pi_X \lesssim 20-30\%$ and similar characteristics for Π_O are predicted as a result of the simultaneous contribution of several active current sheets with different orientations. Additionally, the evolution of the instability results in significant variations, and ψ_X can show variability on a daily timescale; (3) in the TEMZ model, $\Pi_X \lesssim 30\%$ and rapid changes in Π_X and ψ_X less than a day timescale are predicted because of the superposition of the emission of several cells, in which the magnetic field is uniform but randomly oriented [39]. In addition, mild chromatic behavior, $\Pi_X/\Pi_O \lesssim 2$, is expected as a large number of cells can radiate at lower frequencies. Thus, the variability is erratic and somewhat driven by the dimension and number of cells. Consequently, expanding multiwavelength polarimetric measurements on synchrotron emission to the X-ray band provides clues about the layout of the magnetic field as particles emitting at different energies can assist in identifying the predominant acceleration process.

Compton scattering of unpolarized seed photons by highly relativistic electrons produces zero polarization in the scattered radiation [55]. Therefore, detecting X-ray polarization from LSP blazars, whose emission component is primarily dominated by Compton scattering in X-rays, would effectively rule out the possibility of EC processes. In contrast, under the conditions of proton synchrotron radiation, the polarization levels could potentially reach around 70%, consistent with the proton spectral indices in Equation (1), where p is approximately 2–3 [55]. A polarization degree significantly higher than that at mm-waves and similar to or greater than that at optical wavelengths, and a slope of the X-ray emission that is sufficiently flat, could be interpreted as an indication of proton synchrotron emission. Furthermore, for cases involving a combination of synchrotron emission at soft X-ray energies and SSC/EC at harder X-rays, detailed SED modeling and decomposition of the polarization properties of each component are required. The model considering multiple emission zones with turbulent magnetic fields predicts $\Pi_{SSC}/\Pi_{Sync} \approx 0.3$ [56].

2.2. IXPE Observations of High Synchrotron Peak Blazars

IXPE observations of HSP blazars have been conducted for Mrk 501 [57], Mrk 421 [58–60], 1ES1959+650 [61], PG1553+113 [62], and 1ES0229+200 [63]. The X-ray polarization of each target was estimated through time-averaged analysis by considering the entire observation period and time- and energy-resolved analyses by dividing the observations into sub-segments using model independent event Stokes parameter analysis [64] and spectropolarimetric analysis [65]. In addition, during all the IXPE blazar observations, a simultaneous multiwavelength polarimetry campaign was conducted, including radio, infrared (IR), and optical wavelengths, to examine the synchrotron emission properties across different wavelengths. In this section, we summarize the reported X-ray and multiwavelength polarimetry features of each observation, as well as the X-ray spectral properties. Table 1 presents a summary of the time-averaged multiwavelength polarization information for all the HSP observations, and Table 2 provides a summary of the X-ray spectral characteristics, including the photon index and flux derived based on the log parabolic model [66]. The log parabolic model represents power laws with a photon index that varies according to a log parabola in energy:

$$N(E) = K(E/E_{pivot})^{(\alpha-\beta\log(E/E_{pivot}))}, \tag{2}$$

where the pivot energy E_{pivot} is a scaling factor, α describes the slope of the photon spectrum at E_{pivot}, β expresses the spectral curvature, and K denotes a normalization constant.

Table 1. Contemporaneous multiwavelength polarization properties of HSPs.

Source	X-ray		Optical & IR [a]		Radio [a]	
	$\Pi(\%)$	$\psi(°)$	$\Pi(\%)$	$\psi(°)$	$\Pi(\%)$	$\psi(°)$
Mrk 501 I [1]	10 ± 2	134 ± 5	4 ± 1	119 ± 9	1.5 ± 0.5	152 ± 10
Mrk 501 II [1]	11 ± 2	115 ± 4	5 ± 1	117 ± 3	–	–
Mrk 421 I [2]	15 ± 2	35 ± 4	2.9 ± 0.5	32 ± 5	3.4 ± 0.4	55 ± 2
Mrk 421 II [3]	10 ± 1	Rotation	4.4 ± 0.4	140 ± 6	2.4 ± 0.1	139 ± 8
Mrk 421 III [3]	10 ± 1	Rotation	5.4 ± 0.4	145 ± 1	–	–
Mrk 421 IV [4]	14 ± 1	107 ± 3	4.6 ± 1.3	206 ± 9	1.8 ± 0.1	167 ± 4
1ES1959+650 I [5]	8 ± 2	123 ± 8	4.5 ± 0.2	159 ± 1	–	–
1ES1959+650 II [5]	<5	-	4.7 ± 0.6	151 ± 19	<1.6	-
PG1553+113 [6]	10 ± 2	86 ± 8	4.2 ± 0.5	Rotation	2.6 ± 0.7	133 ± 7
1ES0229+200 [7]	18 ± 3	25 ± 5	3.2 ± 0.7	-5 ± 9	<7	-

[1–8] Results compiled from the following references: [57–63]; [a] median polarization properties during the IXPE observation. Especially for optical and IR polarization, only corrected polarization values, accounting for the dilution of polarization by unpolarized starlight from the host galaxy, were considered for calculation; [b] at the lowest radio frequency (4.85 GHz).

Table 2. X-ray spectral properties of HSPs during IXPE observation.

Source	Photon Index	$\text{Flux}_{2-8\text{ keV}}$ ($\times 10^{-11}$ erg s^{-1} cm^{-2})	Telescopes
Mrk 501 I [1]	2.27 ± 0.01	10.0 ± 0.5	IXPE + Swift-XRT+NuSTAR
Mrk 501 II [1]	2.05 ± 0.02	21.0 ± 0.6	IXPE + Swift-XRT
Mrk 421 I [2]	2.97 ± 0.01	8.67 ± 0.03	IXPE + XMM-Newton+NuSTAR
Mrk 421 II [3]	2.32 ± 0.01 [a]	15.7 ± 0.1	Swift-XRT
Mrk 421 III [3]	2.13 ± 0.01 [a]	30.2 ± 0.2	Swift-XRT
Mrk 421 IV [4]	2.82 ± 0.01	22.5 ± 0.4	IXPE + XMM-Newton
1ES1959+650 I [5]	2.50 ± 0.02	12.4 ± 0.2	IXPE + XMM-Newton
1ES1959+650 II [5]	2.29 ± 0.01	14.7 ± 0.1	IXPE + XMM-Newton
PG1553+113 [6]	2.49 ± 0.01	2.55 ± 0.03	IXPE + XMM-Newton
1ES0229+200 [7]	1.82 ± 0.01	$0.84 \pm -$	IXPE + XMM-Newton

[1–8] Results compiled from the following references: [57–63]; [a] average value between variation over time.

Mrk 501 was the first observed HSP blazar. This target was observed at two different epochs (Mrk 501 I: 8–10 March 2022 and Mrk 501 II: 26–28 March 2022). The time-averaged X-ray polarization in the 2–8 keV range of these observations was consistent within a 3σ error range, as shown in Table 1. There was no significant variability within the duration of the individual IXPE observations over time and energy, and the X-ray activity was in an average state. The polarization in the optical, IR, and radio bands was significantly lower than in the X-rays by a factor of $\sim$2, indicating strong chromatic behavior (Figure 2). Additionally, ψ is aligned with the jet direction, $\psi_{43\text{GHz}} = 119.7 \pm 11.8$, obtained from the 43 GHz Very Long Baseline Array (VLBA) observation [57]. These multiwavelength polarimetric results suggest, among the various proposed models, the "energy-stratified shock acceleration model" in which particles accelerated by shocks experience increasing turbulence as they move away from the accelerating site, resulting in a decrease in the polarization degree. Moreover, shocks can push the particles far away, causing a decrease in Π_X. In addition, the energy-stratified shock acceleration model predicts ψ_X along the jet axis. Thereby, the different polarization angles at different wavelengths suggest a scenario where particles emitting at different energies experience different magnetic field conditions.

Mrk 421 has been observed four times (Mrk 421 I: 4–6 May 2022; Mrk 421 II: 4–6 June 2022; Mrk 421 III: 7–9 June 2022; and Mrk 421 IV: 6–8 December 2022). Mrk 421 I exhibited a slightly higher Π_X of 15% compared to Mrk 501 within a statistical difference of less than 2σ and a Π_X/Π_O ratio of $\sim$5, indicating a stronger chromatic behavior relative to Mrk 501. These results are consistent with the energy-stratified shock acceleration model based on the Π_X/Π_O , where a smaller value was predicted in magnetic reconnection, and the TEMZ models. Subsequent observations of Mrk 421 II and III did not reveal significant polarization information in the time-averaged polarization during the observation periods. However, the time-resolved results, as shown in Figure 2, revealed a smooth rotation of ψ_X by more than 360° between two consecutive observations (Mrk 421 II: from $9 \pm 1°$ to $130 \pm 11°$ and Mrk 421 III: from $222 \pm 7°$ to $360 \pm 20°$). The rotation rates of ψ_X were measured as $80 \pm 9°$ per day and $91 \pm 8°$ per day for the two observations, respectively, while the polarization degrees remained consistent within uncertainty as $\Pi_X = 10 \pm 1\%$. During the observed period, the X-ray flux increased by approximately a factor of two between the two observations and then decreased to a level similar to that at the beginning of the observation, and the spectrum properties showed a decrease in the photon index and an increase in the hardness ratio (i.e., the spectrum became harder and flatter), although we cannot be sure whether this is directly related to the rotation in the polarization angle. The multiwavelength polarimetry results showed chromatic trends conistent with previous findings. Therefore, the results of Mrk 421 II and III suggest a geometric structure where the energy-stratified shock propagates linearly or radially along the helical magnetic field within the jet [59]. Additionally, the Mrk 421 IV showed similar characteristics to those of Mrk 421 I in terms of Π_X, but ψ_X exhibited a different value compared to previous observations (Mrk 421 I, II, and III) and varied across all the observations. In particular, Mrk 421 IV confirmed temporal episodic variability in polarization during this observation period. In particular, Ref. [60] showed a possible correlation between the rotation of ψ_X and the observation of a highly polarized knot emerging from the jet core and moving away from the "core" based on the 43 GHz VLBA observations performed during the observation periods of Mrk 421 II, III, and IV. In addition, multiwavelength polarization monitoring of Mrk 421 I, II, III, and IV showed a rotation of ψ in the optical and radio bands over a longer period compared to the X-ray band, with a direction opposite to the rotation of ψ_X. This result suggests the possibility that emission regions at longer wavelengths (millimeter, infrared, and optical) were separate from, or only partially coincident with, those of the X-ray emission but also contain a helical magnetic field that is presumably a continuation of the helical field manifested in the X-ray polarization observations.

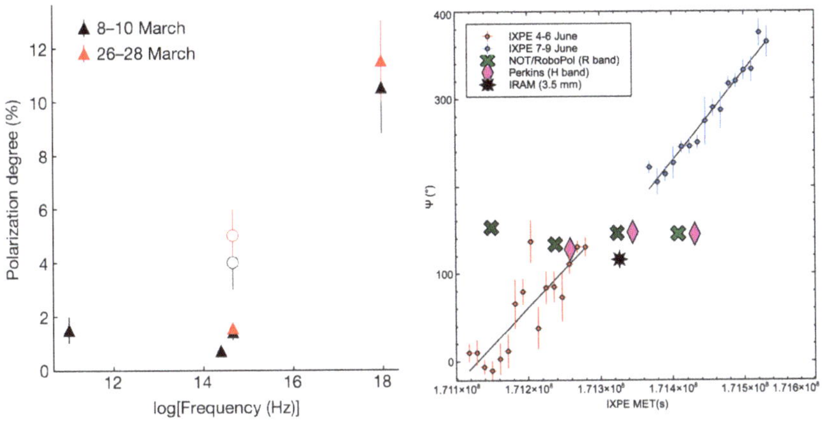

Figure 2. IXPE observation results of Mrk 501 and Mrk 421. (**Left**) Multiwavelength polarization degree of Mrk 501 from radio rays to X-rays. The black symbols represent observations conducted between March 8th and 10th, while the red symbols represent observations from March 26th to 28th. The open symbols indicate the intrinsic optical polarization degree corrected for the host galaxy. The figure was reproduced from Figure 3 in [57]. (**Right**) X-ray polarization angle rotation in Mrk 421. The symbols identify multiwavelength polarimetry measurements obtained from telescopes as labeled. The figure was reproduced from Figure 2 in [59].

In addition, the observations of 1ES1959+650 [61], PG1553+113 [62], and 1ES0229+200 [63] exhibited significant chromatic behavior on Π across multiwavelengths, consistent with the energy-stratified shock acceleration model. Furthermore, concerning energy stratification and cooling inside the shock, the Π_X/Π_Oratio from all the IXPE HSP observations has typically been found to be around ∼2–6, with a trend between the Π_X/Π_O ratio and Π_X , as shown in the left panel of Figure 3. Consequently, as Π_X/Π_O increases, X-ray emission occurs in smaller volumes due to a relatively shorter cooling timescale, indicating well-ordered magnetic fields closer to the shock front and increased turbulence in the downstream direction [67]. Moreover, the comparison between the Π_X/Π_O ratio and the time-averaged photon index in the right panel of Figure 3 suggests a possibility of relatively higher Π_X/Π_O ratios when the spectrum is softer and steeper, but excluding the observations of 1ES0229+200. Especially in the case of 1ES0229+200, an E_{pivot} scale factor of 1 was used in the log parabolic model instead of the common 5 used in other observations [57,59,60]. Therefore, it is possible that this factor may have influenced the observed results compared to other observations, and this could be improved through analysis under similar conditions in the future. Additionally, this characteristic can be interpreted based on the same physical interpretation as when the X-ray emission originates further away from the peak frequency on the spectrum, indicating that the emission is from well-ordered magnetic fields in a smaller volume with a shorter cooling timescale closer to the shock front.

Regarding the variability in polarization properties, smooth rotation on ψ_X and transient ψ_X variation have been observed only in Mrk 421. Hence, it is speculated that the polarization variability observed exclusively in specific targets may be linked to the inherent physical properties of jets. However, it is imperative to scrutinize statistically significant results through concurrent observations based on a larger sample of targets in future studies. Finally, concerning the variability identified through the multiwavelength polarimetry observations of Mrk 421 and an orphan polarization rotation exhibited by PG1553+113 supports the scenario that the X-ray emission region is separate from that at longer wavelengths.

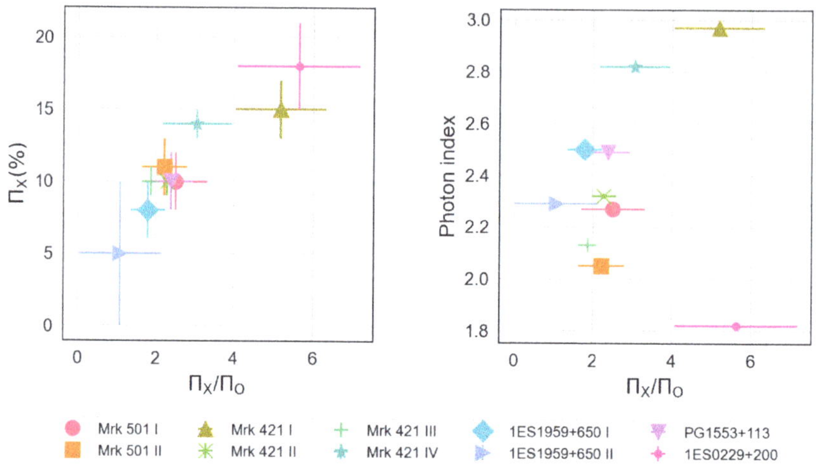

Figure 3. Physical properties of energy-stratified shock emissions from HSP blazars. (**Left**) X-ray polarization vs. the Π_X/Π_Oratio; (**Right**) X-ray power law photon index vs. the Π_X/Π_O ratio of HSP blazars observed by IXPE . Each colored point indicates different IXPE observations labeled in the legend.

2.3. IXPE Observations of Intermediate and Low Synchrotron Peak Blazars

In the case of ISP and LSP blazars, IXPE observations have been carried out for BL Lac [67,68], 3C 273, 3C 279, 3C 454.3, and S5 0716+714 [69]. BL Lac is typically classified as an LSP or ISP, where the Compton hump dominates the soft X-ray band. The X-ray flux and photon index of this target are typically $< 2 \times 10^{-11}$erg s^{-1}and < 2, respectively. BL Lac has been observed by the IXPE three times: during an average flux state (BL Lac I and II; [62]) and during an outburst (BL Lac III; [67]). From BL Lac I and II, the upper limits of $\Pi_X \leq$ 14.2% and $\Pi_X \leq$ 12.6% were estimated, respectively, and the median optical polarization values corresponding to Π_O = 6.8%, ψ_O = 107° (during BL Lac I), and Π_O = 14.2% and ψ_O = 42° (during BL Lac II) were measured from contemporaneous multiwavelength polarimetry observations. These results were interpreted as follows: (1) In the leptonic scenario, considering that emission originating from SSC dominates over EC emission in the *IXPE* energy band according to SED modeling, approximately$\sim$3% of Π_X is predicted from given polarization in radio band, Π_R(BL Lac I$\sim$4% and BL Lac II$\sim$6%) [56]. Therefore, the measured X-ray upper limit is consistent with the leptonic scenario. (2) In the hadronic scenario, Π_X similar to or higher than Π_O is predicted, but Π_O was observed to be high. Therefore, the hadronic scenario is disfavored. In conclusion, this disfavors a significant contribution of proton synchrotron radiation to the X-ray emission at these epochs. Instead, it supports a leptonic origin for the X-ray emission in BL Lac.

Furthermore, in the case of BL Lac III, a distinct synchrotron contribution was observed alongside the outburst, with $F_{2-8\text{keV}} = 2.77 \pm 0.21 \times 10^{-11}$erg s^{-1} and a photon index of 2.10 ± 0.09. Subsequently, Ref. [67] investigated the flux and polarization contributions of the synchrotron emissions across both the soft-energy range (2–4 keV) and the entire *IXPE* band (2–8 keV) using two power law components. Through time-resolved polarization analysis, dividing the total observation period into three equal sub-periods, a significant X-ray polarization, exceeding 99% significance, with $\Pi_X = 21.7^{+5.6}_{-7.9}$ and $\psi_X = -28.7 \pm 8.7$, was detected only in the first sub-period (upper limits were observed for the remaining sub-periods). This was accompanied by a decrease in the flux contribution of the synchrotron emission component. Moreover, the Π_X value was higher than the concurrently measured $\Pi_O = 13.1 \pm 2.4\%$. These results were explained by the influence of turbulence in the magnetic field, closer to the particle acceleration site in a smaller volume with well-ordered magnetic fields, akin to the characteristics of synchrotron emission observed in HSPs, leading to the higher Π_X observed in the high-energy regime. Consequently, BL Lac III was concluded to probe the high-frequency end of the synchrotron component, which diminished during the observation.

For the remaining LSP and ISP sources (3C 273, 3C 279, 3C 454.3, and S5 0716+714), only upper limits could be estimated from *IXPE* observations in the 2–8 keV band (3C 273: $\Pi_X < 9.0\%$, 3C 279: $\Pi_X < 12.7\%$, 3C 454.3: $\Pi_X < 28.0\%$, and S5 0716+714: $\Pi_X < 26.0\%$). These observational results, when compared with polarization in the optical and radio domains, suggest that $\Pi_X < \sim 10\%$ and is likely to be less than or comparable to contemporaneous Π_O. Therefore, although hadronic models are not completely eliminated, these X-ray polarization measurements seem to favor the leptonic process, wherein photons are upscattered by relativistic electrons in the jets of blazars. Lastly, Ref. [69] emphasizes that further IXPE observations can provide additional insights into the X-ray emission mechanism in LSP and ISP blazars.

3. X-ray Polarization from Hot Coronae

3.1. Theoretical Expectations

The geometry and size of the X-ray corona serve as indicators for exploring its physical origin. The corona could manifest as a slab-like or wedge-like structure originating from the accretion disk, e.g., [30,70], and it could assume a conical shape if it forms the base of a jet, e.g., [29,71,72], or a failed jet, e.g., [28]. Often, it has been assumed for simplicity as a point- or sphere-like configuration located above and below the black hole rotation axis, called lamppost geometry, e.g., [73–75].

The X-ray polarization from the corona is expected to depend on its size, geometry, and position in the system. Recent simulation results using the Monte Carlo Comptonization code (MONK) [76,77] yield the following polarizations for different corona geometries: (1) the slab geometry is predicted to exhibit a maximum polarization, Π_X up to 12%, with ψ_X oriented parallel to the disk axis, i.e., parallel to the direction of the jet; (2) conical geometry is expected to result in ψ_X parallel to the major axis of the projection of the disk on the sky with Π_X up to 7%; (3) in the case of spherical lamppost geometry, Π_X of 1–3% and the same ψ_X as conical geometry are predicted (see Figure 4).

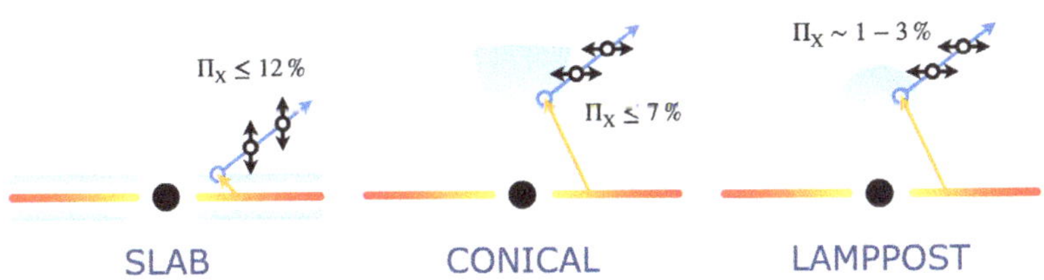

Figure 4. Schematic view of different X-ray corona geometries: slab, conical, and lamppost. The bluish and reddish areas represent the corona and disk emission regions, respectively. Blue and orange arrows indicate upscattered X-ray and UV/optical disk photons, respectively. The black arrow represents the predicted polarization properties (degree in number and angle in direction of the arrow), estimated from [77].

3.2. IXPE Observations of Unobscured Radio-Quiet AGN

Since the beginning of IXPE operation, three unobscured radio-quiet Seyfert 1 galaxies have been observed (NGC 4151: [78], IC 4329A: [79], and MCG-05-23-16: [80,81]). The X-ray polarization analysis for RQAGN was conducted using both model-independent and spectropolarimetric analysis methods. These methods were employed to examine the time-averaged polarization and the variability in polarization over time and energy. In particular, the relatively more complex spectral features of RQAGN were examined based on simultaneous observations from XMM-Newton and NuSTAR, which cover a wider energy range with higher spectral resolutions. This allows for a more detailed spectropolarimetric analysis, resulting in the measurement of X-ray polarization characteristics for each emission component. Finally, the polarization properties measured by the IXPE were used to constrain the geometric characteristics of the corona by comparing the expected X-ray polarization results predicted under the different geometrical conditions of the corona with the MONK simulation results.

First, NGC 4151 exhibited X-ray polarization in the 2–8 keV energy range, with $\Pi_X = 4.9 \pm 1.1\%$ and $\psi_X = 86 \pm 7^\circ$ (as shown in the left panel of Figure 5), corresponding to $\sim 4.4\sigma$ significance, as determined by a model-independent polarization analysis. Moreover, NGC 4151 revealed statistically significant variations in the energy-resolved polarization analysis; i.e., we found $\Pi_X = 4.3 \pm 1.6\%$ and $\psi_X = 42 \pm 11^\circ$ in the soft-energy range of 2–3.5 keV, deviating from the measurements in other energy ranges. Additionally, spectropolarimetric analysis (middle panel in Figure 5) takes into account the polarization dilution effect due to the soft-energy regime and sets the polarization angles of the primary emission and reflection emission components perpendicular to each other. As a result, for the primary emission component, $\Pi_X = 7.7 \pm 1.5\%$ and $\psi_X = 87 \pm 6^\circ$ were derived, while, for the reflection component, an upper limit of $\Pi_X < 27\%$ was obtained. Additionally, NGC 4151 exhibits a radio jet feature in VLBI observations [82], and the measured ψ_X shows that it aligns with the direction of the radio jet emission, $\sim 83^\circ$ [83,84]. This indicates that the corona emission originates from the equatorial plane. These polarization results ruled out lamppost and conical geometry scenarios among the proposed coronal geometries. Furthermore, a comparison with the MONK simulations (right panel in Figure 5) confirmed that the IXPE observational results agree well with the predicted polarization results from the slab or wedge geometry. Moreover, the observed ψ_X in the soft-energy band is similar to that of the extended narrow-line region (NLR), with a confirmed orientation by HST [85,86], supported by a Chandra observation [87] value of $45 \pm 5^\circ$, suggesting that a new emission component may influence the X-ray polarization, which is to be confirmed through additional observations.

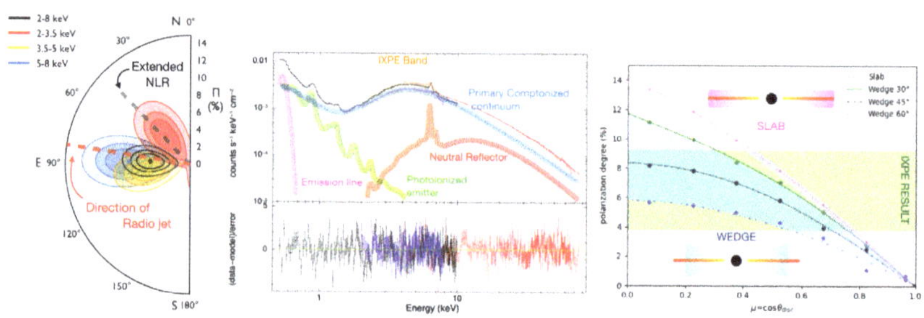

Figure 5. IXPE observation result of NGC 4151. (**Left**) Polarization contours (68%, 90%, and 99% detection significance levels); (**middle**) X-ray spectrum analysis result; (**right**) comparison of IXPE measurements and expected X-ray polarization in slab and wedge geometry coronae calculated from MONK simulations. The figure was reproduced with kind permission from Oxford University Press and the Royal Astronomical Society from [78].

Next, MCG-05-23-16 underwent two observations (MCG-05-23-16 I: May 2022 and MCG-05-23-16 II: November 2022), yielding upper limits of $\Pi_X \leq 4.7\%$ and $\Pi_X \leq 3.3\%$, respectively. The integrated analysis of MCG-05-23-16 I and II resulted in an upper limit of $\Pi_X \leq 3.2\%$, and Figure 6 illustrates the comparison of the X-ray polarization contour plot with the predicted polarization characteristics from the MONK simulations of the proposed corona geometries. Thus, despite ψ being estimated at a $\sim 1\sigma$ confidence level, considering the direction of the NLR estimated by the HST [O III] emission observations to be $\sim 40^\circ$ [88], and taking into account the perpendicular orientation of the NLR and the accretion disk, the results hint at a coronal geometry in the form of a slab or wedge rather than a lamppost or conical shape. Additionally, the disk inclination estimated in [89] of ∼30–50° supports the results of a coronal structure like a slab or wedge.

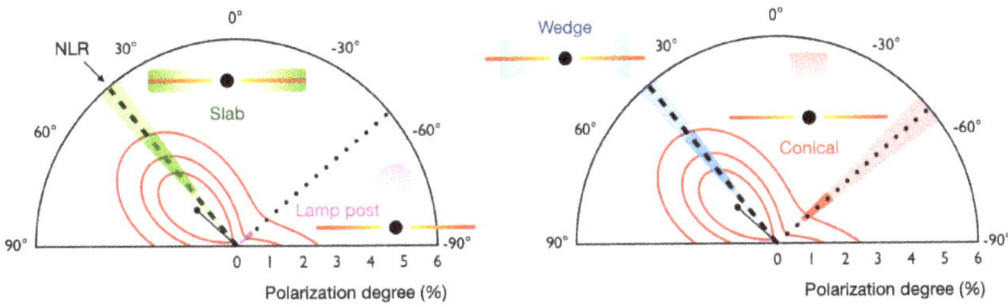

Figure 6. X-ray polarization contours of the integrated analysis of MCG-05-23-16 I and II (red; 68%, 90%, and 99% detection significance levels). The black dashed line indicates the direction of the NLR, while the dotted line denotes the perpendicular direction, implying an angle parallel to the disk. Each colored area represents the expected polarization properties calculated by MONK, with relatively saturated regions indicating the expected degree of polarizaton for inclinations in the range of approximately 30–50°. The figure was reproduced with kind permission from Oxford University Press and the Royal Astronomical Society from Figure 8 in [80].

Finally, IC 4329A has a detected X-ray polarization of the hot corona of $3.3 \pm 1.1\%$ for Π_X and $78 \pm 10°$ for ψ_X in the 2–8 keV band, utilizing simultaneous observational data from IXPE , XMM-Newton, and NuSTAR for the spectropolarimetric analysis. For the spectral model, a Comptonization and a reflection component have been applied. The ψ_X was roughly aligned with the position angle of the extended radio emission at ∼90°, e.g., [90]. This result, consistent with NGC 4151 and MCG-05-23-16, supports the geometric properties of a radially extended corona within the plane of the disk. Additionally, in [79], through a more detailed examination using MONK simulations, a test of the slab and wedge geometry corona system has been conducted. This tentatively suggests, at a 90% confidence level, a lower limit for Π_X, favoring asymmetric configurations possibly characterized by outflowing coronal geometries extended within the disk plane, which would generate highly polarized emissions aligned with the jet.

4. Summary

The IXPE mission has pioneered X-ray polarimetry, achieving the first-ever detections of X-ray polarization from AGN. The diverse physical characteristics of AGN observed in the X-ray range, particularly in the RLAGN population represented by blazars and the RQAGN population represented by unobscured RQAGN, have provided crucial insights into estimating the particle acceleration processes within relativistic jets, the physical and geometrical properties of magnetic fields, and the intrinsic geometry of hot coronae. The IXPE observations of blazars, notably the HSP subclass starting with the first observation of Mrk 501, are consistent with the possibility that the synchrotron emission within the jet originates from energy-stratified shocks. Subsequent observations of Mrk 421 have further supported this scenario and indicated the presence of helical magnetic fields by discovering the smooth rotation of the polarization angle for the first time in the X-ray range. Additionally, other HSP observations, including 1ES1959+650, PG1553+113, and 1ES0229+200, showed values of the Π_X/Π_O ratio around ∼2–6, supporting the conclusions drawn from Mrk501 and Mrk421. Furthermore, extensive comparisons of all the HSP observations have led to a physical interpretation suggesting that synchrotron jet emission at high frequencies takes place in environments with smaller volumes of well-ordered magnetic fields close to shock fronts. The polarization information at longer wavelengths is diluted by magnetic field turbulence. This interpretation could gain further solidity through additional IXPE observations focusing on spectral properties and polarization angles. Furthermore, temporal variations in polarization properties have been solidly observed only in Mrk 421 to date. Hence, the characteristic of significant polarization variability observed only in specific targets should be meticulously investigated in future studies, considering additional parameters such as the physical characteristics of the jet and utilizing more extensive observational data. Lastly, the individual variations in polarization observed across multiple wavelengths imply spatial segregation between X-ray emissions and emissions at other wavelengths.

IXPE observations of other subclasses of blazars, including LSPs and ISPs, were carried out to explore the particle acceleration mechanisms of the Inverse Compton emission component situated in the jet's higher-energy band. The observations of BL Lac objects confirmed that the polarization characteristics predicted by the leptonic-based SSC acceleration mechanisms better explain the IXPE observations than hadronic scenarios. However, the relatively low X-ray flux compared to HSPs poses challenges in achieving statistically significant observational results for many targets. Thus, further exploration through observations during higher-source activity states such as future outbursts or longer exposure times should be considered.

The IXPE observations of RQAGN explore the geometry of the X-ray corona centered around the unobscured central engine. Observations of NGC4151, MCG-05-23-16, and IC4329A with IXPE detected polarization within 3–5% from the coronal emission, aligned with the inferred jet direction. These results suggest that the corona, instead of being located on the black hole axis as expected if it is the base or possibly failed jet, is better explained by a slab-like or wedge-like shape, believed to originate from Comptonization around the accretion disk, such as a disk outflow. The IXPE findings were further validated through a comparison with the predicted X-ray polarization characteristics calculated via MONK simulations based on various geometrical structures. Furthermore, observations of NGC 4151 suggested the possibility of other emission components in the soft-energy band influencing the X-ray polarization, a characteristic that will be further explored in depth through future additional IXPE observations. Finally, future IXPE observations aimed at probing the different states of the corona emission process and reflection emission in the accretion disk will provide additional evidence to expand our understanding of the physical emission processes occurring inside AGN.

Author Contributions: Conceptualization, formal analysis, visualization, and writing—original draft preparation, D.E.K.; writing—review and editing, L.D.G., F.M., A.P.M., G.M., P.S., F.T., E.C. and I.D. All authors have read and agreed to the published version of the manuscript.

Funding: The Imaging X-ray Polarimetry Explorer (IXPE) is a joint US and Italian mission. The US contribution is supported by the National Aeronautics and Space Administration (NASA) and led and managed by its Marshall Space Flight Center (MSFC), with industry partner Ball Aerospace (contract NNM15AA18C). The Italian contribution is supported by the Italian Space Agency (Agenzia Spaziale Italiana, ASI) through contract ASI-OHBI-2017-12-I.0, agreements ASI-INAF-2017-12-H0 and ASI-INFN-2017.13-H0, and its Space Science Data Center (SSDC), and by the Istituto Nazionale di Astrofisica (INAF) and the Istituto Nazionale di Fisica Nucleare (INFN) in Italy. This research used data products provided by the IXPE Team (MSFC, SSDC, INAF, and INFN) and distributed with additional software tools by the High-Energy Astrophysics Science Archive Research Center (HEASARC) at NASA Goddard Space Flight Center (GSFC). The research of APM was partially supported by National Science Foundation grant AST-2108622, NASA Swift grant 80NSSC23K1145, and NASA NuSTAR grant 80NSSC22K0537.

Data Availability Statement: IXPE data are publicly available at https://heasarc.gsfc.nasa.gov/docs/heasarc/missions/ixpe.html (accessed on 17 March 2024).

Conflicts of Interest: The authors declare no conflict of interest.

References

1. Woo, J.H.; Urry, C.M. Active Galactic Nucleus Black Hole Masses and Bolometric Luminosities. *Astrophys. J.* **2002**, *579*, 530–544. [CrossRef]
2. Perna, M.; Lanzuisi, G.; Brusa, M.; Mignoli, M.; Cresci, G. An X-ray/SDSS sample. I. Multi-phase outflow incidence and dependence on AGN luminosity. *Astron. Astrophys.* **2017**, *603*, A99. [CrossRef]
3. Bambi, C.; Sangangelo, A. *Handbook of X-ray and GAMMA-ray Astrophysics*; Springer: Singapore, 2022.
4. Rybicki, G.B.; Lightman, A.P. *Radiative Processes in Astrophysics*; John Wiley & Sons: Hoboken, NJ, USA, 1979.
5. Weisskopf, M.C.; Soffitta, P.; Baldini, L.; Ramsey, B.D.; O'Dell, S.L.; Romani, R.W.; Matt, G.; Deininger, W.D.; Baumgartner, W.H.; Bellazzini, R.; et al. The Imaging X-ray Polarimetry Explorer (IXPE): Pre-Launch. *J. Astron. Telesc. Instrum. Syst.* **2022**, *8*, 026002. [CrossRef]
6. Costa, E.; Soffitta, P.; Bellazzini, R.; Brez, A.; Lumb, N.; Spandre, G. An efficient photoelectric X-ray polarimeter for the study of black holes and neutron stars. *Nature* **2001**, *411*, 662–665. [CrossRef] [PubMed]
7. Urry, C.M.; Padovani, P. Unified Schemes for Radio-Loud Active Galactic Nuclei. *Publ. Astron. Soc. Pac.* **1995**, *107*, 803. [CrossRef]
8. Padovani, P.; Alexander, D.M.; Assef, R.J.; De Marco, B.; Giommi, P.; Hickox, R.C.; Richards, G.T.; Smolčić, V.; Hatziminaoglou, E.; Mainieri, V.; et al. Active galactic nuclei: What's in a name? *Astron. Astrophys. Rev.* **2017**, *25*, 2. [CrossRef]
9. Padovani, P. On the two main classes of active galactic nuclei. *Nat. Astron.* **2017**, *1*, 0194. [CrossRef]
10. Hovatta, T.; Lindfors, E.; Blinov, D.; Pavlidou, V.; Nilsson, K.; Kiehlmann, S.; Angelakis, E.; Fallah Ramazani, V.; Liodakis, I.; Myserlis, I.; et al. Optical polarization of high-energy BL Lacertae objects. *Astron. Astrophys.* **2016**, *596*, A78. [CrossRef]
11. Jones, T.W.; O'Dell, S.L.; Stein, W.A. Physics of Compact Nonthermal Sources. I. Theory of Radiation Processes. *Astrophys. J.* **1974**, *188*, 353–368. [CrossRef]
12. Maraschi, L.; Ghisellini, G.; Celotti, A. A Jet Model for the Gamma-Ray–emitting Blazar 3C 279. *Astrophys. J. Lett.* **1992**, *397*, L5. [CrossRef]
13. Dermer, C.D.; Schlickeiser, R. Model for the High-Energy Emission from Blazars. *Astrophys. J.* **1993**, *416*, 458. [CrossRef]
14. Sikora, M.; Begelman, M.C.; Rees, M.J. Comptonization of Diffuse Ambient Radiation by a Relativistic Jet: The Source of Gamma Rays from Blazars? *Astrophys. J.* **1994**, *421*, 153. [CrossRef]
15. Begelman, M.C.; Sikora, M. Inverse Compton Scattering of Ambient Radiation by a Cold Relativistic Jet: A Source of Beamed, Polarized X-ray and Optical Observations of X-ray–selected BL Lacertae Objects. *Astrophys. J.* **1987**, *322*, 650. [CrossRef]
16. Mannheim, K. The proton blazar. *Astron. Astrophys.* **1993**, *269*, 67–76. [CrossRef]
17. Aharonian, F.A. TeV gamma rays from BL Lac objects due to synchrotron radiation of extremely high energy protons. *New Astron.* **2000**, *5*, 377–395. [CrossRef]
18. Böttcher, M.; Reimer, A.; Sweeney, K.; Prakash, A. Leptonic and Hadronic Modeling of Fermi-detected Blazars. *Astrophys. J.* **2013**, *768*, 54, [CrossRef]
19. Blandford, R.; Meier, D.; Readhead, A. Relativistic Jets from Active Galactic Nuclei. *Annu. Rev. Astron. Astrophys.* **2019**, *57*, 467–509. [CrossRef]
20. Abdo, A.A.; Ackermann, M.; Agudo, I.; Ajello, M.; Aller, H.D.; Aller, M.F.; Angelakis, E.; Arkharov, A.A.; Axelsson, M.; Bach, U.; et al. The Spectral Energy Distribution of Fermi Bright Blazars. *Astrophys. J.* **2010**, *716*, 30–70. [CrossRef]
21. Haardt, F.; Maraschi, L. A Two-Phase Model for the X-ray Emission from Seyfert Galaxies. *Astrophys. J. Lett.* **1991**, *380*, L51. [CrossRef]
22. Marin, F.; Dovčiak, M.; Kammoun, E.S. Contribution of parsec-scale material on to the polarized X-ray spectrum of type 1 Seyfert galaxies. *Mon. Not. R. Astron. Soc.* **2018**, *478*, 950–960. [CrossRef]
23. Antonucci, R. Unified models for active galactic nuclei and quasars. *Annu. Rev. Astron. Astrophys.* **1993**, *31*, 473–521. [CrossRef]
24. Wu, X.B.; Han, J.L. Inclinations and Black Hole Masses of Seyfert 1 Galaxies. *Astrophys. J. Lett.* **2001**, *561*, L59–L62. [CrossRef]
25. Singha, M.; Husemann, B.; Urrutia, T.; O'Dea, C.P.; Scharwächter, J.; Gaspari, M.; Combes, F.; Nevin, R.; Terrazas, B.A.; Pérez-Torres, M.; et al. The Close AGN Reference Survey (CARS). Locating the [O III] wing component in luminous local Type 1 AGN. *Astron. Astrophys.* **2022**, *659*, A123. [CrossRef]
26. Sunyaev, R.A.; Titarchuk, L.G. Comptonization of X-rays in Plasma Clouds - Typical Radiation Spectra. *Astron. Astrophys.* **1980**, *86*, 121.
27. Di Matteo, T. Magnetic reconnection: Flares and coronal heating in active galactic nuclei. *Mon. Not. R. Astron. Soc.* **1998**, *299*, L15–L20. [CrossRef]
28. Ghisellini, G.; Haardt, F.; Matt, G. Aborted jets and the X-ray emission of radio-quiet AGNs. *Astron. Astrophys.* **2004**, *413*, 535–545. [CrossRef]
29. Markoff, S.; Nowak, M.A.; Wilms, J. Going with the Flow: Can the Base of Jets Subsume the Role of Compact Accretion Disk Coronae? *Astrophys. J.* **2005**, *635*, 1203–1216. [CrossRef]
30. Haardt, F.; Matt, G. X-ray polarization in the two-phase model for AGN and X-ray binaries. *Mon. Not. R. Astron. Soc.* **1993**, *261*, 346–352. [CrossRef]
31. Matt, G.; Perola, G.C.; Costa, E.; Piro, L. X-ray polarization of the reprocessed emission from accretion disk in Seyfert galaxies. In *Two Topics in X-ray Astronomy, Volume 1: X Ray Binaries. Volume 2: AGN and the X Ray Background*; Hunt, J., Battrick, B., Eds.; ESA Special Publication: Paris, France, 1989; Volume 296, pp. 991–993.

32. Matt, G.; Fabian, A.C.; Ross, R.R. X-ray photoionized accretion discs: UV and X-ray continuum spectra and polarization. *Mon. Not. R. Astron. Soc.* **1993**, *264*, 839–852. [CrossRef]
33. Jorstad, S.G.; Marscher, A.P.; Lister, M.L.; Stirling, A.M.; Cawthorne, T.V.; Gear, W.K.; Gómez, J.L.; Stevens, J.A.; Smith, P.S.; Forster, J.R.; et al. Polarimetric Observations of 15 Active Galactic Nuclei at High Frequencies: Jet Kinematics from Bimonthly Monitoring with the Very Long Baseline Array. *Astron. J.* **2005**, *130*, 1418–1465. [CrossRef]
34. Kiehlmann, S.; Blinov, D.; Pearson, T.J.; Liodakis, I. Optical EVPA rotations in blazars: Testing a stochastic variability model with RoboPol data. *Mon. Not. R. Astron. Soc.* **2017**, *472*, 3589–3604. [CrossRef]
35. Blinov, D.; Pavlidou, V.; Papadakis, I.; Kiehlmann, S.; Liodakis, I.; Panopoulou, G.V.; Angelakis, E.; Baloković, M.; Hovatta, T.; King, O.G.; et al. RoboPol: Connection between optical polarization plane rotations and gamma-ray flares in blazars. *Mon. Not. R. Astron. Soc.* **2018**, *474*, 1296–1306. [CrossRef]
36. Larionov, V.M.; Jorstad, S.G.; Marscher, A.P.; Morozova, D.A.; Blinov, D.A.; Hagen-Thorn, V.A.; Konstantinova, T.S.; Kopatskaya, E.N.; Larionova, L.V.; Larionova, E.G.; et al. The Outburst of the Blazar S5 0716+71 in 2011 October: Shock in a Helical Jet. *Astrophys. J.* **2013**, *768*, 40. [CrossRef]
37. Marscher, A.P.; Jorstad, S.G.; D'Arcangelo, F.D.; Smith, P.S.; Williams, G.G.; Larionov, V.M.; Oh, H.; Olmstead, A.R.; Aller, M.F.; Aller, H.D.; et al. The inner jet of an active galactic nucleus as revealed by a radio-to-γ-ray outburst. *Nature* **2008**, *452*, 966–969. [CrossRef] [PubMed]
38. Lyutikov, M.; Kravchenko, E.V. Polarization swings in blazars. *Mon. Not. R. Astron. Soc.* **2017**, *467*, 3876–3886. [CrossRef]
39. Marscher, A.P. Turbulent, Extreme Multi-zone Model for Simulating Flux and Polarization Variability in Blazars. *Astrophys. J.* **2014**, *780*, 87. [CrossRef]
40. Marscher, A.P.; Jorstad, S.G.; Williamson, K.E. Modeling the Time-Dependent Polarization of Blazars. *Galaxies* **2017**, *5*, 63. [CrossRef]
41. Marscher, A.P.; Jorstad, S.G. Frequency and Time Dependence of Linear Polarization in Turbulent Jets of Blazars. *Galaxies* **2021**, *9*, 27. [CrossRef]
42. Achterberg, A.; Gallant, Y.A.; Kirk, J.G.; Guthmann, A.W. Particle acceleration by ultrarelativistic shocks: Theory and simulations. *Mon. Not. R. Astron. Soc.* **2001**, *328*, 393–408. [CrossRef]
43. Spitkovsky, A. Particle Acceleration in Relativistic Collisionless Shocks: Fermi Process at Last? *Astrophys. J.* **2008**, *682*, L5, [CrossRef]
44. Mizuno, Y.; Lyubarsky, Y.; Nishikawa, K.I.; Hardee, P.E. Three-Dimensional Relativistic Magnetohydrodynamic Simulations of Current-Driven Instability. I. Instability of a Static Column. *Astrophys. J.* **2009**, *700*, 684–693. [CrossRef]
45. Alves, E.P.; Zrake, J.; Fiuza, F. Efficient Nonthermal Particle Acceleration by the Kink Instability in Relativistic Jets. *Phys. Rev. Lett.* **2018**, *121*, 245101. [CrossRef]
46. Sironi, L.; Spitkovsky, A. Relativistic Reconnection: An Efficient Source of Non-thermal Particles. *Astrophys. J. Lett.* **2014**, *783*, L21, [CrossRef]
47. Guo, F.; Li, H.; Daughton, W.; Liu, Y.H. Formation of Hard Power Laws in the Energetic Particle Spectra Resulting from Relativistic Magnetic Reconnection. *Phys. Rev. Lett.* **2014**, *113*, 155005. [CrossRef]
48. Zhang, H.; Li, X.; Guo, F.; Giannios, D. Large-amplitude Blazar Polarization Angle Swing as a Signature of Magnetic Reconnection. *Astrophys. J. Lett.* **2018**, *862*, L25. [CrossRef]
49. Zhang, H.; Li, X.; Giannios, D.; Guo, F.; Thiersen, H.; Böttcher, M.; Lewis, T.; Venters, T. Radiation and Polarization Signatures from Magnetic Reconnection in Relativistic Jets. II. Connection with γ-Rays. *Astrophys. J.* **2022**, *924*, 90. [CrossRef]
50. Tavecchio, F. Probing Magnetic Fields and Acceleration Mechanisms in Blazar Jets with X-ray Polarimetry. *Galaxies* **2021**, *9*, 37. [CrossRef]
51. Di Gesu, L.; Tavecchio, F.; Donnarumma, I.; Marscher, A.; Pesce-Rollins, M.; Landoni, M. Testing particle acceleration models for BL Lac jets with the Imaging X-ray Polarimetry Explorer. *Astron. Astrophys.* **2022**, *662*, A83. [CrossRef]
52. Tavecchio, F.; Landoni, M.; Sironi, L.; Coppi, P. Probing shock acceleration in BL Lac jets through X-ray polarimetry: the time-dependent view. *Mon. Not. R. Astron. Soc.* **2020**, *498*, 599–608. [CrossRef]
53. Tavecchio, F.; Landoni, M.; Sironi, L.; Coppi, P. Probing dissipation mechanisms in BL Lac jets through X-ray polarimetry. *Mon. Not. R. Astron. Soc.* **2018**, *480*, 2872–2880. [CrossRef]
54. Guo, F.; Liu, Y.H.; Daughton, W.; Li, H. Particle Acceleration and Plasma Dynamics during Magnetic Reconnection in the Magnetically Dominated Regime. *Astrophys. J.* **2015**, *806*, 167. [CrossRef]
55. Zhang, H.; Böttcher, M. X-ray and Gamma-Ray Polarization in Leptonic and Hadronic Jet Models of Blazars. *Astrophys. J.* **2013**, *774*, 18. [CrossRef]
56. Peirson, A.L.; Romani, R.W. The Polarization Behavior of Relativistic Synchrotron Self-Compton Jets. *Astrophys. J.* **2019**, *885*, 76. [CrossRef]
57. Liodakis, I.; Marscher, A.P.; Agudo, I.; Berdyugin, A.V.; Bernardos, M.I.; Bonnoli, G.; Borman, G.A.; Casadio, C.; Casanova, V.; Cavazzuti, E.; et al. Polarized blazar X-rays imply particle acceleration in shocks. *Nature* **2022**, *611*, 677–681. [CrossRef]
58. Di Gesu, L.; Donnarumma, I.; Tavecchio, F.; Agudo, I.; Barnounin, T.; Cibrario, N.; Di Lalla, N.; Di Marco, A.; Escudero, J.; Errando, M.; et al. The X-ray Polarization View of Mrk 421 in an Average Flux State as Observed by the Imaging X-ray Polarimetry Explorer. *Astrophys. J. Lett.* **2022**, *938*, L7. [CrossRef]

59. Di Gesu, L.; Marshall, H.L.; Ehlert, S.R.; Kim, D.E.; Donnarumma, I.; Tavecchio, F.; Liodakis, I.; Kiehlmann, S.; Agudo, I.; Jorstad, S.G.; et al. Discovery of X-ray polarization angle rotation in the jet from blazar Mrk 421. *Nat. Astron.* **2023**, *7*, 1245–1258. [CrossRef]
60. Kim, D.E.; Di Gesu, L.; Liodakis, I.; Marscher, A.P.; Jorstad, S.G.; Middei, R.; Marshall, H.L.; Pacciani, L.; Agudo, I.; Tavecchio, F.; et al. Magnetic field properties inside the jet of Mrk 421. Multiwavelength polarimetry, including the Imaging X-ray Polarimetry Explorer. *Astron. Astrophys.* **2024**, *681*, A12. [CrossRef]
61. Errando, M.; Liodakis, I.; Marscher, A.P.; Marshall, H.L.; Middei, R.; Negro, M.; Peirson, A.L.; Perri, M.; Puccetti, S.; Rabinowitz, P.L.; et al. Detection of X-ray Polarization from the Blazar 1ES 1959+650 with the Imaging X-ray Polarimetry Explorer. *arXiv* **2024**, arXiv:2401.04420. [CrossRef]
62. Middei, R.; Perri, M.; Puccetti, S.; Liodakis, I.; Di Gesu, L.; Marscher, A.P.; Rodriguez Cavero, N.; Tavecchio, F.; Donnarumma, I.; Laurenti, M.; et al. IXPE and Multiwavelength Observations of Blazar PG 1553+113 Reveal an Orphan Optical Polarization Swing. *Astrophys. J. Lett.* **2023**, *953*, L28. [CrossRef]
63. Ehlert, S.R.; Liodakis, I.; Middei, R.; Marscher, A.P.; Tavecchio, F.; Agudo, I.; Kouch, P.M.; Lindfors, E.; Nilsson, K.; Myserlis, I.; et al. X-ray Polarization of the BL Lacertae Type Blazar 1ES 0229+200. *Astrophys. J.* **2023**, *959*, 61. [CrossRef]
64. Kislat, F.; Clark, B.; Beilicke, M.; Krawczynski, H. Analyzing the data from X-ray polarimeters with Stokes parameters. *Astroparticle Physics* **2015**, *68*, 45–51. [arXiv:astro-ph.IM/1409.6214]. [CrossRef]
65. Strohmayer, T.E. X-ray Spectro-polarimetry with Photoelectric Polarimeters. *Astrophys. J.* **2017**, *838*, 72. [CrossRef]
66. Massaro, E.; Perri, M.; Giommi, P.; Nesci, R. Log-parabolic spectra and particle acceleration in the BL Lac object Mkn 421: Spectral analysis of the complete BeppoSAX wide band X-ray data set. *Astron. Astrophys.* **2004**, *413*, 489–503. [CrossRef]
67. Peirson, A.L.; Negro, M.; Liodakis, I.; Middei, R.; Kim, D.E.; Marscher, A.P.; Marshall, H.L.; Pacciani, L.; Romani, R.W.; Wu, K.; et al. X-ray Polarization of BL Lacertae in Outburst. *Astrophys. J. Lett.* **2023**, *948*, L25. [CrossRef]
68. Middei, R.; Liodakis, I.; Perri, M.; Puccetti, S.; Cavazzuti, E.; Di Gesu, L.; Ehlert, S.R.; Madejski, G.; Marscher, A.P.; Marshall, H.L.; et al. X-ray Polarization Observations of BL Lacertae. *Astrophys. J. Lett.* **2023**, *942*, L10. [CrossRef]
69. Marshall, H.L.; Liodakis, I.; Marscher, A.P.; Di Lalla, N.; Jorstad, S.G.; Kim, D.E.; Middei, R.; Negro, M.; Omodei, N.; Peirson, A.L.; et al. Observations of Low and Intermediate Spectral Peak Blazars with the Imaging X-ray Polarimetry Explorer. *arXiv* **2023**, arXiv:2310.11510. [CrossRef]
70. Poutanen, J.; Veledina, A.; Beloborodov, A.M. Polarized X-rays from Windy Accretion in Cygnus X-1. *Astrophys. J. Lett.* **2023**, *949*, L10, [CrossRef]
71. Henri, G.; Pelletier, G. Relativistic Electron-Positron Beam Formation in the Framework of the Two-Flow Model for Active Galactic Nuclei. *Astrophys. J. Lett.* **1991**, *383*, L7. [CrossRef]
72. Henri, G.; Petrucci, P.O. Anisotropic illumination of AGN's accretion disk by a non thermal source. I. General theory and application to the Newtonian geometry. *Astron. Astrophys.* **1997**, *326*, 87–98. [CrossRef]
73. Martocchia, A.; Matt, G. Iron Kalpha line intensity from accretion discs around rotating black holes. *Mon. Not. R. Astron. Soc.* **1996**, *282*, L53–L57. [CrossRef]
74. Miniutti, G.; Fabian, A.C. A light bending model for the X-ray temporal and spectral properties of accreting black holes. *Mon. Not. R. Astron. Soc.* **2004**, *349*, 1435–1448. [CrossRef]
75. Wilkins, D.R.; Fabian, A.C. Understanding X-ray reflection emissivity profiles in AGN: Locating the X-ray source. *Mon. Not. R. Astron. Soc.* **2012**, *424*, 1284–1296. [CrossRef]
76. Zhang, W.; Dovčiak, M.; Bursa, M. Constraining the Size of the Corona with Fully Relativistic Calculations of Spectra of Extended Coronae. I. The Monte Carlo Radiative Transfer Code. *Astrophys. J.* **2019**, *875*, 148. [CrossRef]
77. Ursini, F.; Matt, G.; Bianchi, S.; Marinucci, A.; Dovčiak, M.; Zhang, W. Prospects for differentiating extended coronal geometries in AGNs with the IXPE mission. *Mon. Not. R. Astron. Soc.* **2022**, *510*, 3674–3687. [CrossRef]
78. Gianolli, V.E.; Kim, D.E.; Bianchi, S.; Agís-González, B.; Madejski, G.; Marin, F.; Marinucci, A.; Matt, G.; Middei, R.; Petrucci, P.O.; et al. Uncovering the geometry of the hot X-ray corona in the Seyfert galaxy NGC 4151 with IXPE. *Mon. Not. R. Astron. Soc.* **2023**, *523*, 4468–4476. [CrossRef]
79. Ingram, A.; Ewing, M.; Marinucci, A.; Tagliacozzo, D.; Rosario, D.J.; Veledina, A.; Kim, D.E.; Marin, F.; Bianchi, S.; Poutanen, J.; et al. The X-ray polarization of the Seyfert 1 galaxy IC 4329A. *Mon. Not. R. Astron. Soc.* **2023**, *525*, 5437–5449. [CrossRef]
80. Tagliacozzo, D.; Marinucci, A.; Ursini, F.; Matt, G.; Bianchi, S.; Baldini, L.; Barnouin, T.; Cavero Rodriguez, N.; De Rosa, A.; Di Gesu, L.; et al. The geometry of the hot corona in MCG-05-23-16 constrained by X-ray polarimetry. *Mon. Not. R. Astron. Soc.* **2023**, *525*, 4735–4743. [CrossRef]
81. Marinucci, A.; Muleri, F.; Dovciak, M.; Bianchi, S.; Marin, F.; Matt, G.; Ursini, F.; Middei, R.; Marshall, H.L.; Baldini, L.; et al. Polarization constraints on the X-ray corona in Seyfert Galaxies: MCG-05-23-16. *Mon. Not. R. Astron. Soc.* **2022**, *516*, 5907–5913. [CrossRef]
82. Williams, D.R.A.; McHardy, I.M.; Baldi, R.D.; Beswick, R.J.; Argo, M.K.; Dullo, B.T.; Knapen, J.H.; Brinks, E.; Fenech, D.M.; Mundell, C.G.; et al. Radio jets in NGC 4151: Where eMERLIN meets HST. *Mon. Not. R. Astron. Soc.* **2017**, *472*, 3842–3853. [CrossRef]
83. Harrison, B.; Pedlar, A.; Unger, S.W.; Burgess, P.; Graham, D.A.; Preuss, E. The parsec-scale structure of the radio nucleus of NGC 4151. *Mon. Not. R. Astron. Soc.* **1986**, *218*, 775–784. [CrossRef]

84. Ulvestad, J.S.; Roy, A.L.; Colbert, E.J.M.; Wilson, A.S. A Subparsec Radio Jet or Disk in NGC 4151. *Astrophys. J.* **1998**, *496*, 196–202. [CrossRef]
85. Evans, I.N.; Tsvetanov, Z.; Kriss, G.A.; Ford, H.C.; Caganoff, S.; Koratkar, A.P. Hubble Space Telescope Imaging of the Narrow-Line Region of NGC 4151. *Astrophys. J.* **1993**, *417*, 82. [CrossRef]
86. Das, V.; Crenshaw, D.M. Mapping the Kinematics of the Narrow-Line Regions in NGC 4151 and NGC 1068. In *American Astronomical Society Meeting Abstracts*; American Astronomical Society: Washington, DC, USA, 2003; Volume 203, p. 56.01.
87. Wang, J.; Fabbiano, G.; Elvis, M.; Risaliti, G.; Karovska, M.; Zezas, A.; Mundell, C.G.; Dumas, G.; Schinnerer, E. A Deep Chandra ACIS Study of NGC 4151. III. The Line Emission and Spectral Analysis of the Ionization Cone. *Astrophys. J.* **2011**, *742*, 23. [CrossRef]
88. Ferruit, P.; Wilson, A.S.; Mulchaey, J. Hubble Space Telescope WFPC2 Imaging of a Sample of Early-Type Seyfert Galaxies. *Astrophys. J. Suppl.* **2000**, *128*, 139–169. [CrossRef]
89. Serafinelli, R.; Marinucci, A.; De Rosa, A.; Bianchi, S.; Middei, R.; Matt, G.; Reeves, J.N.; Braito, V.; Tombesi, F.; Gianolli, V.E.; et al. A remarkably stable accretion disc in the Seyfert galaxy MCG-5-23-16. *Mon. Not. R. Astron. Soc.* **2023**, *526*, 3540–3547. [CrossRef]
90. Unger, S.W.; Lawrence, A.; Wilson, A.S.; Elvis, M.; Wright, A.E. Radio observations of a hard X-ray selected sample of active galaxies. *Mon. Not. R. Astron. Soc.* **1987**, *228*, 521–531. [CrossRef]

galaxies MDPI

Article

Galaxy Groups as the Ultimate Probe of AGN Feedback

Dominique Eckert [1,*], Fabio Gastaldello [2], Ewan O'Sullivan [3], Alexis Finoguenov [4], Marisa Brienza [5,6] and the X-GAP Collaboration [†]

1 Department of Astronomy, University of Geneva, Ch. d'Ecogia 16, CH-1290 Versoix, Switzerland
2 INAF—IASF Milano, Via A. Corti 12, I-20133 Milano, Italy; fabio.gastaldello@inaf.it
3 Center for Astrophysics | Harvard & Smithsonian, 60 Garden Street, Cambridge, MA 02138, USA; eosullivan@cfa.harvard.edu
4 Department of Physics, University of Helsinki, Gustaf Hällströmin katu 2, FI-00014 Helsinki, Finland; alexis.finoguenov@helsinki.fi
5 INAF—Osservatorio di Astrofisica e Scienza dello Spazio di Bologna, via Gobetti 93/3, I-40129 Bologna, Italy; marisa.brienza@inaf.it
6 Dipartimento di Fisica e Astronomia, Università di Bologna, Via P. Gobetti 93/2, I-40129 Bologna, Italy
* Correspondence: dominique.eckert@unige.ch
† https://www.astro.unige.ch/xgap/

Abstract: The co-evolution between supermassive black holes and their environment is most directly traced by the hot atmospheres of dark matter halos. The cooling of the hot atmosphere supplies the central regions with fresh gas, igniting active galactic nuclei (AGN) with long duty cycles. The outflows from the central engine tightly couple with the surrounding gaseous medium and provide the dominant heating source, preventing runaway cooling. Every major modern hydrodynamical simulation suite now includes a prescription for AGN feedback to reproduce the realistic populations of galaxies. However, the mechanisms governing the feeding/feedback cycle between the central black holes and their surrounding galaxies and halos are still poorly understood. Galaxy groups are uniquely suited to constrain the mechanisms governing the cooling–heating balance, as the energy supplied by the central AGN can exceed the gravitational binding energy of halo gas particles. Here, we provide a brief overview of our knowledge of the impact of AGN on the hot atmospheres of galaxy groups, with a specific focus on the thermodynamic profiles of the groups. We then present our on-going efforts to improve on the implementation of AGN feedback in galaxy evolution models by providing precise measurements of the properties of galaxy groups. We introduce the XMM-*Newton* Group AGN Project (X-GAP), a large program on XMM-*Newton* targeting a sample of 49 galaxy groups out to R_{500c}.

Keywords: black holes; galaxy groups; elliptical galaxies; intragroup medium/plasma; active nuclei; X-ray observations; hydrodynamical and cosmological simulations

1. Introduction

The overarching goal of galaxy evolution models is to reproduce as closely as possible the properties of the baryonic content of the Universe and its evolution. In the past decade, feedback from active galactic nuclei (AGN), i.e., the self-regulated feedback cycle between central supermassive black holes (SMBH) and their host galaxies and halos, has emerged as the most likely solution to a wide range of issues in galaxy evolution [1,2]. For instance, AGN feedback is necessary to reproduce the cut-off in the galaxy stellar mass function [3], explain the origin of the scaling relations between SMBH mass and galaxy properties [4,5], interpret the co-evolution between star formation rate and SMBH activity [6], and quench catastrophic cooling flows [7]. Modern hydrodynamical galaxy evolution models, such as EAGLE [8], BAHAMAS [9], and IllustrisTNG [10], all include a prescription for AGN feedback. The implemented feedback model ranges from pure thermal feedback to mechanical, directional feedback, e.g., [11]. While the inclusion of AGN feedback into hydrodynamical simulations allowed, for the first time, the reproduction of a wide range of properties of the galaxy populations, the choice of the feedback scheme in state-of-the-art hydrodynamical simulations vastly differs from one simulation to the other. The parameters of the feedback model can then be tuned to reproduce a set of observables, which, in most cases, is the galaxy stellar mass function.

As stated above, the properties of the galaxy population are not sufficient on their own to create a unified AGN feedback model, as the parameters of multiple feedback models can be tuned to match the galaxy stellar mass function equally well. On the other hand, the hot gas content of galaxy groups, i.e., halos in the range $10^{13} < M_{500c} < 10^{14} M_{\odot}$, is highly sensitive to the implemented feedback scheme. At the current epoch, galaxy groups represent the peak of the halo mass density. They occupy a key regime in the evolution of galaxies, as typical $L_{\star}$ galaxies exist within groups of 5–20 members rather than within isolated halos [12]. In terms of the sensitivity to AGN feedback, galaxy groups occupy a transitional regime between isolated galaxies and massive galaxy clusters, as the total feedback energy is comparable to the gravitational binding energy of the gas. Their gravitational potential well is strong enough to retain a substantial hot gaseous atmosphere (the intragroup medium, hereafter, IGrM), whereas the outflows generated by the central SMBH are energetic enough to produce clearly discernible effects in the surrounding medium. Deep observations of nearby galaxy groups such as NGC 5813 [13] and NGC 5044 [14], reveal a wealth of feedback-induced features in the IGrM. Bubbles of outflowing material associated with successive outbursts of the central SMBH expand into the surrounding medium, producing pairs of cavities in the hot gas distribution [15]. The supersonic nature of the ejecta also induces shock waves propagating through the medium perpendicular to the main direction of the outflow [16]. These phenomena inject a large amount of energy into the IGrM, which offsets the radiative losses and prevents the gas from cooling. The low cooling rate eventually stops the supply of molecular gas to the central galaxy and quenches star formation [7].

In a recent paper (Eckert et al. [17]), we provided a detailed review of the AGN feedback processes in the specific context of galaxy groups. This paper provides a summary of an invited review talk given at the "AGN on the beach" conference, which took place in Tropea, Italy, from 10 to 15 September 2023. We also introduce the XMM-*Newton* Group AGN Project (X-GAP), a newly approved large program on XMM-*Newton* that aims at measuring the impact of AGN feedback on the hot atmospheres of galaxy groups in a carefully selected sample of 49 groups.

2. Galaxy Groups as Probes of AGN Feedback

The hot gaseous atmospheres of galaxy groups constitute a privileged environment for the study of AGN-induced feedback processes. To illustrate this point, in the left-hand panel of Figure 1 we show a composite image of the galaxy IC 2476. IC 2476 is a massive ($\log M_{\star} = 11.39 M_{\odot}$), quenched ($\log SFR = -2.124 M_{\odot}/\mathrm{yr}$) elliptical galaxy at $z = 0.0472$ [18]. The blue background image is an SDSS *i*-band image, which highlights the position of the galaxy at the center of the image. While the galaxy lives in a rather poor optical environment, spectroscopic data tell us that it is the dominant galaxy of a group of 12 spectroscopic members [19]. X-ray observations of this system with XMM-*Newton* (red component in Figure 1) reveal the existence of an IGrM extending over several hundred kpc centered on IC2476. The gas temperature of $\sim$1.2 keV is typical of the mass range populated by galaxy groups. On top of that, the image shows radio emission contours from the LOFAR Two-meter sky survey DR2 (LoTSS, [20]). The bright remnant radio galaxy B2 0924+30 [21,22], which extends over >100 kpc from the nucleus, is associated with IC 2476. A spectral aging analysis shows that the radio jets were active for a period of $\sim$100 Myr and switched off $\sim$50 Myr ago [22]. The radio data highlight the presence of radio-emitting electrons well beyond the size of the central galaxy, such that the bulk of the AGN energy is injected at large distances from the central galaxy within the IGrM. The outflows deposit energy within the IGrM, which reheats the surrounding medium and eventually quenches star formation [7].

It is now well established that there exists a relation between radio-loud AGN and rich environments. Indeed, radio-loud AGN are ubiquitous among the central galaxies of cool-core clusters [23,24]. The fraction of radio-loud AGN increases steeply with galaxy stellar mass [25]. The most massive galaxies, corresponding to the central galaxies of the most massive clusters hosting the most massive black holes [26], all exhibit some level of radio emissions [25]. Within systems hosting detectable X-ray cavities [27], we observe a correlation between radio power and total jet power [28], with the total energy output exceeding the radiative output by $\sim$2 orders of magnitude. Moreover, cavity power correlates with the cooling luminosity of the X-ray emitting gas [15]. The energy input from the central AGN is at least sufficient to balance the radiated energy and quench catastrophic cooling flows. Finally, the blueshifted CO absorption lines reveal the existence of radially infalling gas clouds at the vicinity of the SMBH [29,30]. Therefore, the link between radio-loud AGN and their environment is now well established. Nonetheless, the total amount of energy injected in the surroundings by AGN feedback is still poorly known, especially at the mass range populated by galaxy groups.

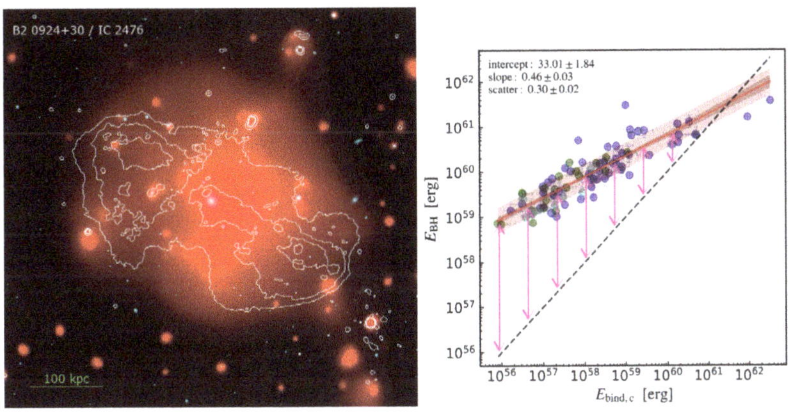

Figure 1. Impact of AGN feedback on galaxy group atmospheres. The left-panel shows an SDSS i-band image of the brightest group galaxy IC 2476 with X-ray observations of the IGrM superimposed in red. The white contours shows radio emission from the associated radio galaxy B2 0924+30 from LoTSS DR2 data [20]. The right-hand panel shows the relation between the binding energy of IGrM gas particles and the energy injection from the central BH (figure reproduced from [17]). The dashed line indicates equality between BH and binding energy.

Since the majority of the feedback energy is dissipated within the IGrM, the structural properties of the gaseous atmospheres of galaxy groups act as fossil records of the total feedback energy integrated over cosmic time. In the right-hand panel of Figure 1, we show the relation between the gas binding energy within group cores, $E_{bind} \approx 2E_{th} \propto M_{\rm gas}k_B T$, and the available SMBH mechanical energy, $E_{BH} = \epsilon_M M_{BH} c^2$, for a sample of galaxy groups with dynamically measured BH masses [31]. The ϵ_M parameter, which represents the fraction of the accreted energy that is converted into heat, was assumed to be constant at the value of $\epsilon_M = 10^{-3}$ [32]. We can see that for low-mass (i.e., low-temperature) systems, the available BH energy largely exceeds the binding energy of gas particles in halo cores, such that the energy supplied by AGN feedback is sufficient to unbind gas particles and eject them from the halo. Therefore, the total baryon fraction of galaxy groups within R_{500c} falls short of the cosmic baryon fraction Ω_b/Ω_m (e.g., [33–36]). While most studies agree on the existence of some level of gas ejection outside of halos, the exact baryon fraction of galaxy groups is still widely debated, and the scatter in the gas fraction at a fixed mass is unknown.

Given the large amount of injected energy, we expect the effect of AGN feedback to extend throughout the entire volume of these systems, and possibly even beyond their virial radius [37–39]. In Figure 2, we show the thermodynamic profiles of simulated galaxy groups extracted from four state-of-the-art codes (EAGLE, Illustris-TNG, SIMBA, and ROMULUS [40]). These simulations all implement a different prescription for AGN feedback. For instance, EAGLE and ROMULUS implement some version of the Booth and Schaye [41] model, where the energy deposition by AGN is fully thermal and isotropic. Conversely, SIMBA and Illustris-TNG implement dual models including a mechanical component to consider the impact of AGN jets and winds. The predicted properties of the IGrM largely differ from one simulation to the other. Models with strong feedback, such as SIMBA [42], predict substantially lower gas densities (Figure 2a) and higher entropies (Figure 2b) than models with relatively weak feedback (e.g., ROMULUS [43]). The discrepancies are echoed in the overall gas and baryon fractions within R_{500c} (see the discussion in Section 5.2 of [17]): while in the galaxy cluster regime ($M_{500} > 10^{14} M_\odot$), the aforementioned codes predict very similar integrated gas fractions, in the group regime, the predictions differ by up to an order of magnitude. In case the implemented feedback is very strong (e.g., Illustris [44]), most of the gas is evacuated from the halo and the measured gas fractions are very low. Conversely, models implementing a more gentle feedback scheme such as EAGLE [8] predict a high gas fraction within the halo. It is worth noting that these two simulation sets predict very similar galaxy stellar mass functions, galaxy morphologies, and star formation rate distributions. Therefore, modern simulation suites have little predictive power on the baryon content of groups, even when the properties of the galaxy population are accurately reproduced.

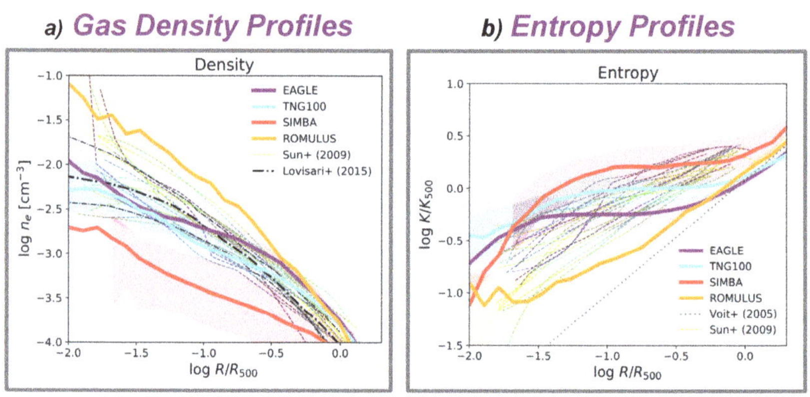

Figure 2. Impact of AGN feedback on the thermodynamic profiles of galaxy groups. The figure shows the profiles of IGrM gas density (**a**) and entropy (**b**) in four state-of-the-art simulations (EAGLE, Illustris-TNG100, SIMBA, and ROMULUS; [40]). Increasing the feedback energy leads to lower gas densities (**left**) and higher gas entropies (**right**), highlighting the sensitivity of the IGrM to the feedback scheme. For comparison, the dashed curves show observed galaxy group thermodynamic profiles from *Chandra* [45] and XMM-*Newton* [46].

On top of that, recent studies have shown that AGN feedback modifies the large-scale matter distribution in the Universe in a complicated way, which represents a leading source of systematic uncertainties for upcoming cosmology experiments. Baryonic processes affect the predicted matter power spectrum at the level of 10–20%, i.e., about an order of magnitude larger than the statistical precision of upcoming cosmic shear measurements [47]. The impact of baryonic processes depends very sensitively on the details of the chosen feedback model [48]. Calibrating the simulations against unbiased measurements of group gas fractions will be crucial for robustly predicting the impact of baryons on cosmological observables and warrant the success of upcoming cosmic shear experiments like ESA's *Euclid* mission [49].

The use of the IGrM as a probe of AGN feedback has, up until this point, been limited by our observational knowledge of its properties, which is relatively primitive compared to our understanding of the intra-cluster medium in more massive halos. While it has been known for 20 years that the scaling relations deviate from self-similarity at galaxy group scale [50,51], little is known on the thermodynamic profiles of the group population as a whole. Previous works based on *Chandra* [45] and XMM-*Newton* [46,52] are based on archival studies and focus on the brightest, most nearby systems. In the vast majority of cases, the selected systems are very nearby and fill the *Chandra* and XMM-*Newton* field of view (FOV), such that direct constraints beyond $0.5R_{500c}$ are available only for a handful of systems. Alternatively, constraints on the gas fraction in galaxy groups have been obtained from X-ray surveys [35,36], such as the XXL survey, which covers an area of 50 square degrees with *XMM-Newton* [53]. However, such observations are very shallow, and the resulting uncertainties are large: the *mean* gas fraction can typically be determined with $\sim$20% uncertainty at $10^{14} M_{\odot}$, and no information on the intrinsic scatter can be obtained. For a more comprehensive review of our knowledge of feedback effects on the IGrM, we refer the reader to Section 3 of Eckert et al. (2021) [17].

3. The XMM-Newton Group AGN Project (X-GAP)

To address the science questions highlighted in Section 2, we[1] initiated a program aiming at measuring the properties of the IGrM out to R_{500c} in a carefully selected galaxy group sample spanning the mass range $10^{13} \lesssim M_{500c} \lesssim 10^{14} M_{\odot}$. However, the question of selecting a pure and unbiased sample of galaxy groups is a complex one.

Historically, samples of galaxy groups have been selected mostly based on their optical or X-ray properties. Given the low richness of galaxy groups, selection algorithms based on photometric data (e.g., redMaPPer, [54]) are strongly affected by projection effects in the mass range of galaxy groups and are thus not well suited for their detection. Conversely, large spectroscopic surveys allow for the detection of groups in three-dimensional space using algorithms such as Friends-of-Friends (FoF). Such algorithms were applied to large spectroscopic surveys (SDSS e.g., [19,55], GAMA [12]). While spectroscopic group catalogues are much cleaner than their photometric equivalent, they can still be affected by projection effects. For instance, testing their algorithm on mock data, Robotham et al. [12] estimate that 20–30% of the groups detected with at least five group members are not bona fide group-scale halos, but are rather unvirialised systems composed of several smaller halos (see also [56]). Conversely, group selection based on X-ray surveys such as the ROSAT all-sky survey (RASS, [50,57]) or eROSITA [58], yields group samples that are very pure, since the selection is based on the presence of a virialised IGrM. However, given the limited sensitivity of these surveys, the detection is limited to the local Universe ($z < 0.1$). On top of that, the structural properties of the IGrM, and, in particular, the presence or absence of a cool core, strongly affect the detectability of groups [59,60], such that X-ray group samples are likely biased towards the most relaxed, X-ray brightest groups.

Given our goal of selecting a highly pure and unbiased sample, following Damsted et al. [56] we attempted to combine the best features of both selection methods by cross-correlating galaxy groups selected from optical spectroscopic surveys with the presence of faint, extended X-ray sources. Starting from the SDSS FoF group catalogue of Tempel et al. [19], we selected galaxy groups with a minimum of eight spectroscopic members, and cross-correlated their position with diffuse sources selected from the RASS data. Our X-ray source detection algorithm [61] removes the central flux of X-ray point-like sources and then performs a wavelet search for core-excised extended sources on scales greater than $12'$. As a result, our approach is sensitive only to large-scale X-ray emissions and is not biased toward centrally peaked systems. This strategy maximises the purity of the sample and is more complete than a pure X-ray selection. Moreover, the cross-correlation with SDSS groups provides a wealth of supporting optical data, in particular, redshifts, velocity dispersions, star formation rates, and stellar masses. For more details on the selection of the parent sample, we refer the reader to Damsted et al. [56].

In Figure 3, we show the X-ray luminosity of the detected systems and their mass estimated from a luminosity–mass relation. The points are color-coded by the apparent size θ_{500} of R_{500c}. For the purpose of this study, we perform a cut according to the following criteria:

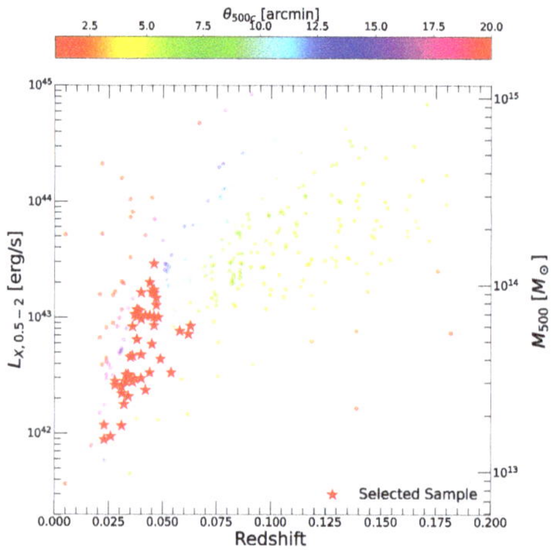

Figure 3. X-GAP group selection cross-matching extended sources detected in RASS with SDSS spectroscopic galaxy groups [56]. The plot shows the RASS X-ray luminosity and the corresponding mass estimated through a luminosity–mass relation as a function of the source redshift. The points are color-coded by the apparent size θ_{500} of R_{500c}. The selected sample is highlighted by the red stars.

- $\theta_{500} < 15'$: This ensures that R_{500c} of the selected system is contained within the XMM-*Newton* FOV;
- $z < 0.06$: Given the sensitivity curve of our sample, the systems located at greater redshifts are almost exclusively clusters with $M_{200} > 10^{14} M_{\odot}$;
- Number of member galaxies >8: This criterion removes loosely bound galaxy systems.

The selected sample comprises 49 systems with an estimated mass $10^{13} \lesssim M_{500} \lesssim 10^{14} M_{\odot}$ in the narrow redshift range $0.025 < z < 0.06$. The properties and size of the sample are sufficient to investigate the dependence of the thermodynamic profiles and the gas fraction on halo mass. The sample size of 49 systems will also allow us to reliably determine the intrinsic scatter of the quantities of interest at a fixed mass. The availability of SDSS optical data and 2MASS and WISE near-IR data ensures that we will be able to estimate the stellar fractions as well, allowing us to determine the total baryon fraction and the relative contribution of gas and stars. Finally, the vast majority of the selected systems fall within the LoTSS footprint, which will allow us to correlate the properties of the IGrM with those of the central radio galaxies.

To demonstrate the capabilities of our observing strategy, we searched the XMM-*Newton* archive and analysed the existing public data for a subset of four groups that were already observed by XMM-*Newton* and that are representative of the systems available in the archive in terms of their mass and redshift. We reduced the available data using `XMMSAS`v19.1 and extracted thermodynamic profiles and hydrostatic mass profiles using the Python package `hydromass` [62]. The resulting thermodynamic and mass profiles are shown in Figure 4. The determined masses lie in the range $5 \times 10^{13} M_{\odot} < M_{500} < 1.5 \times 10^{14} M_{\odot}$, and the recovered profiles extend to R_{500c}, which shows that our selection criteria are adequate. This analysis shows that we are able to determine f_{gas} inside R_{500c} *without extrapolation*. The availability of velocity dispersion measurements for all the systems provides an additional, independent estimate of the group's mass, which is important to verify that our mass measurements are not severely biased by the hydrostatic assumption.

Interestingly, all four systems show very flat surface brightness profiles and a strong deficit of gas within their central regions. The corresponding entropy profiles show a striking entropy excess extending all the way out to R_{500c} and high central cooling times. The selected systems also have low pressure and do not show any central temperature drop. Our selection process may thus unveil a population of low surface brightness groups that was missed by the standard RASS detection pipelines and is absent from previous galaxy group samples [45,46]. If confirmed, this finding will have profound consequences for AGN feedback models (see Figure 2), as it would imply that the properties of the IGrM are more diverse than previously thought.

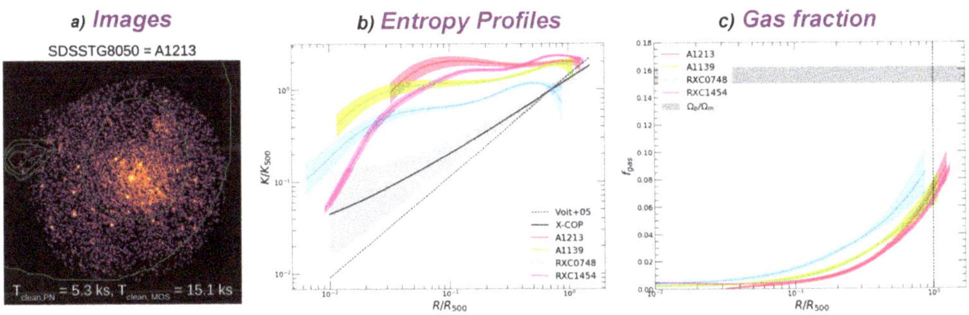

Figure 4. Analysis results for four groups matching the X-GAP selection criteria with available observations in the XMM-*Newton* archive. The selected groups exhibit a flat brightness distribution, as highlighted in the (**a**), where we show the combined XMM-*Newton*/EPIC image of one of the systems (SDSSTG 8050 = A1213) with RASS contours overlaid in green. The low surface brightness implies a strong entropy excess (**b**) extending all the way out to R_{500c}. The gas is evacuated towards the outskirts, where the gas fraction rises sharply (**c**). Given our selection process, we are able to determine the gas fraction at R_{500c} without requiring any extrapolation.

The X-GAP program aims at providing observations of similar quality over the entire sample of 49 groups presented in Figure 3. The program was accepted into A priority during XMM-*Newton* AO-19 for a total observing time of 852ks, which constitutes the largest observing program awarded that year. The X-GAP program will yield at least 20,000 source photons for each target, which is similar to the four archival observations presented in Figure 4. We note, however, that the observations previously available in the archive are skewed towards the bright end of the full sample, such that the full sample covers a substantially wider mass and luminosity range. In Appendix A, we provide a master table with the description of the 49 groups selected for X-ray follow-up with XMM-*Newton*. Out of the 49 groups, 11 were already observed previously, whereas the remaining 38 are new observations. We also provide an image gallery of all the observed systems. All the observations were reduced using `XMMSAS`v20.0 and the X-COP analysis pipeline [63]. The images provided in the gallery are adaptively smoothed, background subtracted, and vignetting corrected maps in the [0.7–1.2] keV band. The location of SDSS FoF member galaxies is also highlighted with the cyan squares. The very high success rate of our observing strategy is apparent in Figures A1–A4. Indeed, the vast majority of the selected groups host bright diffuse X-ray emission extending over 10 arcmin or more, which confirms the efficiency of our selection process. The data quality is sufficient to obtain results of similar quality as those presented in Figure 4 for a large sample of galaxy groups. Follow-up campaigns in the radio and optical are now being undertaken, which will eventually bring us a more comprehensive view of the properties of galaxy groups.

4. Conclusions

In this paper, we gave an overview of an invited talk on the impact of AGN on the hot atmospheres of galaxy groups given at the "AGN on the beach" conference in Tropea, Italy. The paper can be summarised in the following way:

- Galaxy groups occupy a key mass regime where the energy injected by AGN feedback is sufficient to affect the baryonic properties of the system over the entire volume, yet not so strong that most of the baryons are evacuated from the halo. The hot atmospheres of galaxy groups can be used as a fossil record of the feedback energy dissipated by AGN over the entire history of these systems;
- The mechanical energy injected by radio AGN is deposited far outside of the central galaxy into the surrounding IGrM (see the left-hand panel of Figure 1), which is eventually responsible for cutting the supply of fresh gas on the central galaxy and quenching star formation;
- The properties of the IGrM predicted by modern hydrodynamical simulations implementing different feedback models strongly vary from one simulation to another (see Figure 2). Calibrating the feedback model against high-fidelity measurements of IGrM properties is therefore a key step toward creating a unified model of energy injection by AGN in galaxy evolution models;
- To advance our understanding of IGrM properties, we were recently awarded the XMM-*Newton* Group AGN Project (X-GAP), a large program on XMM-*Newton* targeting a sample of 49 galaxy groups selected by cross-matching SDSS FoF catalogues with weak RASS extended sources (see Figure 3; [56]). In Appendix A, we provide a master table describing the selected sources and an XMM-*Newton* image gallery.

Author Contributions: D.E.: lead author and X-GAP PI. F.G. and E.O.: X-GAP co-PIs. A.F.: X-GAP sample selection. M.B.: X-GAP LOFAR images. Conceptualization: all, Methodology: D.E. and A.F., Software: D.E., Validation: D.E., Formal Analysis: D.E., A.F. and M.B., Investigation: E.O. and F.G., Resources: all, Data curation: D.E., Writing—original draft: D.E., Writing—review & editing: D.E., Visualization: D.E., Supervision: D.E., Project Administration: D.E., E.O. and F.G., Funding acquisition: all. All authors have read and agreed to the published version of the manuscript.

Funding: D.E. is supported by the Swiss National Science Foundation (SNSF) through grant agreement 200021_212576. M.B. acknowledges support from the agreement ASI-INAF n. 2017-14-H.O and from the PRIN MIUR 2017PH3WAT "Blackout".

Data Availability Statement: X-GAP data will be publicly released on the project website: https://www.astro.unige.ch/xgap/ (accessed on 5 May 2024).

Acknowledgments: D.E. thanks Ilaria Ruffa for the organisation of a beautiful conference and the SOC for the kind invitation.

Conflicts of Interest: The authors declare no conflicts of interest.

Appendix A. X-GAP Master Table and Image Gallery

Table A1. Master table of X-GAP galaxy groups.

ID	RA (deg)	Dec (deg)	z	$F_{[0.1-2.4]}$ (10^{-13} ergs/cm^2/s)	N_{gal}	σ_v (km/s)	M_{200c} ($10^{13} M_\odot$)	BGG Name	BGG z
828	153.41	−0.93	0.046	54.22 ± 4.70	55	750	19.6 ± 1.8	UGC 5515	0.0453
885	117.05	18.55	0.047	26.43 ± 3.84	62	626	2.9 ± 0.6	MCG+03-20-013	0.0225
1011	223.60	16.36	0.046	42.15 ± 6.30	11	374	2.4 ± 0.5	IC 4516	0.0241
1162	232.37	7.57	0.044	41.03 ± 7.68	18	316	4.9 ± 0.7	NGC 5931	0.0267
1398	228.20	7.43	0.046	30.26 ± 4.54	74	617	4.3 ± 0.8	UGC 9767	0.0279
1601	161.09	14.08	0.034	7.24 ± 3.39	23	355	5.3 ± 1.0	NGC 3357	0.0323

Table A1. *Cont.*

ID	RA (deg)	Dec (deg)	z	$F_{[0.1-2.4]}$ (10^{-13} ergs/cm^2/s)	N_{gal}	σ_v (km/s)	M_{200c} ($10^{13} M_\odot$)	BGG Name	BGG z
1695	155.42	23.92	0.04	40.96 ± 4.25	44	441	4.0 ± 1.1	NGC 3216	0.0326
2424	241.61	15.69	0.04	7.44 ± 2.70	85	627	5.3 ± 0.9	MCG+03-41-123	0.0335
2473	174.80	55.67	0.063	8.24 ± 2.30	19	368	5.5 ± 0.8	MCG+09-19-143	0.0335
2620	233.13	4.68	0.039	29.99 ± 4.77	43	498	6.9 ± 0.9	UGC 9886	0.0351
3128	243.15	29.48	0.032	7.00 ± 1.73	27	313	10.1 ± 1.0	NGC 6086	0.0352
3460	170.74	34.11	0.044	21.28 ± 4.31	26	359	5.0 ± 0.7	UGC 6394	0.0358
3513	252.57	23.58	0.036	8.59 ± 3.13	24	362	11.7 ± 0.8	NGC 6233	0.0361
3669	162.19	22.22	0.048	17.21 ± 3.22	30	426	12.5 ± 1.1	MCG+04-26-010	0.0379
4047	174.01	55.08	0.058	8.88 ± 2.42	35	413	12.5 ± 1.2	MCG+09-19-131	0.0389
4436	159.30	50.12	0.046	16.02 ± 2.61	30	562	11.0 ± 1.0	NGC 3298	0.0392
4654	188.92	26.52	0.023	6.87 ± 2.36	24	297	7.0 ± 1.1	NGC 4555	0.0390
4936	145.89	39.42	0.042	5.33 ± 1.72	18	442	15.7 ± 1.0	UGC 5193	0.0392
5742	176.59	33.16	0.034	11.29 ± 2.65	42	425	5.2 ± 1.1	NGC 3880	0.0402
6058	156.09	41.71	0.045	11.54 ± 2.63	17	403	11.5 ± 1.1	MCG+07-22-001	0.0413
6159	203.99	33.43	0.026	5.69 ± 1.81	23	321	11.4 ± 1.4	IC 4305	0.0427
8050	169.09	29.25	0.046	32.34 ± 5.90	68	582	17.7 ± 2.1	MCG+05-27-037	0.0437
8102	238.76	41.58	0.033	10.37 ± 3.61	26	492	15.7 ± 1.4	MCG+07-33-011	0.0444
9178	162.50	0.32	0.04	11.97 ± 3.12	18	281	8.1 ± 1.1	MCG+00-28-017	0.0445
9370	196.24	43.55	0.038	17.94 ± 2.63	29	340	21.9 ± 1.2	MCG+07-27-026	0.0457
9399	140.85	22.31	0.035	15.02 ± 2.88	35	561	10.1 ± 1.0	UGC 4991	0.0451
9647	138.41	29.99	0.023	9.12 ± 3.31	29	347	11.3 ± 1.1	NGC 2783	0.0452
9695	165.24	10.51	0.038	32.36 ± 4.54	47	577	16.1 ± 1.8	NGC 3492	0.0453
9771	205.60	29.82	0.044	6.89 ± 2.19	33	482	13.2 ± 1.3	NGC 5275	0.0461
10094	151.72	14.37	0.031	10.61 ± 2.70	35	364	14.6 ± 1.3	NGC 3121	0.0468
10159	206.32	23.22	0.031	9.28 ± 2.71	31	406	11.2 ± 1.3	LEDA 48750	0.0466
10842	164.55	1.60	0.04	24.23 ± 3.49	51	439	6.8 ± 0.8	UGC 6057	0.0340
11320	146.72	54.45	0.045	32.34 ± 4.54	45	492	15.5 ± 1.5	MCG+09-16-044	0.0458
11631	239.59	18.08	0.046	19.02 ± 3.00	41	504	2.8 ± 0.5	2MASX J15582067+1804512	0.0580
11844	216.17	26.63	0.038	28.27 ± 4.44	9	304	2.4 ± 0.5	MCG+05-34-033	0.0222
12349	200.06	33.14	0.037	31.14 ± 3.41	41	406	5.2 ± 1.1	NGC 5098	0.0336
15354	181.10	42.56	0.054	4.50 ± 1.96	13	372	5.8 ± 1.1	2MASX J12042469+4233432	0.0427
15641	141.97	29.99	0.028	14.63 ± 3.20	12	258	6.9 ± 1.1	IC 2476	0.0472
15776	164.43	37.65	0.036	14.51 ± 3.01	14	291	4.5 ± 0.9	MCG+06-24-039	0.0405
16150	123.65	55.14	0.033	11.97 ± 3.60	24	346	5.0 ± 1.1	MCG+09-14-020	0.0361
16386	249.32	44.42	0.031	4.91 ± 1.43	10	267	8.9 ± 0.8	2MASX J16370588+4416111	0.0389
16393	152.71	54.21	0.047	22.84 ± 3.59	46	295	4.8 ± 0.7	MCG+09-17-036	0.0298
21128	197.18	13.81	0.062	7.27 ± 2.43	13	376	4.8 ± 0.6	2MASX J13084384+1348248	0.0265
22635	230.05	25.72	0.034	10.93 ± 3.10	30	438	3.8 ± 0.6	MCG+04-36-038	0.0325
28674	203.24	32.61	0.037	8.66 ± 1.85	20	282	11.9 ± 1.2	MCG+06-30-029	0.0371
35976	136.98	49.60	0.036	25.88 ± 4.33	31	391	9.3 ± 1.6	MCG+08-17-034	0.0571
39344	184.91	28.50	0.028	13.51 ± 2.73	21	270	9.9 ± 1.7	LEDA 39736	0.0612
40241	239.17	20.17	0.049	7.23 ± 1.92	34	439	5.7 ± 1.5	2MASX J15564131+2010172	0.0555
46701	123.19	54.14	0.042	23.45 ± 3.74	17	369	9.4 ± 1.9	2MASX J08124599+5408228	0.0607

Column description: 1: Group ID [1]. 2: Right ascension [1]. 3: Declination [1]. 4: ROSAT all-sky survey flux in the [0.1–2.4] keV band [2]. 5: Number of FoF galaxies with spectroscopic redshift [1]. 6: Gapper velocity dispersion in km/s [2]. 7: Halo mass within an overdensity of 200 times the critical density estimated from a mass-luminosity relation [2]. 8: Name of brightest group galaxy[3]. 9: Redshift of brightest group galaxy [3]. References: [1] [19]; [2] [56]; [3] This work.

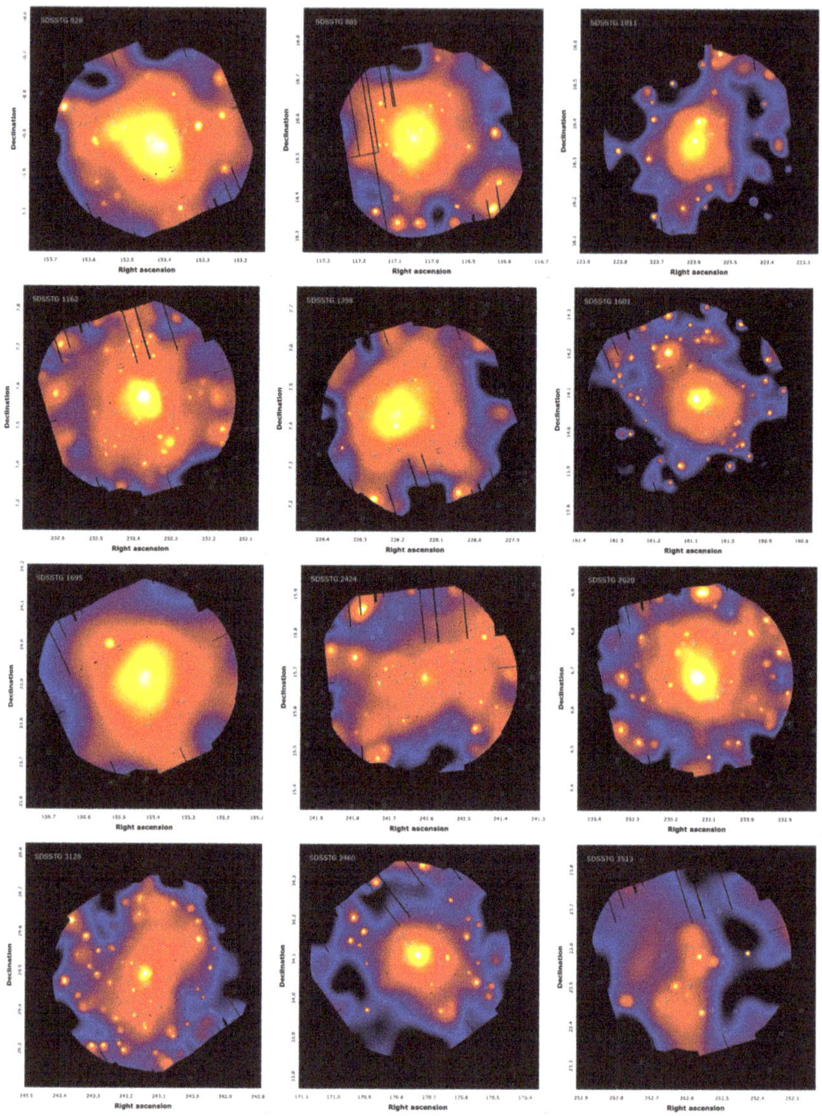

Figure A1. Background subtracted, vignetting corrected, and adaptively smoothed *XMM-Newton*/EPIC maps of X-GAP groups in the [0.7–1.2] keV band. Shown are SDSSTG 828, SDSSTG 885, SDSSTG 1011 (top row), SDSSTG 1162, SDSSTG 1398, SDSSTG 1601 (second row), SDSSTG 1695, SDSSTG 2424, SDSSTG 2620 (third row), SDSSTG 3128, SDSSTG 3460, SDSSTG 3513 (bottom row). The magenta squares show the position of SDSS member galaxies selected using the FoF algorithm [19]. In all cases, the axes are right ascension and declination in unit of degrees.

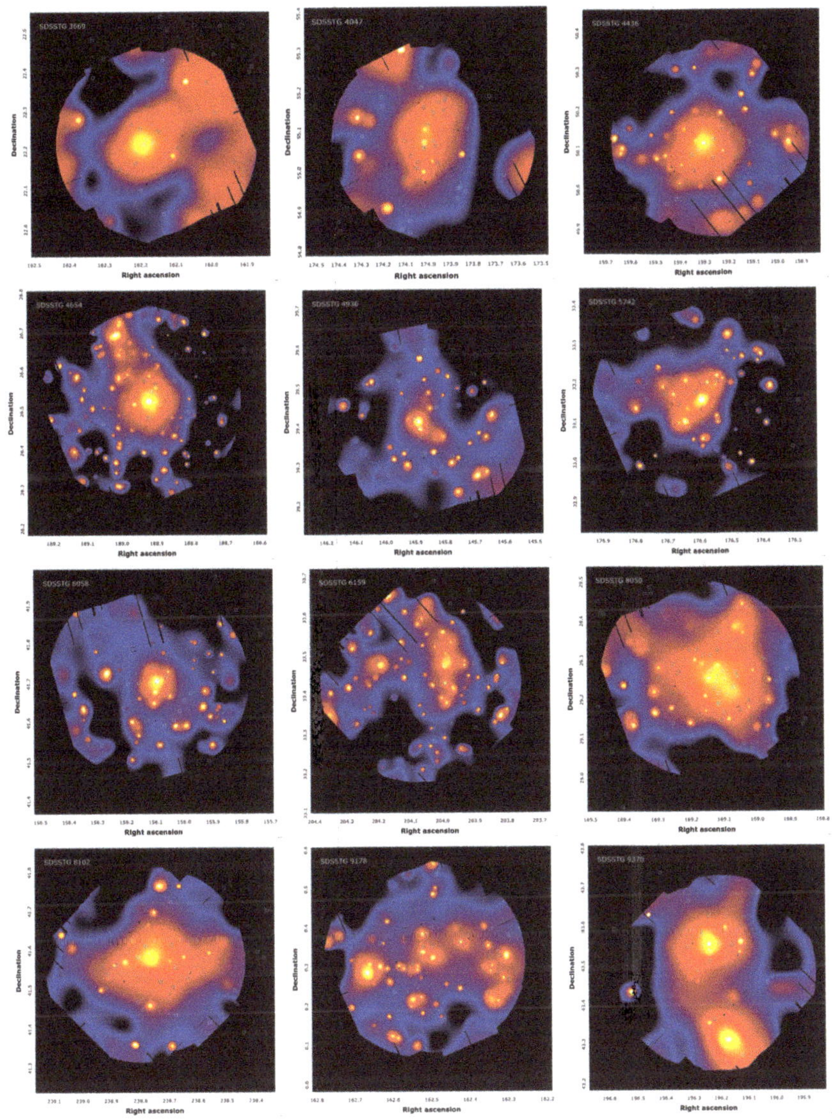

Figure A2. Same as Figure A1 for SDSSTG 3669, SDSSTG 4047, SDSSTG 4436 (top row), SDSSTG 4654, SDSSTG 4936, SDSSTG 5742 (second row), SDSSTG 6058, SDSSTG 6159, SDSSTG 8050 (third row), SDSSTG 8102, SDSSTG 9178, SDSSTG 9370 (bottom row).

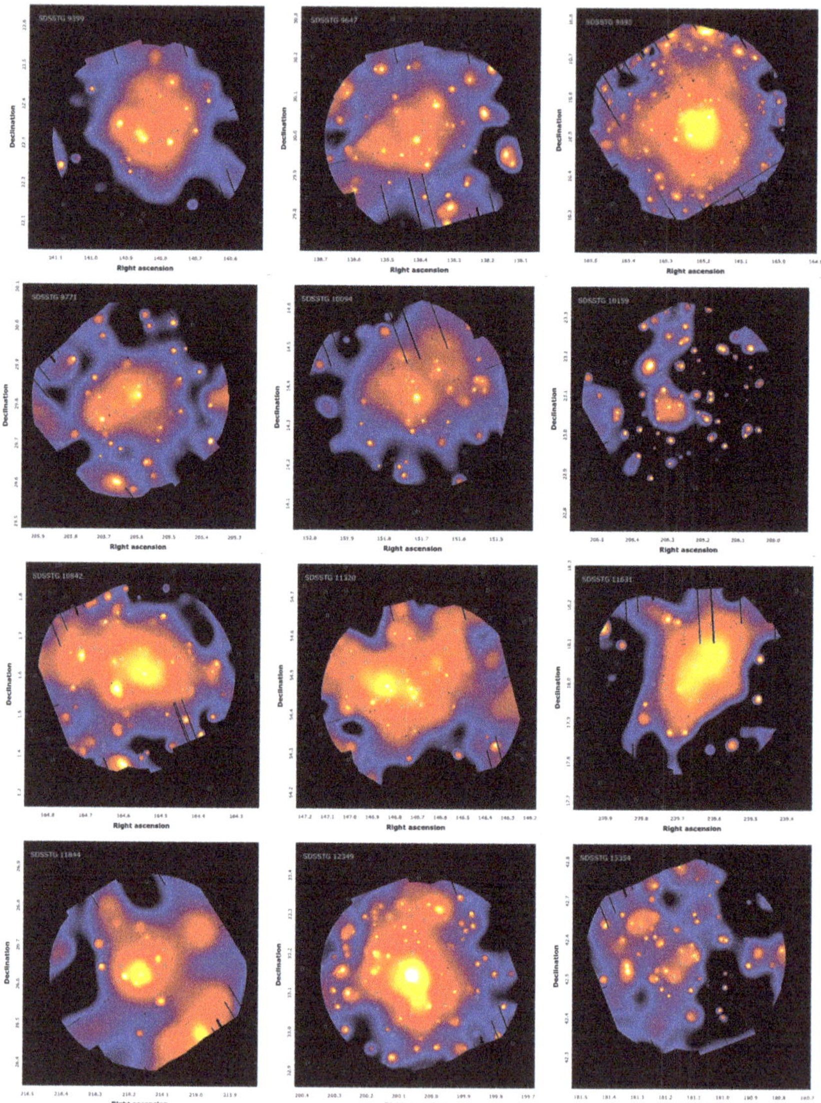

Figure A3. Same as Figure A1 for SDSSTG 9399, SDSSTG 9647, SDSSTG 9695 (top row), SDSSTG 9771, SDSSTG 10094, SDSSTG 10159 (second row), SDSSTG 10842, SDSSTG 11320, SDSSTG 11631 (third row), SDSSTG 11844, SDSSTG 12349, SDSSTG 15354 (bottom row).

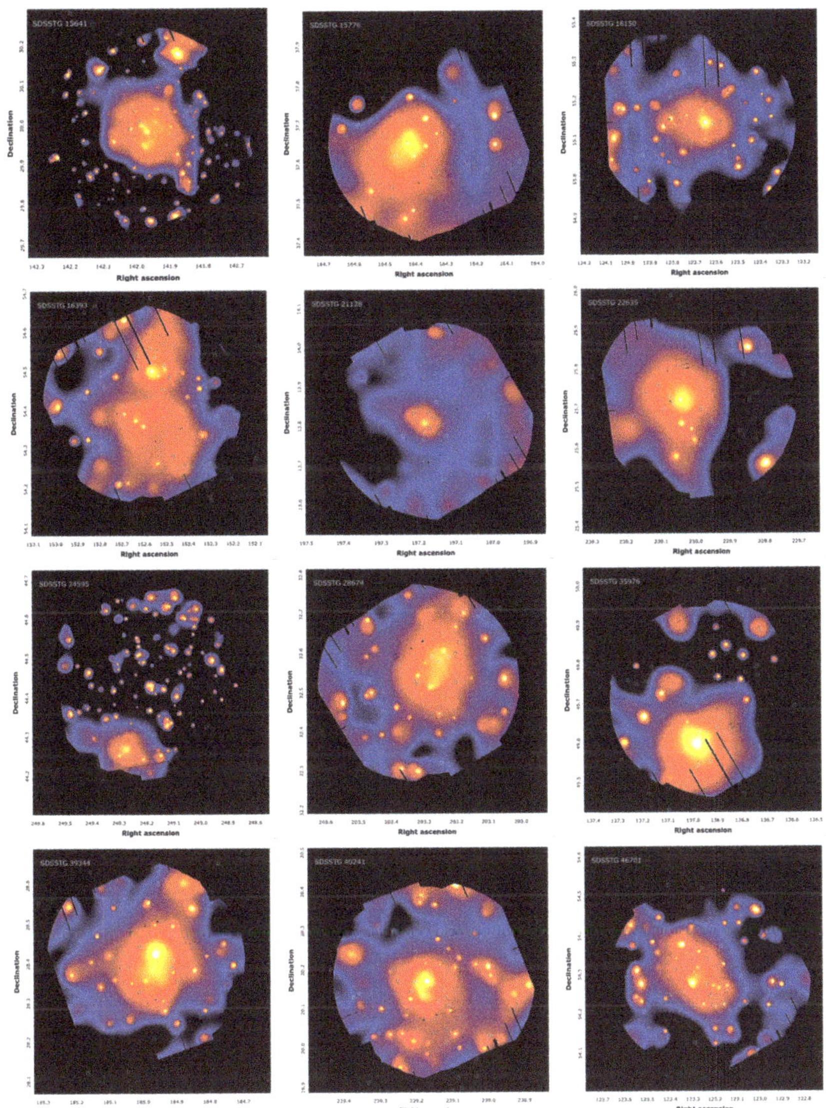

Figure A4. Same as Figure A1 for SDSSTG 15641, SDSSTG 15776, SDSSTG 16150 (top row), SDSSTG 16393, SDSSTG 21128, SDSSTG 22635 (second row), SDSSTG 24595, SDSSTG 28674, SDSSTG 35976 (third row), SDSSTG 39344, SDSSTG 40241, SDSSTG 46701 (bottom row).

Notes

1 https://www.astro.unige.ch/xgap/blog/people, accessed on 5 May 2024.

References

1. Granato, G.L.; De Zotti, G.; Silva, L.; Bressan, A.; Danese, L. A Physical Model for the Coevolution of QSOs and Their Spheroidal Hosts. *Astrophys. J.* **2004**, *600*, 580–594. [CrossRef]
2. Croton, D.J.; Springel, V.; White, S.D.M.; De Lucia, G.; Frenk, C.S.; Gao, L.; Jenkins, A.; Kauffmann, G.; Navarro, J.F.; Yoshida, N. The many lives of active galactic nuclei: cooling flows, black holes and the luminosities and colours of galaxies. *Mon. Not. R. Astron. Soc.* **2006**, *365*, 11–28. [CrossRef]
3. Silk, J.; Rees, M.J. Quasars and galaxy formation. *Astron. Astrophys.* **1998**, *331*, L1–L4.
4. Cattaneo, A.; Haehnelt, M.G.; Rees, M.J. The distribution of supermassive black holes in the nuclei of nearby galaxies. *Mon. Not. R. Astron. Soc.* **1999**, *308*, 77–81. [CrossRef]
5. Kauffmann, G.; Haehnelt, M. A unified model for the evolution of galaxies and quasars. *Mon. Not. R. Astron. Soc.* **2000**, *311*, 576–588. [CrossRef]
6. Madau, P.; Ferguson, H.C.; Dickinson, M.E.; Giavalisco, M.; Steidel, C.C.; Fruchter, A. High-redshift galaxies in the Hubble Deep Field: colour selection and star formation history to $z\sim4$. *Mon. Not. R. Astron. Soc.* **1996**, *283*, 1388–1404. [CrossRef]
7. McNamara, B.R.; Nulsen, P.E.J. Heating Hot Atmospheres with Active Galactic Nuclei. *Annu. Rev. Astron. Astrophys.* **2007**, *45*, 117–175. [CrossRef]
8. Schaye, J.; Crain, R.A.; Bower, R.G.; Furlong, M.; Schaller, M.; Theuns, T.; Dalla Vecchia, C.; Frenk, C.S.; McCarthy, I.G.; Helly, J.C.; et al. The EAGLE project: simulating the evolution and assembly of galaxies and their environments. *Mon. Not. R. Astron. Soc.* **2015**, *446*, 521–554. [CrossRef]
9. McCarthy, I.G.; Schaye, J.; Bird, S.; Le Brun, A.M.C. The BAHAMAS project: calibrated hydrodynamical simulations for large-scale structure cosmology. *Mon. Not. R. Astron. Soc.* **2017**, *465*, 2936–2965. [CrossRef]
10. Springel, V.; Pakmor, R.; Pillepich, A.; Weinberger, R.; Nelson, D.; Hernquist, L.; Vogelsberger, M.; Genel, S.; Torrey, P.; Marinacci, F.; et al. First results from the IllustrisTNG simulations: matter and galaxy clustering. *Mon. Not. R. Astron. Soc.* **2018**, *475*, 676–698. [CrossRef]
11. Sijacki, D.; Springel, V.; Di Matteo, T.; Hernquist, L. A unified model for AGN feedback in cosmological simulations of structure formation. *Mon. Not. R. Astron. Soc.* **2007**, *380*, 877–900. [CrossRef]
12. Robotham, A.S.G.; Norberg, P.; Driver, S.P.; Baldry, I.K.; Bamford, S.P.; Hopkins, A.M.; Liske, J.; Loveday, J.; Merson, A.; Peacock, J.A.; et al. Galaxy and Mass Assembly (GAMA): the GAMA galaxy group catalogue (G^3Cv1). *Mon. Not. R. Astron. Soc.* **2011**, *416*, 2640–2668. [CrossRef]
13. Randall, S.W.; Nulsen, P.E.J.; Jones, C.; Forman, W.R.; Bulbul, E.; Clarke, T.E.; Kraft, R.; Blanton, E.L.; David, L.; Werner, N.; et al. A Very Deep Chandra Observation of the Galaxy Group NGC 5813: AGN Shocks, Feedback, and Outburst History. *Astrophys. J.* **2015**, *805*, 112. [CrossRef]
14. Gastaldello, F.; Buote, D.A.; Temi, P.; Brighenti, F.; Mathews, W.G.; Ettori, S. X-Ray Cavities, Filaments, and Cold Fronts in the Core of the Galaxy Group NGC 5044. *Astrophys. J.* **2009**, *693*, 43–55. [CrossRef]
15. Bîrzan, L.; McNamara, B.R.; Nulsen, P.E.J.; Carilli, C.L.; Wise, M.W. Radiative Efficiency and Content of Extragalactic Radio Sources: Toward a Universal Scaling Relation between Jet Power and Radio Power. *Astrophys. J.* **2008**, *686*, 859. [CrossRef]
16. Liu, W.; Sun, M.; Nulsen, P.; Clarke, T.; Sarazin, C.; Forman, W.; Gaspari, M.; Giacintucci, S.; Lal, D.V.; Edge, T. AGN feedback in galaxy group 3C 88: cavities, shock, and jet reorientation. *Mon. Not. R. Astron. Soc.* **2019**, *484*, 3376–3392. [CrossRef]
17. Eckert, D.; Gaspari, M.; Gastaldello, F.; Le Brun, A.M.C.; O'Sullivan, E. Feedback from Active Galactic Nuclei in Galaxy Groups. *Universe* **2021**, *7*, 142. [CrossRef]
18. Salim, S.; Lee, J.C.; Janowiecki, S.; da Cunha, E.; Dickinson, M.; Boquien, M.; Burgarella, D.; Salzer, J.J.; Charlot, S. GALEX-SDSS-WISE Legacy Catalog (GSWLC): Star Formation Rates, Stellar Masses, and Dust Attenuations of 700,000 Low-redshift Galaxies. *Astrophys. J. Suppl. Ser.* **2016**, *227*, 2. [CrossRef]
19. Tempel, E.; Tuvikene, T.; Kipper, R.; Libeskind, N.I. Merging groups and clusters of galaxies from the SDSS data. The catalogue of groups and potentially merging systems. *Astron. Astrophys.* **2017**, *602*, A100. [CrossRef]
20. Shimwell, T.W.; Hardcastle, M.J.; Tasse, C.; Best, P.N.; Röttgering, H.J.A.; Williams, W.L.; Botteon, A.; Drabent, A.; Mechev, A.; Shulevski, A.; et al. The LOFAR Two-metre Sky Survey. V. Second data release. *Astron. Astrophys.* **2022**, *659*, A1. [CrossRef]
21. Jamrozy, M.; Klein, U.; Mack, K.H.; Gregorini, L.; Parma, P. Spectral ageing in the relic radio galaxy B2 0924+30. *Astron. Astrophys.* **2004**, *427*, 79–86. https://doi.org/10.1051/0004-6361:20048056.
22. Shulevski, A.; Morganti, R.; Harwood, J.J.; Barthel, P.D.; Jamrozy, M.; Brienza, M.; Brunetti, G.; Röttgering, H.J.A.; Murgia, M.; White, G.J.; et al. Radiative age mapping of the remnant radio galaxy B2 0924+30: the LOFAR perspective. *Astron. Astrophys.* **2017**, *600*, A65. [CrossRef]
23. Burns, J.O. The radio properties of cD galaxies in Abell clusters. I - an X-ray selected sample. *Astron. J.* **1990**, *99*, 14–30. https://doi.org/10.1086/115307.
24. Mittal, R.; Hudson, D.S.; Reiprich, T.H.; Clarke, T. AGN heating and ICM cooling in the HIFLUGCS sample of galaxy clusters. *Astron. Astrophys.* **2009**, *501*, 835–850. [CrossRef]
25. Sabater, J.; Best, P.N.; Hardcastle, M.J.; Shimwell, T.W.; Tasse, C.; Williams, W.L.; Brüggen, M.; Cochrane, R.K.; Croston, J.H.; de Gasperin, F.; et al. The LoTSS view of radio AGN in the local Universe. The most massive galaxies are always switched on. *Astron. Astrophys.* **2019**, *622*, A17. [CrossRef]

26. Kormendy, J.; Ho, L.C. Coevolution (Or Not) of Supermassive Black Holes and Host Galaxies. *Annu. Rev. Astron. Astrophys.* **2013**, *51*, 511–653. https://doi.org/10.1146/annurev-astro-082708-101811.
27. Dunn, R.J.H.; Fabian, A.C. Investigating AGN heating in a sample of nearby clusters. *Mon. Not. R. Astron. Soc.* **2006**, *373*, 959–971. [CrossRef]
28. Cavagnolo, K.W.; McNamara, B.R.; Nulsen, P.E.J.; Carilli, C.L.; Jones, C.; Birzan, L. A relationship between AGN jet power and radio power. *Astrophys. J.* **2010**, *720*, 1066–1072. [CrossRef]
29. Tremblay, G.R.; Oonk, J.B.R.; Combes, F.; Salomé, P.; O'Dea, C.P.; Baum, S.A.; Voit, G.M.; Donahue, M.; McNamara, B.R.; Davis, T.A.; et al. Cold, clumpy accretion onto an active supermassive black hole. *Nature* **2016**, *534*, 218–221. [CrossRef]
30. Oosterloo, T.; Morganti, R.; Murthy, S. Closing the feedback-feeding loop of the radio galaxy 3C 84. *Nat. Astron.* **2024**, *8*, 256–262. [CrossRef]
31. Gaspari, M.; Eckert, D.; Ettori, S.; Tozzi, P.; Bassini, L.; Rasia, E.; Brighenti, F.; Sun, M.; Borgani, S.; Johnson, S.D.; et al. The X-Ray Halo Scaling Relations of Supermassive Black Holes. *Astrophys. J.* **2019**, *884*, 169. [CrossRef]
32. Gaspari, M.; Sądowski, A. Unifying the Micro and Macro Properties of AGN Feeding and Feedback. *Astrophys. J.* **2017**, *837*, 149. [CrossRef]
33. Gonzalez, A.H.; Sivanandam, S.; Zabludoff, A.I.; Zaritsky, D. Galaxy Cluster Baryon Fractions Revisited. *Astrophys. J.* **2013**, *778*, 14. [CrossRef]
34. Laganá, T.F.; de Souza, R.S.; Keller, G.R. On the influence of non-thermal pressure on the mass determination of galaxy clusters. *Astron. Astrophys.* **2010**, *510*, A76. [CrossRef]
35. Eckert, D.; Ettori, S.; Coupon, J.; Gastaldello, F.; Pierre, M.; Melin, J.B.; Le Brun, A.M.C.; McCarthy, I.G.; Adami, C.; Chiappetti, L.; et al. The XXL Survey. XIII. Baryon content of the bright cluster sample. *Astron. Astrophys.* **2016**, *592*, A12. [CrossRef]
36. Akino, D.; Eckert, D.; Okabe, N.; Sereno, M.; Umetsu, K.; Oguri, M.; Gastaldello, F.; Chiu, I.N.; Ettori, S.; Evrard, A.E.; et al. HSC-XXL: Baryon budget of the 136 XXL groups and clusters. *Publ. Astron. Soc. Jpn.* **2022**, *74*, 175–208. [CrossRef]
37. Ayromlou, M.; Nelson, D.; Pillepich, A. Feedback reshapes the baryon distribution within haloes, in halo outskirts, and beyond: the closure radius from dwarfs to massive clusters. *Mon. Not. R. Astron. Soc.* **2023**, *524*, 5391–5410. [CrossRef]
38. Davies, J.J.; Crain, R.A.; McCarthy, I.G.; Oppenheimer, B.D.; Schaye, J.; Schaller, M.; McAlpine, S. The gas fractions of dark matter haloes hosting simulated $\sim L^\star$ galaxies are governed by the feedback history of their black holes. *Mon. Not. R. Astron. Soc.* **2019**, *485*, 3783–3793. [CrossRef]
39. Voit, G.M.; Oppenheimer, B.D.; Bell, E.F.; Terrazas, B.; Donahue, M. Black Hole Growth, Baryon Lifting, Star Formation, and IllustrisTNG. *Astrophys. J.* **2024**, *960*, 28. [CrossRef]
40. Oppenheimer, B.D.; Babul, A.; Bahé, Y.; Butsky, I.S.; McCarthy, I.G. Simulating Groups and the IntraGroup Medium: The Surprisingly Complex and Rich Middle Ground between Clusters and Galaxies. *Universe* **2021**, *7*, 209. [CrossRef]
41. Booth, C.M.; Schaye, J. Cosmological simulations of the growth of supermassive black holes and feedback from active galactic nuclei: method and tests. *Mon. Not. R. Astron. Soc.* **2009**, *398*, 53–74. [CrossRef]
42. Davé, R.; Anglés-Alcázar, D.; Narayanan, D.; Li, Q.; Rafieferantsoa, M.H.; Appleby, S. SIMBA: Cosmological simulations with black hole growth and feedback. *Mon. Not. R. Astron. Soc.* **2019**, *486*, 2827–2849. [CrossRef]
43. Tremmel, M.; Karcher, M.; Governato, F.; Volonteri, M.; Quinn, T.R.; Pontzen, A.; Anderson, L.; Bellovary, J. The Romulus cosmological simulations: a physical approach to the formation, dynamics and accretion models of SMBHs. *Mon. Not. R. Astron. Soc.* **2017**, *470*, 1121–1139. [CrossRef]
44. Vogelsberger, M.; Genel, S.; Springel, V.; Torrey, P.; Sijacki, D.; Xu, D.; Snyder, G.; Nelson, D.; Hernquist, L. Introducing the Illustris Project: simulating the coevolution of dark and visible matter in the Universe. *Mon. Not. R. Astron. Soc.* **2014**, *444*, 1518–1547. [CrossRef]
45. Sun, M.; Voit, G.M.; Donahue, M.; Jones, C.; Forman, W.; Vikhlinin, A. Chandra Studies of the X-Ray Gas Properties of Galaxy Groups. *Astrophys. J.* **2009**, *693*, 1142–1172. [CrossRef]
46. Lovisari, L.; Reiprich, T.H.; Schellenberger, G. Scaling properties of a complete X-ray selected galaxy group sample. *Astron. Astrophys.* **2015**, *573*, A118. [CrossRef]
47. Schneider, A.; Teyssier, R.; Stadel, J.; Chisari, N.E.; Le Brun, A.M.C.; Amara, A.; Refregier, A. Quantifying baryon effects on the matter power spectrum and the weak lensing shear correlation. *J. Cosmol. Astropart. Phys.* **2019**, *2019*, 020. [CrossRef]
48. Chisari, N.E.; Mead, A.J.; Joudaki, S.; Ferreira, P.G.; Schneider, A.; Mohr, J.; Tröster, T.; Alonso, D.; McCarthy, I.G.; Martin-Alvarez, S.; et al. Modelling baryonic feedback for survey cosmology. *Open J. Astrophys.* **2019**, *2*, 4. [CrossRef]
49. van Daalen, M.P.; McCarthy, I.G.; Schaye, J. Exploring the effects of galaxy formation on matter clustering through a library of simulation power spectra. *Mon. Not. R. Astron. Soc.* **2020**, *491*, 2424–2446. [CrossRef]
50. Ponman, T.J.; Cannon, D.B.; Navarro, J.F. The thermal imprint of galaxy formation on X-ray clusters. *Nature* **1999**, *397*, 135–137. [CrossRef]
51. Finoguenov, A.; Jones, C.; Böhringer, H.; Ponman, T.J. ASCA Observations of Groups at Radii of Low Overdensity: Implications for the Cosmic Preheating. *Astrophys. J.* **2002**, *578*, 74–89. [CrossRef]
52. Gastaldello, F.; Buote, D.A.; Humphrey, P.J.; Zappacosta, L.; Bullock, J.S.; Brighenti, F.; Mathews, W.G. Probing the Dark Matter and Gas Fraction in Relaxed Galaxy Groups with X-Ray Observations from Chandra and XMM-Newton. *Astrophys. J.* **2007**, *669*, 158–183. [CrossRef]

53. Pierre, M.; Pacaud, F.; Adami, C.; Alis, S.; Altieri, B.; Baran, N.; Benoist, C.; Birkinshaw, M.; Bongiorno, A.; Bremer, M.N.; et al. The XXL Survey. I. Scientific motivations - XMM-Newton observing plan—Follow-up observations and simulation programme. *Astron. Astrophys.* **2016**, *592*, A1. [CrossRef]
54. Rykoff, E.S.; Rozo, E.; Busha, M.T.; Cunha, C.E.; Finoguenov, A.; Evrard, A.; Hao, J.; Koester, B.P.; Leauthaud, A.; Nord, B.; et al. redMaPPer. I. Algorithm and SDSS DR8 Catalog. *Astrophys. J.* **2014**, *785*, 104. [CrossRef]
55. Tinker, J.L. A Self-Calibrating Halo-Based Group Finder: Application to SDSS. *Astrophys. J.* **2021**, *923*, 154. [CrossRef]
56. Damsted, S.; Finoguenov, A.; Lietzen, H.; Mamon, G.A.; Comparat, J.; Tempel, E.; Dmitrieva, I.; Clerc, N.; Collins, C.; Gozaliasl, G.; et al. AXES-SDSS: comparison of SDSS galaxy groups with All-sky X-ray Extended Sources. *arXiv* **2024**, arXiv:2403.17055.
57. Osmond, J.P.F.; Ponman, T.J. The GEMS project: X-ray analysis and statistical properties of the group sample. *Mon. Not. R. Astron. Soc.* **2004**, *350*, 1511–1535. [CrossRef]
58. Bahar, Y.E.; Bulbul, E.; Ghirardini, V.; Sanders, J.S.; Zhang, X.; Liu, A.; Clerc, N.; Artis, E.; Balzer, F.; Biffi, V.; et al. The SRG/eROSITA All-Sky Survey: Constraints on AGN Feedback in Galaxy Groups. *arXiv* **2024**, arXiv:2401.17276. https://doi.org/10.48550/arXiv.2401.17276.
59. Eckert, D.; Molendi, S.; Paltani, S. The cool-core bias in X-ray galaxy cluster samples. I. Method and application to HIFLUGCS. *A&A* **2011**, *526*, A79. [CrossRef]
60. Xu, W.; Ramos-Ceja, M.E.; Pacaud, F.; Reiprich, T.H.; Erben, T. A new X-ray-selected sample of very extended galaxy groups from the ROSAT All-Sky Survey. *Astron. Astrophys.* **2018**, *619*, A162. [CrossRef]
61. Käfer, F.; Finoguenov, A.; Eckert, D.; Clerc, N.; Ramos-Ceja, M.E.; Sanders, J.S.; Ghirardini, V. Toward the low-scatter selection of X-ray clusters. Galaxy cluster detection with eROSITA through cluster outskirts. *Astron. Astrophys.* **2020**, *634*, A8. [CrossRef]
62. Eckert, D.; Ettori, S.; Pointecouteau, E.; van der Burg, R.F.J.; Loubser, S.I. The gravitational field of X-COP galaxy clusters. *Astron. Astrophys.* **2022**, *662*, A123. [CrossRef]
63. Ghirardini, V.; Eckert, D.; Ettori, S.; Pointecouteau, E.; Molendi, S.; Gaspari, M.; Rossetti, M.; De Grandi, S.; Roncarelli, M.; Bourdin, H.; et al. Universal thermodynamic properties of the intracluster medium over two decades in radius in the X-COP sample. *Astron. Astrophys.* **2019**, *621*, A41. [CrossRef]

galaxies

MDPI

Review

The Seeding of Cosmic Ray Electrons by Cluster Radio Galaxies: A Review

Franco Vazza [1,2,3,*] and Andrea Botteon [2]

[1] Dipartimento di Fisica e Astronomia, Universitá di Bologna, Via Gobetti 93/2, 40129 Bologna, Italy
[2] INAF-Istituto di Radio Astronomia, Via Gobetti 101, 40129 Bologna, Italy; andrea.botteon@inaf.it
[3] Hamburger Sternwarte, Universität Hamburg, Gojenbergsweg 112, 41029 Hamburg, Germany
* Correspondence: franco.vazza2@unibo.it

Abstract: Radio galaxies in clusters of galaxies are a prominent reservoir of magnetic fields and of non-thermal particles, which become mixed with the intracluster medium. We review the observational and theoretical knowledge of the role of these crucial ingredients for the formation of diffuse radio emission in clusters (radio halos, relics, mini halos) and outline the open questions in this field.

Keywords: AGN feedback; radio jets; galaxy evolution; clusters of galaxies

1. Introduction

Jets from radio galaxies can store the majority of their internal energy in the form of non-thermal components (relativistic particles and magnetic fields), although the balance between thermal and non-thermal components may depend on the radio galaxy type, as well as on the interaction with the surrounding environment, e.g., [1,2]. Given the very active dynamics of the intracluster medium (ICM) under the effect of the episodic accretion of gas and dark matter, e.g., [3–7], as well as of the powerful feedback from radio galaxies themselves, e.g., [8–12], it appears unavoidable that such interactions result in the fueling of a prominent and potentially visible reservoir of magnetic fields and non-thermal particles in the ICM, e.g., [13,14].

Recent deep radio observations have detected complex morphologies of remnant plasma, injected by radio galaxies and in different stages of mixing with the surrounding environment, e.g., [15–19], and often in conjunction with diffuse $\sim$Mpc-sized cluster radio sources.

In the case of "radio relics" (i.e., elongated and polarized radio sources that are typically co-located with merger shock waves; see, e.g., [20] for a review), a pre-existing population of low-energy relativistic electrons is often required to explain the brightness of radio relics associated with weak $\mathcal{M} \leq 2.5$ shocks, in which case shock re-acceleration, rather than the injection of freshly accelerated electrons from the thermal pool, can be hypothesized, e.g., [21–24]. In "radio halos" (i.e., centrally located, roughly spherically symmetric and unpolarized sources believed to trace turbulence in clusters with recent mergers; see, e.g., [25] for a review), fossil electrons are required to allow Fermi II turbulent re-acceleration to produce the observed level of radio emission, which also shows hints of curvature at high energies. Similar requirements have been proposed to explain "radio mini-halos" (smaller than radio halos and preferentially found in cool-core clusters, e.g., [26,27]) and also the pervasive emission recently detected at the extreme periphery of clusters and in between pairs of interacting clusters of galaxies, e.g., [28–36].

In this review, we wish to present the state of the art of the existing observational evidences for the mixing of plasma ejected by radio galaxies with the ICM (Section 2), and to give the status of our theoretical and numerical understanding of this process, which requires the understanding of the dynamics of the ICM across a wide range of spatial scales (Section 3).

2. Observations

The electrons emitted in diffuse cluster radio sources are, in most cases, expected not to be accelerated in situ. Radio phoenixes and disturbed radio tails, such as those shown in Figure 1, represent the clearest evidence of how the relativistic plasma ejected by cluster AGN can seed the ICM with non-thermal components while being mixed and distributed in the surrounding environment by thermal gas motions.

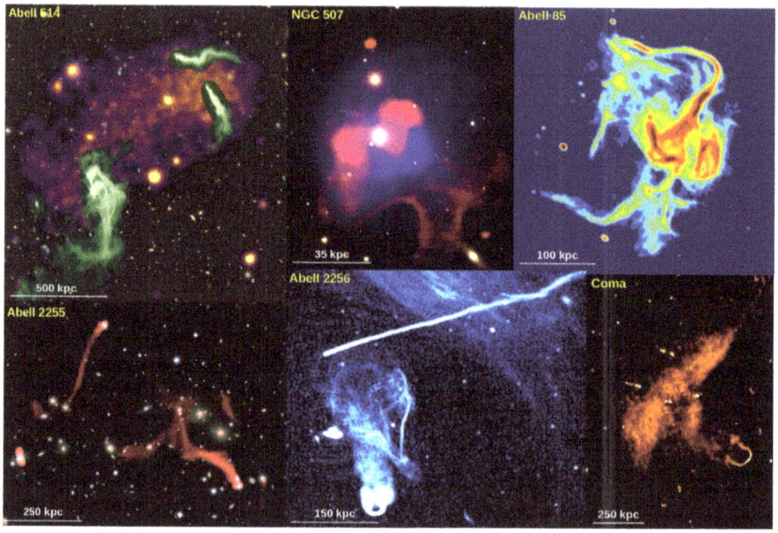

Figure 1. Examples of ongoing cosmic ray seeding and interaction between radio galaxies and thermal gas. From top left to bottom right: AGN with bent radio tails (green) due to the interaction with the X-ray gas (purple) in Abell 514 [37]; radio plasma (red) transported by thermal gas (blue) sloshing motions in the NGC 507 group [19]; the filamentary radio phoenix in Abell 85 [38]; the complex system of filamentary-tailed radio AGN (red) at the center of Abell 2255 [39]; the radio phoenix and the long head–tail radio galaxy in Abell 2256 [40]; and a radio galaxy feeding the radio relic in the Coma cluster [41].

2.1. Radio Phoenixes

Radio phoenixes are diffuse radio sources with sizes of a few hundred kpc at most, characterized by ultra-steep spectra ($\alpha > 1.8$, where $I(\nu) \propto \nu^{-\alpha}$ is the radio emission spectrum) showing high-frequency spectral steepening. They trace fossil radio plasma ejected from past episodes of AGN activity (also called "AGN remnants"), revived by the dynamical motions within the ICM. Consequently, such types of emission are also referred to as "revived fossil plasma". As a result of interactions with the thermal gas, phoenixes typically show amorphous morphologies, with an often unclear connection with the host galaxy that originally ejected the radio plasma (cf., Figure 1).

Despite radio phoenixes (before the term "phoenix" was introduced in Ref. [42] these sources were called "relics"; nowadays, the term "radio relic" is used to classify large and arc-shaped diffuse emission associated with shock fronts in cluster outskirts) have been known for about three decades, e.g., [43–47], with the advent of high-quality observations, and particularly at $\lesssim$1 GHz frequencies (where the radio emission appears brighter due to its steep spectral index), in recent years the number of these sources has significantly increased, e.g., [17,39,48–55]. Instances of phoenixes with extreme spectral indexes of >2.5 have also been found, e.g., [45–47,50,56]. High-resolution images have started to reveal their complex morphologies, often characterized by filamentary structures [17,38,57–59]; see, for example, the remarkable cases of the phoenixes in Abell 85 and Abell 2256 reported in Figure 1. The existence of these phoenixes indicates the presence of extended patches of old, relativistic electrons in the sub-GeV energy range in the ICM. Although these electrons are fully relativistic (their Lorentz factor is $\gamma\sim100$), they are usually considered as low-energy or even "mildly relativistic" electrons in this context, considering that the electrons emitting detectable radio emission have a $\geq$GeV energy (corresponding to $\gamma \gg 10^3$). While statistical studies on the properties of radio phoenixes, such as their occurrence with cluster mass and dynamical state, are still lacking, the recent observations of large samples of clusters, e.g., [60–66] suggest that sources of fossil plasma may be very common, if not ubiquitous, in all clusters of galaxies, especially at lower frequencies.

2.2. Disturbed Radio Tails

Jets originating from supermassive black holes residing in cluster AGN are deflected by the ram pressure while their host galaxy moves through the ICM ($P_{\rm ram} \propto \rho v^2$, where ρ is the ICM density and v is the relative velocity between jets, or the galaxy, and the ICM), generating the so-called tailed radio sources [67]. Depending on the degree of jet bending, these sources can be classified as wide-angle tails (WATs), narrow-angle tails (NATs), or head–tails (HTs). The most spectacular tailed radio galaxies can extend beyond the Mpc scale, leaving a long trail of non-thermal plasma behind their path through the ICM. This material is eventually dispersed in the environment. Relativistic electrons in the magnetized ICM primarily lose their energy via synchrotron and inverse Compton losses on the short timescale given by $t_{\rm cool}\sim\gamma/\dot{\gamma}\sim10$ Myr (using $\gamma = 10^4$, $B = 1$ μG and $z = 0$ as reference values for the ICM), e.g., [68]. The cooling timescale approximately scales inversely with the energy of the electrons ($t_{\rm cool} \propto 1/\gamma$); hence, the least energetic electrons in tails are found at larger distances from the AGN, and steeper synchrotron spectra are progressively found along the tail. As a result, radio tails appear longer at lower frequencies.

However, new highly sensitive radio observations (again, mainly at low frequency) are revealing that tails (i) often exhibit longer extensions than expected from the radiative lifetime of relativistic electrons, e.g., [69]; (ii) may feature regions of surface brightness and spectral index flattening (contrary to the gradual steepening expected by particle aging) [15,70,71]; and (iii) frequently display disturbed morphology, especially at their terminal ends, where the tail structure "breaks", e.g., [16,49,60,72–76]; see Figure 1. These properties suggest a non-trivial ongoing interplay between the non-thermal components in tails and the surrounding thermal gas, leading to processes that can sustain particle lifetimes for periods of time and distances longer than usually expected. Radio tails, thus, naturally fuel the cluster environment with seed mildly relativistic electrons, and the latter can be re-energized in this first interaction with the ICM, i.e., even before they are re-accelerated again by the processes leading to diffuse radio emissions (see next section).

2.3. Evidence of Interaction with the ICM and Connection with Large-Scale Diffuse Radio Emission

Shock, cold fronts, turbulence, and sloshing in the ICM can significantly influence the fate and morphology of non-thermal plasma ejected by cluster radio galaxies. Nowadays, radio and X-ray observations are providing clear evidence of this kind of interaction in many clusters (Figure 1). Shocks or cold fronts co-located with sudden direction changes of radio galaxy tails have been reported in Abell 3411 [77], Coma [78], Abell 3376 [79], Abell 1775 [49], Abell 3562 [80], Abell 514 [37], and are invoked in other cases to explain the disturbed morphology of the radio emission [81–84]. Shocks can re-energize cosmic ray electrons both via re-acceleration (Fermi I processes) and adiabatic compression, e.g., [85–87]. Shear flows inside cold fronts stretch and amplify magnetic fields, e.g., [79,88], possibly dragging the relativistic plasma with them in tangential directions, e.g., [19,37,79,80,84,89]. Turbulence, either induced by cluster–cluster mergers or by sloshing or by the wake of a fast-moving radio galaxy, can trigger Fermi II re-acceleration mechanisms [90–94]. In the extreme case of the so-called "gently re-energized tail" (GReET) in Abell 1033, a turbulent re-acceleration mechanism acting on a timescale comparable to that of relativistic electrons, barely balancing their cooling, has been proposed to explain the observed radio proper-tics [15,71]. Adiabatic compression of fossil plasma due to a recent shock passage is also the favored formation scenario for radio phoenixes [85]. However, as of now, the detection of a shock front at the location of a radio phoenix has been reported only in Abell 2443 [95], leaving the formation scenario for these sources still uncertain (although the detection of shock-heated gas co-located with phoenixes has also been claimed in some cases [96–99]).

Evidence of the seeding of radio halos and relics by radio galaxies has also been reported. Radio galaxies embedded into radio relics suggest the ongoing seeding of cosmic ray electrons in the environment [41,77,100–104]. In the case of Abell 3411, a clear connection between a radio galaxy compressed by a shock front and a radio relic has been used to support the scenario in which particles are re-accelerated by the weak shock [77], and to test the adiabatic compression scenario [105]. Radio galaxies whose tails fade into radio halos or that demonstrate a complex mixing with central non-thermal diffuse emission/turbulent motions have been also associated with the seeding of cosmic rays on cluster scales, e.g., [16,30,32,39,54,84,106]. Recently, a connection between a radio phoenix and central diffuse emission in Abell 85 has also been reported [107].

Although here we focus on systems with strong indications of interaction between AGN radio plasma and diffuse radio emission on larger scales, it is worth noting that several interesting observations have detected different stages of the likely interplay between radio galaxies and the surrounding medium, e.g., [108–112], to cite a few. These interactions are often characterized by filamentary radio structures, which can, in principle, yield precious information about the life cycle of relativistic fossil electrons in the ICM.

Despite the observational evidence of the seeding of cosmic ray electrons by cluster radio galaxies, a number of questions still need to be addressed. What is the contribution of the single dynamical processes (shocks and turbulence) in the redistribution and re-acceleration of cosmic ray electrons in clusters? How many radio galaxies are needed to fuel the ICM with non-thermal particles? Is a direct connection between radio galaxies and diffuse sources always needed to explain the emission from radio halos and relics? What about diffuse sources that do not show such a connection with cluster AGN?

3. Theoretical and Numerical Models

One of the very first seminal works on this subject [113] noted that the contribution from a few radio galaxies hosted in the Coma cluster may explain the normalization of the observed radio halo emission, provided that the magnetic field in the entire region was of order $\sim\mu G$. Similar estimates were later attempted for different systems, e.g., in A3562 [114], or more recently for A2163 [31], just to cite a few.

A rough estimate can be obtained assuming energy equipartition between the relativistic electrons—also including those not emitting at the frequency of radio observations—the ICM magnetic field, and possibly the additional cosmic ray protons in the system (whose energy ratio with respect to electrons is K), assuming a power-law distribution of cosmic ray electrons between $\gamma_{\rm min}$ and $\gamma_{\rm max}$ ($N(\gamma) = N_0\gamma^{-\delta}$). Here, we follow the formalism in [115], where the equipartition magnetic field, $B_{\rm eq}$, was computed, also considering the contribution from cosmic ray electrons with energies below the observing frequencies as

$$B_{\rm eq} = [C(\alpha)\frac{P(\nu)\nu^{\alpha}}{\phi V}(1+K)]^{\frac{1}{\alpha+3}}\gamma_{\rm min}{}^{\frac{1-2\alpha}{\alpha+3}} \tag{1}$$

where $P(\nu)$ is the radio emission (in units erg/s/Hz), ν is the observing frequency (in Hz), $\alpha = (\delta-1)/2$ is the radio spectral index, V is the volume of the emitting region (in cm^3), and ϕ is the filling factor. $C(\alpha)$ is a function of the spectral index, whose expression is given by Equation (A3) in [115]. The total energy of cosmic ray electrons (in units of erg) across their entire energy range is thus given by

$$U_{\rm CRe} = \frac{2\phi V}{\alpha+1}\frac{B_{\rm eq}^2}{(1+K)8\pi} = m_e c^2 N_0 \int_{\gamma_{\rm min}}^{\gamma_{\rm max}} \gamma^{-\delta+1} d\gamma \tag{2}$$

From the above relation, we derive the total number of electrons, which has the advantage of being a conserved quantity within the cluster volume (regardless of radiative losses or re-acceleration by turbulence). It should be noted that the above estimate neglects the possibility of a continuous production of secondary electrons by the hadronic process [116], which, however, cannot be the dominant process to explain radio halos given the very tight constraints imposed by the non-detection of hadronic γ-ray emission by high-energy observations, e.g., [117–119].

The total number of cosmic ray electrons can thus be computed with

$$N_{\rm CRe} = \frac{U_{\rm CRe}}{m_e c^2} \cdot \left(\frac{-\delta+2}{-\delta+1}\right) \cdot \left(\frac{\gamma_{\rm min}^{-\delta+1} - \gamma_{\rm max}^{-\delta+1}}{\gamma_{\rm min}^{-\delta+2} - \gamma_{\rm max}^{-\delta+2}}\right). \quad (3)$$

One can use the above relation, after specifying the minimum and maximum energies of the cosmic ray electron population, to estimate the number of electrons required by an observed radio structure for which energy equipartition can be reasonably hypothesized.

The left panel of Figure 2 gives the total number of cosmic ray electrons under different model variations (e.g., for different values of K, $\gamma_{\rm min}$) for a radio halo with volume $V = 1^3$ Mpc^3 at $z = 0$ and with a typical radio power of 10^{32} erg/s/Hz at 144 MHz, for different values of the radio spectral index. The right panel of the same figure shows the number of electrons as a function of the radio power instead, by fixing the spectrum to a reference value of $\alpha = 1.3$.

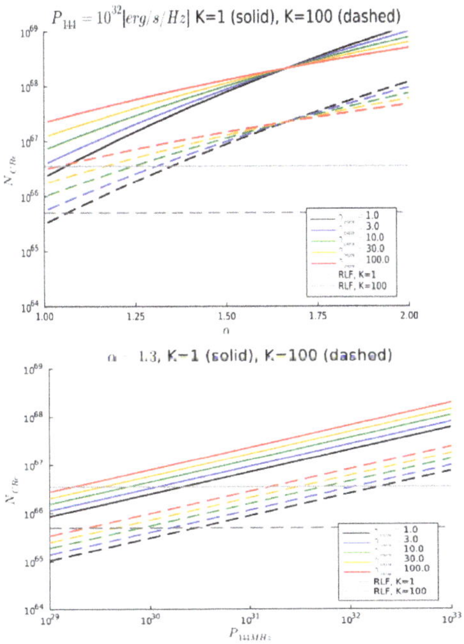

Figure 2. Top panel: Number of cosmic ray electrons to explain a radio halo emitting 10^{32} erg/s/Hz at 144 MHz, as a function of the observed radio spectrum (α) and for different choices of the ratio between cosmic ray protons and electrons (k) and for different choices of the minimum energy of electrons. Bottom panel: Number of cosmic ray electrons required to explain radio halo emissions of different powers at 144 MHz, assuming $\alpha = 1.3$ and $\gamma_{\rm min} = 10$ and for $K = 1$ or $K = 100$. In both panels, we assume that the radio halo has a volume of 1^3 Mpc^3 and that the host cluster has a 10^{15} $M_\odot$ mass. The horizontal lines give the number of electrons injected by the entire population of radio galaxies in a cluster and only using a single activity burst, assuming a $\alpha = 0.8$ spectrum for all radio galaxies and either $K = 1$ or $K = 100$.

The two additional horizontal lines in each panel mark the total number of cosmic ray electrons injected by all radio galaxies present in a $10^{15} M_\odot$ cluster (again for the $K = 1$ or $K = 100$ extreme scenarios, and always considering $\gamma_{\rm min} = 10$). We computed this quantity based on the the same formalism outlined above, i.e., after assuming equiparition between magnetic fields and cosmic rays, but this time using the distribution of radio luminosities of cluster radio galaxies derived at 843 MHz in [120], and assuming a power-size relation from [121] to obtain the largest linear scale (LLS). From the latter, we estimate the volume occupied by radio lobes following the recent theoretical results in [122], in which the evolution of simulated FRI radio sources in idealized cluster atmospheres was studied, reporting an approximate scaling of $V_g \sim LLS^3 \cdot (10~\text{kpc}/\text{LLS})$ between the size of expanding lobes and the volume they occupy.

In essence, this simplistic model gives an idea of the number of cosmic ray electrons that can be available as a result of a single generation of radio lobes by all radio galaxies in a typical cluster. Note that since it is reasonable that the K ratio in lobes and in radio halos is the same, to a first approximation, only the lines referred to as equal K should be compared here. Using as a reference the case of a $\alpha = 1.3$ radio halo in a $10^{15}~M_\odot$ cluster, the left panel shows that a single episode of activity by all radio galaxies in a cluster can easily account for all electrons required by a $\sim 5 \times 10^{30}$–10^{31} erg/s/Hz halo at 144 MHz; while for more powerful halos, multiple burst episodes (up to ~ 10) are required instead. The actual number of bursts from radio galaxies can be slightly lowered if the additional injection by cosmic rays by galactic winds or structure formation shocks is included, e.g., [123,124]. This simple formalism can also be extended to other kinds of diffuse sources like radio relics, yet the assumption of equipartition between cosmic rays and magnetic fields is much more questionable, owing to the much shorter acceleration timescale that is supposed to be at work in shock-accelerated cosmic rays.

Although this analysis suggests that the present and past activity by a typical population of radio galaxies in the ICM can explain the total number of electrons required to power diffuse radio sources, the actual ICM dynamics leading to the transport of cosmic rays has to be properly modeled in order to test whether the injection by radio galaxies is a viable seeding scenario to quantitatively explain observations.

Numerical Simulations

The actual transport of electrons injected by radio jets mixing with the (often turbulent) ICM can only be followed with fluid simulations, where also larger scale mixing motions induced by accretions, or by the repeated activity of AGN feedback, are included. While accounting for the full multi-scale complexity of this process is a challenge, even for modern numerical simulations, several studies have at least captured the most important pattern associated with large-scale circulation of the material ejected by jets.

Starting from the pioneering work in [125], in which the mixing of buoyant bubbles in a typical cluster environment was simulated and their radio detectability as a function of time was studied, numerical simulations have kept refining the state of the art of our understanding of how AGN feedback can affect the ICM; see, e.g., [12] for a recent review. Relatively fewer works have focused on the interplay between the remnant plasma ejected by AGN and the cluster environment, often resorting to a combination of different numerical methods in order to best capture the multi-scale nature of this process. For example, Ref. [126] used a Eulerian magneto-hydrodynamical approach to account for "cluster weather" on radio lobes expanding into the perturbed environment of a cluster of galaxies produced by a smoothed-particle hydrodynamics simulation. Likewise, the semi-analytic modeling of radio jets in [127] has been recently coupled to the Eulerian hydrodynamical re-simulation of jets crossing the ICM extracted from a large suite of SPH cosmological simulations in [128], which allowed them to probe the impressive resolution of 0.05 kpc/cell, and even to explore the potential connection with the production of "odd radio circles" in follow-up work [129].

A few studies have explored the possibility of a more direct connection between observed radio relics and the fueling by single, nearby radio galaxies, by simulating the evolution of re-accelerated fossil electrons. In a series of works, Refs. [21,130,131] modeled the formation of radio relics using 1D diffusive convection simulations with the goal of reproducing, first, the integrated radio emission spectra and the profile of radio emission across the width of real relics (e.g., the "toothbrush" in the merging cluster 1RXS J060303.3, and the "sausage" in cluster CIZA J2242.8 + 5301). In particular, they tested direct shock acceleration models, based on diffusive shock acceleration, versus models in which the radio-emitting particles were re-accelerated by weak shocks ($\mathcal{M} \leq 2$) running over a pre-existing population of fossil electrons with a low-energy cutoff ($\gamma \sim 8 \times 10^4$). The latter scenario was shown to reproduce reasonably well the observed integrated spectrum and the profile of radio surface brightness for the toothbrush, as well as the downstream curvature of the spectrum. A necessary ingredient for the model was, therefore, the presence of a wide and rather uniform distribution of "fossil" relativistic electrons, which they suggest could have been seeded by the previous activity of AGN in this sector of the cluster (e.g., Figure 3).

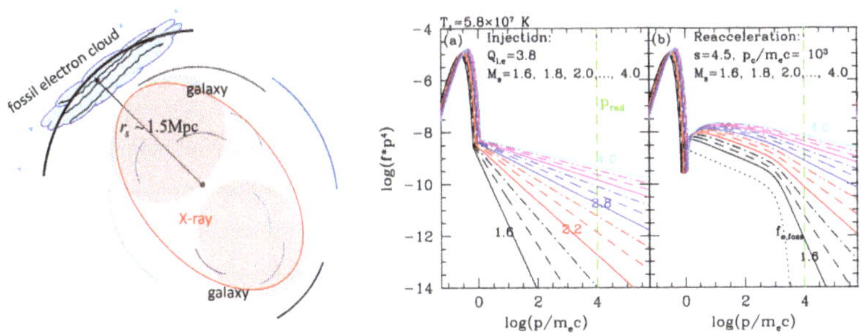

Figure 3. Left: Cartoon sketch of the bubble of fossil relativistic electrons at the location of the "sausage" radio relic in cluster CIZA J2242.8 + 5301. Centrl and right panel: electron spectra resulting from the injection of fresh electrons via diffusive shock acceleration, as a function of Mach number (central panel), and from the re-acceleration of fossil relativistic electrons by the same shocks (right panel); taken from [131].

The formation and long-term evolution of disturbed radio tails in clusters have been extensively investigated in a series of works [132–135] in which non-cosmological high-resolution MHD simulations were coupled to the simulation of the cosmic ray fluid associated with remnant plasma.

Refs. [24,136] investigated whether the morphology of some radio relics can reflect the distribution of fossil relativistic electrons previously injected by central AGN in clusters, and later dispersed into the ICM by sloshing gas motions generated by a previous merger event. They found that the transport of electrons along a spiral path, and its stretching tangentially to the spiral pattern, could produce a filamentary emission pattern compatible with real radio relics (see, e.g., Figure 4). Following this approach, Ref. [89] proposed a similar scenario to model the filamentary diffuse radio emission in Abel 2657 out of a shredded AGN bubble.

Recently, Ref. [79] produced high-resolution MHD simulations to model the complex radio structure in A3376, in which they accounted for the interaction between an AGN jet and a curved ICM magnetic field. They showed that the simulated jet propagation is quenched by the magnetic tension, which induces a later escape of the flow and of the relativistic particles carried with it, which can reasonably mimic the sharp "double-scythe" shape of the real observation. The magnetic field which best reproduces the synchrotron morphology is of order $\sim$10 μG, and the modeling suggests that magnetic reconnection might additionally be at work to produce an efficient injection of cosmic rays in the region.

However, whether a single burst from AGN combined with turbulence and mixing motions in clusters can provide a sufficient seeding of relativistic electrons to cover the entire extent of observed radio relics is not a fully settled issue.

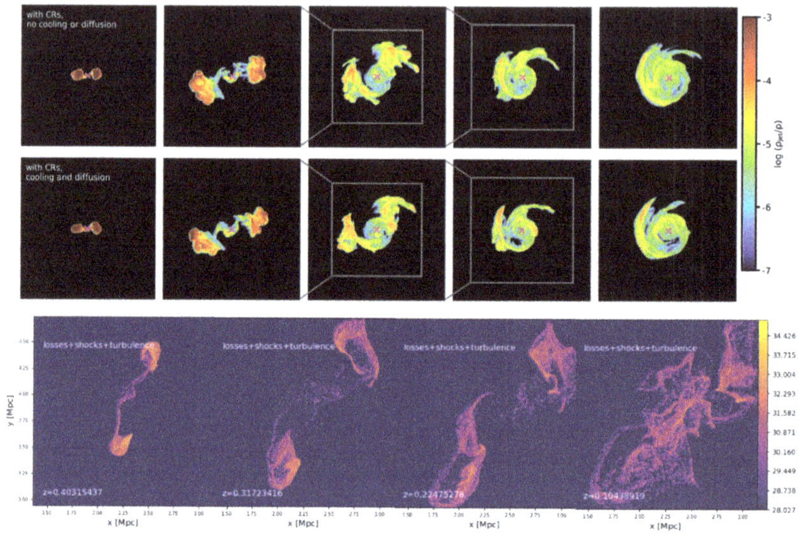

Figure 4. Top panels: Evolution of the projected density of fluid tracers ejected by AGN-inflated bubbles in simulations in [136]. The top row shows the evolution in the case in which cosmic rays are passively advected with the fluid, while in the lower row the effects of cosmic ray diffusion and Alfvén losses are included. Last row: Projected synchrotron radio emission at 150 MHz and for four different evolutionary steps, for a simulated pair of bubbles inflated by an AGN in the cosmological simulation in [137], in which electrons could age by radiative processes, as well as be re-energized by shocks and turbulent re-acceleration.

To explore this question, Refs. [137,138] modeled the dynamics of radio jets mixing with the turbulent ICM in cosmological simulations, while also following the spatial and spectral energy evolution of fossil electrons injected by AGN sources (see bottom panels of Figure 4). Their simulations showed that right after the active stage of radio jets, the buoyant stage of mixing of relativistic electrons is soon followed by a longer stage of large-scale mixing driven by turbulent motions in the ICM. During that phase, the remnant radio plasma can remain detectable at low frequencies out to $\leq$0.5 Mpc distance from sources, and even up to $\sim$0.5–1 Gyr since its first injection, provided that re-acceleration from weak shocks or turbulence is triggered by the cluster accretion dynamics. The emission from re-accelerated fossil electrons can produce detectable emission at any time, but it can typically produce only small ($\leq$100 kpc) and often filamentary radio features with steep radio spectra ($\alpha \geq 1.5$–2). With the same approach, Ref. [139] studied the mixing of electrons injected by eight realistic radio sources all activated at once in the same cluster, and compared their efficiency in seeding the ICM with cosmic rays to shocks driven by mergers in the same cluster, finding that the latter more naturally generate a volume-filling distribution of fossil electrons on the scales typically covered by radio relics, as also reported by other complementary analyses [140,141].

A common prediction from all of the aforementioned numerical approaches is that the detectable structures associated with plasma tails or remnant plasma from radio galaxies can only probe "the tip of the iceberg" of a broader and more volume-filling distribution of fossil electrons. To visualize this, we show in Figure 5 the measured evolution of the area-filling factor, $\Psi(2D)$, of the radio emission from the same simulation by [139], where eight radio galaxies release magnetic jets in a $\sim 10^{14}$ $M_{\odot}$ galaxy cluster, all at $z = 0.5$ and with a range of realistic jet powers. In the figure, we show the fraction of area of radial shells from the cluster center which is visible at different radio frequencies (50, 140, 610, and 1400 MHz), assuming for simplicity a detection threshold which scales with frequency as $\sigma(\nu) = \sigma_{LBA}[\nu/(50\ \mathrm{MHz})]^{-1}$, where $\sigma_{LBA} = \sigma = 5.7 \cdot 10^{-4}$ Jy/beam, as in, e.g., [32], assuming a beam of $\theta = 12.5''$ and a (non-evolving) luminosity distance of the cluster of $d_L = 132$ Mpc. The additional gray dotted lines in each panel show instead the area-covering factor of the entire distribution of cosmic ray electrons injected by simulated radio galaxies, projected along the same line of sight. Of course, while the area-filling factor of electrons mixing in the ICM can be large, the filling factor of the detectable radio emission can be much smaller, because it depends on the fraction of relativistic electrons in the high-energy tail of the distribution, as well as on the observing frequency, and it varies with the available sensitivity of each radio observation. In particular, given the same volume-filling distribution of relativistic electrons released by radio galaxies (or other mechanisms), different scenarios for the energy evolution of such electrons can vary their observable area-filling factor at different radio wavelengths. As can be seen, the area-filling factor of the observable emission rapidly drops with time, at an increased rate for higher frequencies. At a time of ~1.2 Gyr after their injection in the cluster, the electrons from all galaxies have spread to large radii, due to the combined effect of the initial jet velocity of the large-scale turbulent motions in the cluster, which can reach $\sigma_v \sim 300$ km/s in this simulated cluster [138]. However, cosmic rays dilute as they expand into the cluster volume, and ~0.25 Gyr after the injection less than ~30% of the projected area at ~0.5 R_{200} is covered by cosmic rays, which further declines to ~20% 1 Gyr later. At this epoch, less than ~5% of the projected area at ~R_{200} is covered by cosmic rays along the line of sight. The area-covering factor of the observable emission drops sharply with frequency and time: already ~0.25 Gyr since the jets' injection, ~30% of the surface at ~0.5 R_{200} is detectable at 50 MHz, but less than ~5% at 1400 MHz. One Gyr later, there is virtually no detectable emission from the remnant plasma from jets beyond ~0.4 R_{200}, and the very steep spectral energy distribution makes a detection only possible for low-frequency ($\lesssim$140 MHz) radio observations within that radius. It should be noted that this modeling cannot capture the additional effect of cosmic ray diffusion in the tangled ICM magnetic field, which can surely increase the volume-filling factor of cosmic rays. However, the timescales of the spatial diffusion of cosmic ray electrons are much longer than the evolutionary range considered here. In the range of energy of interest here, the diffusion coefficient should be in the $D \leq 10^{30} \sim 10^{31}$ cm^2/s ballpark, e.g., [25] and discussion therein, which means that the diffusion timescale over a $L \approx 1$ Mpc scale is $\tau_{\mathrm{diff}} \sim L^2/(4D) \gtrsim 7.5$–75 Gyr, and hence, is a a very slow and sub-dominant mechanism.

In summary, in line with the simplistic theoretical derivation of the previous section, the repeated activity of a few radio galaxies in clusters ($\lesssim$10), together with the typical turbulent gas motions expected from numerical simulations, appears sufficient to fill most of the area covered by observed radio halos ($\lesssim$0.1 R_{200}) with relativistic electrons.

On the other hand, the external regions where radio relics typically form ($\geq$0.5 R_{200}, e.g., [142,143]) are hardly replenished by a significant amount of fossil cosmic ray electrons, even by multiple radio galaxies, and probably additional cluster wide motions (e.g., sloshing) are required to conveniently distribute the remnant plasma in a more uniform way, before shocks revive it, as already suggested, e.g., [24]. How common the appearance of such configurations is in the lifetime of clusters of galaxies remains to be assessed. Complementary to this, the injection of cosmic rays by merger shocks can naturally lead to large and correlated populations of fossil electrons that subsequent re-acceleration events can illuminate on ~Mpc scales [22,139–141], and moreover, the ubiquitous presence of turbulent motions, even in the outskirts of the cluster, can maintain electrons at a higher energy than expected only considering the effect of radiative losses, i.e., up to $\gamma \sim 10^3$ [35]. This would imply the existence of a second and rather uniform population of fossil electrons on larger cluster scales, in addition to the outcome of evolving radio cluster galaxies.

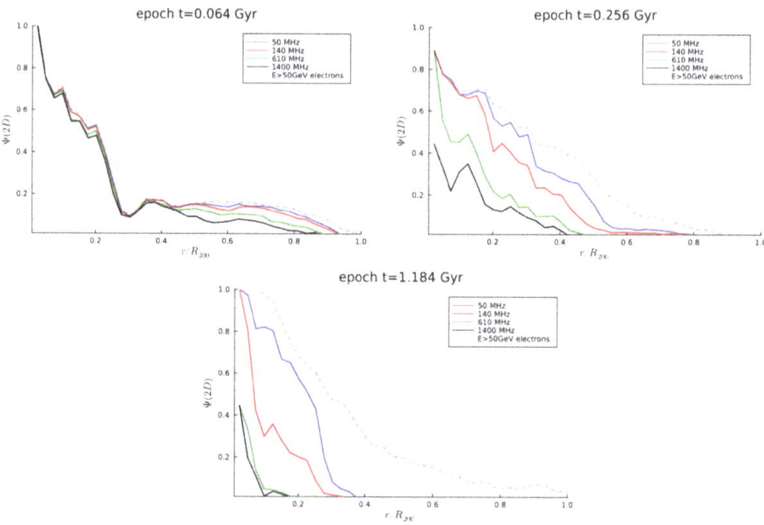

Figure 5. Area-filling factors of detectable radio emission as a function of frequency (different colors) and of relativistic electrons (dotted gray lines) for three different epochs since the injection of radio jets in the cosmological simulation by [139].

4. Conclusions and Future Perspectives

In summary, both the direct observations of remnant radio plasma and of disturbed radio tails, and the more indirect indications of a connection between the activity of radio galaxies and large diffuse radio emissions in clusters of galaxies (i.e., radio halos and radio relics) support the idea that the fueling of relativistic electrons from cluster radio galaxies is a key process to regulate the non-thermal energy budget of the ICM.

From the theoretical viewpoint, the time-integrated activity of the population of radio galaxies in clusters appear sufficient to fuel the entire extent of diffuse radio emissions with the necessary amount of fossil electrons, while the exact timescales and efficiency with which relativistic electrons spread across the ICM depend on the multi-scale gas dynamics, and its investigation with modern numerical methods has just begun. Observations at very low frequencies (<100 MHz) performed with LOFAR2.0 and SKA-LOW will soon make it possible to recover the emission from older populations of cosmic ray electrons in the ICM. In parallel, with the excellent polarization capabilities of the new generation of radio telescopes (e.g., MeerKAT+, SKA-MID), these instruments will allow the investigation of highly inefficient re-energization processes sustaining particle lifetimes in the ICM.

Among the main processes to re-accelerate fossil electrons in the ICM, shocks and turbulence are the dominant mechanisms. While the detection of shock fronts has been already enabled by instruments such as *Chandra*, *XMM-Newton*, and *Suzaku*, the direct observation of turbulent motions is limited to the case of Perseus by Hitomi [144]. The deployment of new X-ray facilities capable of measuring the gas turbulent motions, either with a large integration area (e.g., *XRISM*) or in a more spatially resolved manner (e.g., *Athena*), promises to enable better constraints to be provided on the multi-scale dynamical interactions between the relativistic plasma dispersed by radio galaxies, and the evolving ICM.

Author Contributions: F.V. and A.B. curated the manuscript, the figures and the interpretation of results in equal parts. All authors have read and agreed to the published version of the manuscript.

Funding: F.V. acknowledges partial financial support from the Cariplo "BREAKTHRU" funds Rif: 2022-2088 CUP J33C22004310003. A.B. acknowledges financial support from the European Union-Next Generation EU.

Acknowledgments: The authors thank Annalisa Bonafede, Marisa Brienza, Hyesung Kang, Wonki Lee, Ramij Raja, Kamlesh Rajpurohit and John ZuHone for the kind permission to reuse the figures of their papers for this work. We thank Lawrence Rudnick for his constructive feedback on the manuscript, as well as our anonymous reviewers.

Conflicts of Interest: The authors declare no conflicts of interest.

Abbreviations

The following abbreviations are used in this manuscript:

AGN	Active galactic nucleus
AMR	Adaptive mesh refinement
ASKAP	Australian Square Kilometre Array Pathfinder
BCG	Brightest central galaxy
DSA	Diffusive shock acceleration
GReET	Gently re-energized tail
HBA	High-band antenna
HT	Head–tail
ICM	Intracluster medium
LBA	Low-band antenna
LLS	Largest linear scale
LOFAR	Low Frequency Array
MHD	Magnetohydro dynamics
MWA	Murchison Widefield Array
NAT	Narrow-angle tail
SMBH	Super massive black hole
SPH	Smoothed-particle Hydrodynamics
WAT	Wide-angle tail
ΛCDM	Lambda cold dark matter

References

1. Croston, J.H.; Hardcastle, M.J.; Birkinshaw, M.; Worrall, D.M.; Laing, R.A. An XMM-Newton study of the environments, particle content and impact of low-power radio galaxies. *Mon. Not. R. Astron. Soc.* **2008**, *386*, 1709–1728. [CrossRef]
2. Croston, J.H.; Ineson, J.; Hardcastle, M.J. Particle content, radio-galaxy morphology, and jet power: All radio-loud AGN are not equal. *Mon. Not. R. Astron. Soc.* **2018**, *476*, 1614–1623. [CrossRef]
3. Norman, M.L.; Bryan, G.L. Cluster Turbulence. In Proceedings of the Radio Galaxy Messier 87,Tegernsee, Germany, 15–19 September 1997; Röser, H.-J., Meisenheimer, K., Eds.; Springer: Berlin/Heidelberg, Germany, 1999; Volume 530, pp. 106–115. [CrossRef]
4. Dolag, K.; Vazza, F.; Brunetti, G.; Tormen, G. Turbulent gas motions in galaxy cluster simulations: The role of smoothed particle hydrodynamics viscosity. *Mon. Not. R. Astron. Soc.* **2005**, *364*, 753–772. [CrossRef]
5. Vazza, F.; Brunetti, G.; Gheller, C.; Brunino, R.; Brüggen, M. Massive and refined. II. The statistical properties of turbulent motions in massive galaxy clusters with high spatial resolution. *Astron. Astrophys.* **2011**, *529*, A17. [CrossRef]
6. Schmidt, W.; Byrohl, C.; Engels, J.F.; Behrens, C.; Niemeyer, J.C. Viscosity, pressure and support of the gas in simulations of merging cool-core clusters. *Mon. Not. R. Astron. Soc.* **2017**, *470*, 142–156. [CrossRef]
7. Ayromlou, M.; Nelson, D.; Pillepich, A.; Rohr, E.; Truong, N.; Li, Y.; Simionescu, A.; Lehle, K.; Lee, W. An Atlas of Gas Motions in the TNG-Cluster Simulation: From Cluster Cores to the Outskirts. *arXiv* **2023**, arXiv:2311.06339.
8. Churazov, E.; Forman, W.; Jones, C.; Sunyaev, R.; Böhringer, H. XMM-Newton observations of the Perseus cluster-II. Evidence for gas motions in the core. *Mon. Not. R. Astron. Soc.* **2004**, *347*, 29–35. [CrossRef]
9. Heinz, S.; Brüggen, M.; Young, A.; Levesque, E. The answer is blowing in the wind: Simulating the interaction of jets with dynamic cluster atmospheres. *Mon. Not. R. Astron. Soc.* **2006**, *373*, L65–L69. [CrossRef]
10. Gaspari, M.; Ruszkowski, M.; Sharma, P. Cause and Effect of Feedback: Multiphase Gas in Cluster Cores Heated by AGN Jets. *Astrophys. J.* **2012**, *746*, 94. [CrossRef]
11. Li, Y.; Bryan, G.L. Modeling Active Galactic Nucleus Feedback in Cool-core Clusters: The Balance between Heating and Cooling. *Astrophys. J.* **2014**, *789*, 54. [CrossRef]
12. Bourne, M.A.; Yang, H.Y.K. Recent Progress in Modeling the Macro- and Micro-Physics of Radio Jet Feedback in Galaxy Clusters. *Galaxies* **2023**, *11*, 73. [CrossRef]
13. Völk, H.J.; Atoyan, A.M. Early Starbursts and Magnetic Field Generation in Galaxy Clusters. *Astrophys. J.* **2000**, *541*, 88–94. [CrossRef]
14. Hardcastle, M.J.; Croston, J.H. Radio galaxies and feedback from AGN jets. *New Astron. Rev.* **2020**, *88*, 101539. [CrossRef]

15. de Gasperin, F.; Intema, H.T.; Shimwell, T.W.; Brunetti, G.; Brüggen, M.; Enßlin, T.A.; van Weeren, R.J.; Bonafede, A.; Röttgering, H.J.A. Gentle reenergization of electrons in merging galaxy clusters . *Sci. Adv.* **2017**, *3*, e1701634. [CrossRef]
16. Wilber, A.; Brüggen, M.; Bonafede, A.; Savini, F.; Shimwell, T.; van Weeren, R.J.; Rafferty, D.; Mechev, A.P.; Intema, H.; Andrade-Santos, F.; et al. LOFAR discovery of an ultra-steep radio halo and giant head-tail radio galaxy in Abell 1132 . *Mon. Not. R. Astron. Soc.* **2018**, *473*, 3536–3546. [CrossRef]
17. Mandal, S.; Intema, H.T.; van Weeren, R.J.; Shimwell, T.W.; Botteon, A.; Brunetti, G.; de Gasperin, F.; Brüggen, M.; Di Gennaro, G.; Kraft, R.; et al. Revived fossil plasma sources in galaxy clusters . *Astron. Astrophys.* **2020**, *634*, A4. [CrossRef]
18. Quici, B.; Turner, R.J.; Seymour, N.; Hurley-Walker, N.; Shabala, S.S.; Ishwara-Chandra, C.H. Selecting and modelling remnant AGNs with limited spectral coverage. *Mon. Not. R. Astron. Soc.* **2022**, *514*, 3466–3484. [CrossRef]
19. Brienza, M.; Lovisari, L.; Rajpurohit, K.; Bonafede, A.; Gastaldello, F.; Murgia, M.; Vazza, F.; Bonnassieux, E.; Botteon, A.; Brunetti, G.; et al. The galaxy group NGC 507: Newly detected AGN remnant plasma transported by sloshing. *Astron. Astrophys.* **2022**, *661*, A92. [CrossRef]
20. van Weeren, R.J.; de Gasperin, F.; Akamatsu, H.; Brüggen, M.; Feretti, L.; Kang, H.; Stroe, A.; Zandanel, F. Diffuse Radio Emission from Galaxy Clusters. *Sci. Space Rev.* **2019**, *215*, 16. [CrossRef]
21. Kang, H.; Ryu, D.; Jones, T.W. Diffusive Shock Acceleration Simulations of Radio Relics. *Astrophys. J.* **2012**, *756*, 97. [CrossRef]
22. Pinzke, A.; Oh, S.P.; Pfrommer, C. Giant radio relics in galaxy clusters: Reacceleration of fossil relativistic electrons? *Mon. Not. R. Astron. Soc.* **2013**, *435*, 1061–1082. [CrossRef]
23. Botteon, A.; Brunetti, G.; Ryu, D.; Roh, S. Shock acceleration efficiency in radio relics. *Astron. Astrophys.* **2020**, *634*, A64. [CrossRef]
24. ZuHone, J.A.; Markevitch, M.; Weinberger, R.; Nulsen, P.; Ehlert, K. How Merger-driven Gas Motions in Galaxy Clusters Can Turn AGN Bubbles into Radio Relics. *Astrophys. J.* **2021**, *914*, 73. [CrossRef]
25. Brunetti, G.; Jones, T.W. Cosmic Rays in Galaxy Clusters and Their Nonthermal Emission. *Int. J. Mod. Phys. D* **2014**, *23*, 1430007. [CrossRef]
26. Cassano, R.; Gitti, M.; Brunetti, G. A morphological comparison between giant radio halos and radio mini-halos in galaxy clusters. *Astron. Astrophys.* **2008**, *486*, L31–L34. [CrossRef]
27. Richard-Laferrière, A.; Hlavacek-Larrondo, J.; Nemmen, R.S.; Rhea, C.L.; Taylor, G.B.; Prasow-Émond, M.; Gendron-Marsolais, M.; Latulippe, M.; Edge, A.C.; Fabian, A.C.; et al. On the relation between mini-halos and AGN feedback in clusters of galaxies. *Mon. Not. R. Astron. Soc.* **2020**, *499*, 2934–2958. [CrossRef]
28. Govoni, F.; Orrù, E.; Bonafede, A.; Iacobelli, M.; Paladino, R.; Vazza, F.; Murgia, M.; Vacca, V.; Giovannini, G.; Feretti, L.; et al. A radio ridge connecting two galaxy clusters in a filament of the cosmic web. *Science* **2019**, *364*, 981–984. [CrossRef]
29. Brunetti, G.; Vazza, F. Second-order Fermi Reacceleration Mechanisms and Large-Scale Synchrotron Radio Emission in Intracluster Bridges. *Phys. Rev. Lett.* **2020**, *124*, 051101. [CrossRef]
30. de Jong, J.; van Weeren, R.J.; Botteon, A.; Oonk, J.; Brunetti, G.; Shimwell, T.W.; Cassano, R.; Röttgering, H.J.; Tasse, C. Deep study of A399-401: Application of a wide-field facet calibration. *Astron. Astrophys.* **2022**, *668*, A107. [CrossRef]
31. Shweta, A.; Athreya, R.M.; Sekhar, S. Reenergization of Radio Halo Electrons in the Merging Galaxy Cluster A2163 . *Astrophys. J.* **2020**, *897*, 115. [CrossRef]
32. Botteon, A.; van Weeren, R.J.; Brunetti, G.; Vazza, F.; Shimwell, T.W.; Brüggen, M.; Röttgering, H.J.; de Gasperin, F.; Akamatsu, H.; Bonafede, A.; et al. Magnetic fields and relativistic electrons fill entire galaxy cluster. *Sci. Adv.* **2022**, *8*, eabq7623. [CrossRef] [PubMed]
33. Cuciti, V.; de Gasperin, F.; Brüggen, M.; Vazza, F.; Brunetti, G.; Shimwell, T.W.; Edler, H.W.; van Weeren, R.J.; Botteon, A.; Cassano, R.; et al. Galaxy clusters enveloped by vast volumes of relativistic electrons. *Nature* **2022**, *609*, 911–914. [CrossRef] [PubMed]
34. Bruno, L.; Botteon, A.; Shimwell, T.W.; Cuciti, V.; de Gasperin, F.; Brunetti, G.; Dallacasa, D.; Gastaldello, F.; Rossetti, M.; van Weeren, R.J.; et al. A three-component giant radio halo: The puzzling case of the galaxy cluster Abell 2142. *Astron. Astrophys.* **2023**, *678*, A133. [CrossRef]
35. Beduzzi, L.; Vazza, F.; Brunetti, G.; Cuciti, V.; Wittor, D.; Corsini, E. Exploring the origins of mega radio halos. *Astron. Astrophys.* **2023**, *678*, L8. [CrossRef]
36. Nishiwaki, K.; Asano, K.; Murase, K. High-energy Neutrino Constraints on Cosmic-Ray Reacceleration in Radio Halos of Massive Galaxy Clusters. *Astrophys. J.* **2023**, *954*, 188. [CrossRef]
37. Lee, W.; ZuHone, J.A.; James Jee, M.; HyeongHan, K.; Kale, R.; Ahn, E. Discovery of A Large-scale Bent Radio Jet in the Merging Cluster A514. *Astrophys. J.* **2023**, *957*, L4. [CrossRef]
38. Raja, R.; Rahaman, M.; Datta, A.; Smirnov, O.M. A Multi-Frequency View of the Radio Phoenix in the Abell 85 Cluster. *arXiv* **2023**, arXiv:2309.14244.
39. Botteon, A.; Brunetti, G.; van Weeren, R.J.; Shimwell, T.W.; Pizzo, R.F.; Cassano, R.; Iacobelli, M.; Gastaldello, F.; Bîrzan, L.; Bonafede, A.; et al. The Beautiful Mess in Abell 2255. *Astrophys. J.* **2020**, *897*, 93. [CrossRef]
40. Rajpurohit, K.; Osinga, E.; Brienza, M.; Botteon, A.; Brunetti, G.; Forman, W.R.; Riseley, C.J.; Vazza, F.; Bonafede, A.; van Weeren, R.J.; et al. Deep low-frequency radio observations of Abell 2256. II. The ultra-steep spectrum radio halo. *Astron. Astrophys.* **2023**, *669*, A1. [CrossRef]
41. Bonafede, A.; Brunetti, G.; Rudnick, L.; Vazza, F.; Bourdin, H.; Giovannini, G.; Shimwell, T.W.; Zhang, X.; Mazzotta, P.; Simionescu, A.; et al. The Coma Cluster at LOFAR Frequencies. II. The Halo, Relic, and a New Accretion Relic. *Astrophys. J.* **2022**, *933*, 218. [CrossRef]

42. Kempner, J.C.; Blanton, E.L.; Clarke, T.E.; Enßlin, T.A.; Johnston-Hollitt, M.; Rudnick, L. Conference Note: A Taxonomy of Extended Radio Sources in Clusters of Galaxies. In Proceedings of the Riddle of Cooling Flows in Galaxies and Clusters of Galaxies, Charlottesville, VA, USA, 31 May–4 June 2003; Reiprich, T., Kempner, J., Soker, N., Eds.; p. 335.
43. Slee, O.; Reynolds, J. Steep-spectrum radio sources in clusters of galaxies - the southern sample. *Publ. Astron. Soc. Aust.* **1984**, *5*, 516–529. [CrossRef]
44. Slee, O.; Roy, A. An extreme example of a radio relic in Abell 4038. *Mon. Not. R. Astron. Soc.* **1998**, *297*, L86–L92. [CrossRef]
45. Slee, O.; Roy, A.; Murgia, M.; Andernach, H.; Ehle, M. Four Extreme Relic Radio Sources in Clusters of Galaxies. *Astron. J.* **2001**, *122*, 1172–1193. [CrossRef]
46. Subrahmanyan, R.; Beasley, A.; Goss, W.; Golap, K.; Hunstead, R.W. PKS B1400-33: An Unusual Radio Relic in a Poor Cluster. *Astron. J.* **2003**, *125*, 1095–1106. [CrossRef]
47. Green, D.A.; Lacy, M.; Bhatnagar, S.; Gates, E.; Warner, P. Radio and near-infrared observations of the steep-spectrum Galactic plane radio source WKB 0314+57.8. *Mon. Not. R. Astron. Soc.* **2004**, *354*, 1159–1164. [CrossRef]
48. Duchesne, S.W.; Johnston-Hollitt, M.; Zhu, Z.H.; Wayth, R.; Line, J. Murchison Widefield Array detection of steep-spectrum, diffuse, non-thermal radio emission within Abell 1127. *Publ. Astron. Soc. Aust.* **2020**, *37*, e037. [CrossRef]
49. Botteon, A.; Giacintucci, S.; Gastaldello, F.; Venturi, T.; Brunetti, G.; van Weeren, R.J.; Shimwell, T.W.; Rossetti, M.; Akamatsu, H.; Brüggen, M.; et al. Nonthermal phenomena in the center of Abell 1775. An 800 kpc head-tail, revived fossil plasma and slingshot radio halo. *Astron. Astrophys.* **2021**, *649*, A37. [CrossRef]
50. Hodgson, T.; Bartalucci, I.; Johnston-Hollitt, M.; McKinley, B.; Vazza, F.; Wittor, D. Ultra-steep-spectrum Radio "Jellyfish" Uncovered in A2877. *Astrophys. J.* **2021**, *909*, 198. [CrossRef]
51. Pandge, M.; Sebastian, B.; Seth, R.; Raychaudhury, S. A detailed study of X-ray cavities in the intracluster environment of the cool core cluster Abell 3017. *Mon. Not. R. Astron. Soc.* **2021**, *504*, 1644–1656. [CrossRef]
52. Pandge, M.; Kale, R.; Dabhade, P.; Mahato, M.; Raychaudhury, S. Giant Metrewave Radio Telescope unveils steep-spectrum antique filaments in the galaxy cluster Abell 725. *Mon. Not. R. Astron. Soc.* **2022**, *509*, 1837–1847. [CrossRef]
53. Pasini, T.; Edler, H.; Brüggen, M.; de Gasperin, F.; Botteon, A.; Rajpurohit, K.; van Weeren, R.J.; Gastaldello, F.; Gaspari, M.; Brunetti, G.; et al. Particle re-acceleration and diffuse radio sources in the galaxy cluster Abell 1550. *Astron. Astrophys.* **2022**, *663*, A105. [CrossRef]
54. Riseley, C.J.; Bonnassieux, E.; Vernstrom, T.; Galvin, T.; Chokshi, A.; Botteon, A.; Rajpurohit, K.; Duchesne, S.W.; Bonafede, A.; Rudnick, L.; et al. Radio fossils, relics, and haloes in Abell 3266: Cluster archaeology with ASKAP-EMU and the ATCA. *Mon. Not. R. Astron. Soc.* **2022**, *515*, 1871–1896. [CrossRef]
55. Groenveld, C.; van Weeren, R.J.; Osinga, E.; Williams, W.L.; Callingham, J.R.; de Gasperin, F.; Botteon, A.; Shimwell, T.; Swejen, F.; de Jong, J.M.G.H.J.; et al. Characterisation of the decameter sky at sub-arcminute resolution. *Nat. Astron.* **2024** , *accepted*.
56. Mandal, S.; Intema, H.T.; Shimwell, T.W.; van Weeren, R.J.; Botteon, A.; Röttgering, H.J.; Hoang, D.N.; Brunetti, G.; de Gasperin, F.; Giacintucci, S.; et al. Ultra-steep spectrum emission in the merging galaxy cluster Abell 1914. *Astron. Astrophys.* **2019**, *622*, A22. [CrossRef]
57. Juett, A.; Sarazin, C.L.; Clarke, T.E.; Andernach, H.; Ehle, M.; Fujita, Y.; Kempner, J.C.; Roy, A.; Rudnick, L.; Slee, O. A Chandra Observation of Abell 13: Investigating the Origin of the Radio Relic. *Astrophys. J.* **2008**, *672*, 138–145. [CrossRef]
58. Owen, F.N.; Rudnick, L.; Eilek, J.A.; Rau, U.; Bhatnagar, S.; Kogan, L. Wideband Very Large Array Observations of A2256. I. Continuum, Rotation Measure, and Spectral Imaging. *Astrophys. J.* **2014**, *794*, 24. [CrossRef]
59. Werner, N.; Zhuravleva, I.; Canning, R.E.; Allen, S.W.; King, A.; Sanders, J.S.; Simionescu, A.; Taylor, G.B.; Morris, R.G.; Fabian, A.C. Deep Chandra study of the truncated cool core of the Ophiuchus cluster. *Mon. Not. R. Astron. Soc.* **2016**, *460*, 2752–2764. [CrossRef]
60. van Weeren, R.J.; Shimwell, T.W.; Botteon, A.; Brunetti, G.; Brüggen, M.; Boxelaar, J.M.; Cassano, R.; Di Gennaro, G.; Andrade-Santos, F.; Bonnassieux, E.; et al. LOFAR observations of galaxy clusters in HETDEX. Extraction and self-calibration of individual LOFAR targets. *Astron. Astrophys.* **2021**, *651*, A115. [CrossRef]
61. Duchesne, S.W.; Johnston-Hollitt, M.; Offringa, A.R.; Pratt, G.W.; Zheng, Q.; Dehghan, S. Diffuse galaxy cluster emission at 168 MHz within the Murchison Widefield Array Epoch of Reionization 0-h field. *Publ. Astron. Soc. Aust.* **2021**, *38*, e010. [CrossRef]
62. Duchesne, S.W.; Johnston-Hollitt, M.; Bartalucci, I. Low-frequency integrated radio spectra of diffuse, steep-spectrum sources in galaxy clusters: Palaeontology with the MWA and ASKAP. *Publ. Astron. Soc. Aust.* **2021**, *38*, e053. [CrossRef]
63. Botteon, A.; Shimwell, T.W.; Cassano, R.; Cuciti, V.; Zhang, X.; Bruno, L.; Camillini, L.; Natale, R.; Jones, A.; Gastaldello, F.; et al. The Planck clusters in the LOFAR sky. I. LoTSS-DR2: New detections and sample overview. *Astron. Astrophys.* **2022**, *660*, A78. [CrossRef]
64. Hoang, D.N.; Brüggen, M.; Botteon, A.; Shimwell, T.W.; Zhang, X.; Bonafede, A.; Bruno, L.; Bonnassieux, E.; Cassano, R.; Cuciti, V.; et al. Diffuse radio emission from non-Planck galaxy clusters in the LoTSS-DR2 fields. *Astron. Astrophys.* **2022**, *665*, A60. [CrossRef]
65. Knowles, K.; Cotton, W.; Rudnick, L.; Camilo, F.; Goedhart, S.; Deane, R.; Ramatsoku, M.; Bietenholz, M.; Brüggen, M.; Button, C.; et al. The MeerKAT Galaxy Cluster Legacy Survey. I. Survey Overview and Highlights. *Astron. Astrophys.* **2022**, *657*, A56. [CrossRef]

66. Duchesne, S.W.; Botteon, A.; Koribalski, B.S.; Loi, F.; Rajpurohit, K.; Riseley, C.J.; Rudnick, L.; Vernstrom, T.; Andernach, H.; Hopkins, A.M.; et al. Evolutionary Map of the Universe (EMU): A pilot search for diffuse, non-thermal radio emission in galaxy clusters with the Australian SKA Pathfinder. *arXiv* **2024**, arXiv:2402.06192.
67. Miley, G.K. The structure of extended extragalactic radio sources. *Annu. Rev. Astron. Astrophys.* **1980**, *18*, 165–218. [CrossRef]
68. Sarazin, C.L. The energy spectrum of primary cosmic-ray electrons in clusters of galaxies and inverse compton emission. *Astrophys. J.* **1999**, *520*, 529–547. [CrossRef]
69. Sebastian, B.; Lal, D.V.; Pramesh Rao, A. Giant Metrewave Radio Telescope Observations of Head-Tail Radio Galaxies. *Astron. J.* **2017**, *154*, 169. [CrossRef]
70. Cuciti, V.; Brunetti, G.; van Weeren, R.J.; Bonafede, A.; Dallacasa, D.; Cassano, R.; Venturi, T.; Kale, R. New giant radio sources and underluminous radio halos in two galaxy clusters. *Astron. Astrophys.* **2018**, *609*, A61. [CrossRef]
71. Edler, H.; de Gasperin, F.; Brunetti, G.; Botteon, A.; Cuciti, V.; van Weeren, R.J.; Cassano, R.; Shimwell, T.W.; Brüggen, M.; Drabent, A. Abell 1033: Radio halo and gently reenergized tail at 54 MHz. *Astron. Astrophys.* **2022**, *666*, A3. [CrossRef]
72. van Weeren, R.J.; Röttgering, H.J.; Rafferty, D.; Pizzo, R.F.; Bonafede, A.; Brüggen, M.; Brunetti, G.; Ferrari, C.; Orrù, E.; Heald, G.; et al. First LOFAR observations at very low frequencies of cluster-scale non-thermal emission: The case of Abell 2256. *Astron. Astrophys.* **2012**, *543*, A43. [CrossRef]
73. Srivastava, S.; Singal, A.K. GMRT observations of IC 711-the longest head-tail radio galaxy known. *Mon. Not. R. Astron. Soc.* **2020**, *493*, 3811–3824. [CrossRef]
74. Venturi, T.; Giacintucci, S.; Merluzzi, P.; Bardelli, S.; Busarello, G.; Dallacasa, D.; Sikhosana, S.P.; Marvil, J.; Smirnov, O.M.; Bourdin, H.; et al. Radio footprints of a minor merger in the Shapley Supercluster: From supercluster down to galactic scales. *Astron. Astrophys.* **2022**, *660*, A81. [CrossRef]
75. Ignesti, A.; Brunetti, G.; Shimwell, T.W.; Gitti, M.; Bîrzan, L.; Botteon, A.; Brüggen, M.; de Gasperin, F.; Di Gennaro, G.; Edge, A.C.; et al. A LOFAR view into the stormy environment of the galaxy cluster 2A0335+096. *Astron. Astrophys.* **2022**, *659*, A20. [CrossRef]
76. Lusetti, G.; de Gasperin, F.; Cuciti, V.; Brüggen, M.; Spinelli, C.; Edler, H.; Brunetti, G.; van Weeren, R.J.; Botteon, A.; Di Gennaro, G.; et al. Re-energization of AGN head-tail radio galaxies in the galaxy cluster ZwCl 0634.1+47474. *Mon. Not. R. Astron. Soc.* **2024**, *528*, 141–159. [CrossRef]
77. van Weeren, R.J.; Andrade-Santos, F.; Dawson, W.A.; Golovich, N.R.; Lal, D.V.; Kang, H.; Ryu, D.; Brüggen, M.; Ogrean, G.; Forman, W.R.; et al. The case for electron re-acceleration at galaxy cluster shocks. *Nat. Astron.* **2017**, *1*, 5. [CrossRef]
78. Lal, D.V. NGC 4869 in the Coma Cluster: Twist, Wrap, Overlap, and Bend. *Astron. J.* **2020**, *160*, 161. [CrossRef]
79. Chibueze, J.O.; Sakemi, H.; Ohmura, T.; Machida, M.; Akamatsu, H.; Akahori, T.; Nakanishi, H.; Parekh, V.; van Rooyen, R.; Takeuchi, T.T. Jets from MRC 0600-399 bent by magnetic fields in the cluster Abell 3376. *Nature* **2021**, *593*, 47–50. [CrossRef] [PubMed]
80. Giacintucci, S.; Venturi, T.; Markevitch, M.; Bourdin, H.; Mazzotta, P.; Merluzzi, P.; Dallacasa, D.; Bardelli, S.; Sikhosana, S.P.; Smirnov, O.M.; et al. A Candle in the Wind: A Radio Filament in the Core of the A3562 Galaxy Cluster. *Astrophys. J.* **2022**, *934*, 49. [CrossRef]
81. Pfrommer, C.; Jones, T.W. Radio Galaxy NGC 1265 Unveils the Accretion Shock Onto the Perseus Galaxy Cluster. *Astrophys. J.* **2011**, *730*, 22. [CrossRef]
82. Botteon, A.; Shimwell, T.W.; Bonafede, A.; Dallacasa, D.; Gastaldello, F.; Eckert, D.; Brunetti, G.; Venturi, T.; van Weeren, R.J.; Mandal, S.; et al. The spectacular cluster chain Abell 781 as observed with LOFAR, GMRT, and XMM-Newton. *Astron. Astrophys.* **2019**, *622*, A19. [CrossRef]
83. Wilber, A.G.; Brüggen, M.; Bonafede, A.; Rafferty, D.; Shimwell, T.W.; van Weeren, R.J.; Akamatsu, H.; Botteon, A.; Savini, F.; Intema, H.T.; et al. Evolutionary phases of merging clusters as seen by LOFAR. *Astron. Astrophys.* **2019**, *622*, A25. [CrossRef]
84. Gendron-Marsolais, M.L.; Hull, C.; Perley, R.A.; Rudnick, L.; Kraft, R.P.; Hlavacek-Larrondo, J.; Fabian, A.C.; Roediger, E.; van Weeren, R.J.; Richard-Laferrière, A.; et al. VLA Resolves Unexpected Radio Structures in the Perseus Cluster of Galaxies. *Astrophys. J.* **2021**, *911*, 56. [CrossRef]
85. Enßlin, T.A.; Gopal-Krishna. Reviving fossil radio plasma in clusters of galaxies by adiabatic compression in environmental shock waves. *Astron. Astrophys.* **2001**, *366*, 26–34. [CrossRef]
86. Enßlin, T.A.; Brüggen, M. On the formation of cluster radio relics. *Mon. Not. R. Astron. Soc.* **2002**, *331*, 1011–1019. [CrossRef]
87. Markevitch, M.; Govoni, F.; Brunetti, G.; Jerius, D. Bow Shock and Radio Halo in the Merging Cluster A520. *Astrophys. J.* **2005**, *627*, 733–738. [CrossRef]
88. ZuHone, J.A.; Markevitch, M.; Lee, D. Sloshing of the Magnetized Cool Gas in the Cores of Galaxy Clusters. *Astrophys. J.* **2011**, *743*, 16. [CrossRef]
89. Botteon, A.; Gastaldello, F.; ZuHone, J.A.; Balboni, M.; Bartalucci, I.; Brunetti, G.; Bonafede, A.; Brüggen, M.; Shimwell, T.W.; van Weeren, R.J. A radio bubble shredded by gas sloshing? *Mon. Not. R. Astron. Soc.* **2024**, *527*, 919–930. [CrossRef]
90. Pacholczyk, A.; Scott, J. In situ particle acceleration and physical conditions in radio tail galaxies. *Astrophys. J.* **1976**, *203*, 313–322. [CrossRef]
91. Wilson, A.; Vallée, J.P. The structures of the head-tail radio galaxies IC 708 and IC 711 at 1.4 GHz. *Astron. Astrophys.* **1977**, *58*, 79–91.
92. Jones, T.W.; Owen, F.N. Hot gas in elliptical galaxies and the formation of head-tail radio sources. *Astrophys. J.* **1979**, *234*, 818–824. [CrossRef]

93. Brunetti, G.; Setti, G.; Feretti, L.; Giovannini, G. Particle reacceleration in the Coma cluster: Radio properties and hard X-ray emission. *Mon. Not. R. Astron. Soc.* **2001**, *320*, 365–378. [CrossRef]
94. Petrosian, V. On the Nonthermal Emission and Acceleration of Electrons in Coma and Other Clusters of Galaxies. *Astrophys. J.* **2001**, *557*, 560–572. [CrossRef]
95. Clarke, T.E.; Randall, S.W.; Sarazin, C.L.; Blanton, E.L.; Giacintucci, S. Chandra View of the Ultra-steep Spectrum Radio Source in A2443: Merger Shock-induced Compression of Fossil Radio Plasma? *Astrophys. J.* **2013**, *772*, 84. [CrossRef]
96. Ogrean, G.; Brüggen, M.; van Weeren, R.J.; Simionescu, A.; Röttgering, H.J.; Croston, J. Evidence for a merger-revived radio phoenix in MaxBCG J217.95869+13.53470. *Mon. Not. R. Astron. Soc.* **2011**, *414*, 1175–1182. [CrossRef]
97. Schellenberger, G.; Giacintucci, S.; Lovisari, L.; O'Sullivan, E.; Vrtilek, J.M.; David, L.P.; Melin, J.B.; Lal, D.V.; Ettori, S.; Kolokythas, K.; et al. The Unusually Weak and Exceptionally Steep Radio Relic in A2108. *Astrophys. J.* **2022**, *925*, 91. [CrossRef]
98. Rahaman, M.; Raja, R.; Datta, A.; Burns, J.O.; Rapetti, D. On the origin of diffuse radio emission in Abell 85-insights from new GMRT observations. *Mon. Not. R. Astron. Soc.* **2022**, *515*, 2245–2255. [CrossRef]
99. Whyley, A.; Randall, S.W.; Clarke, T.E.; van Weeren, R.J.; Rajpurohit, K.; Forman, W.R.; Edge, A.C.; Blanton, E.L.; Lovisari, L.; Intema, H.T. Understanding the Nature of the Ultra-Steep Spectrum Diffuse Radio Source in the Galaxy Cluster Abell 272. *arXiv* **2024**, arXiv:2402.04876.
100. Bonafede, A.; Intema, H.T.; Brüggen, M.; Girardi, M.; Nonino, M.; Kantharia, N.G.; van Weeren, R.J.; Röttgering, H.J. Evidence for Particle Re-acceleration in the Radio Relic in the Galaxy Cluster PLCKG287.0+32.9. *Astrophys. J.* **2014**, *785*, 1. [CrossRef]
101. Shimwell, T.W.; Markevitch, M.; Brown, S.; Feretti, L.; Gaensler, B.M.; Johnston-Hollitt, M.; Lage, C.; Srinivasan, R. Another shock for the Bullet cluster, and the source of seed electrons for radio relics. *Mon. Not. R. Astron. Soc.* **2015**, *449*, 1486–1494. [CrossRef]
102. Botteon, A.; Gastaldello, F.; Brunetti, G.; Dallacasa, D. A shock at the radio relic position in Abell 115. *Mon. Not. R. Astron. Soc.* **2016**, *460*, L84–L88. [CrossRef]
103. Di Gennaro, G.; van Weeren, R.J.; Hoeft, M.; Kang, H.; Ryu, D.; Rudnick, L.; Forman, W.R.; Röttgering, H.J.; Brüggen, M.; Dawson, W.A.; et al. Deep Very Large Array Observations of the Merging Cluster CIZA J2242.8+5301: Continuum and Spectral Imaging. *Astrophys. J.* **2018**, *865*, 24. [CrossRef]
104. Stuardi, C.; Bonafede, A.; Wittor, D.; Vazza, F.; Botteon, A.; Locatelli, N.; Dallacasa, D.; Golovich, N.R.; Hoeft, M.; van Weeren, R.J.; et al. Particle re-acceleration and Faraday-complex structures in the RXC J1314.4-2515 galaxy cluster. *Mon. Not. R. Astron. Soc.* **2019**, *489*, 3905–3926. [CrossRef]
105. Button, C.; Marchegiani, P. The application of the adiabatic compression scenario to the radio relic in the galaxy cluster Abell 3411–3412. *Mon. Not. R. Astron. Soc.* **2020**, *499*, 864–872. [CrossRef]
106. Velović, V.; Cotton, W.D.; Filipović, M.D.; Norris, R.P.; Barnes, L.A.; Condon, J.J. MeerKAT view of the dancing ghosts-peculiar galaxy pair PKS 2130-538 in Abell 3785. *Mon. Not. R. Astron. Soc.* **2023**, *523*, 1933–1945. [CrossRef]
107. Raja, R.; Rahaman, M.; Datta, A.; Smirnov, O.M. A radio bridge connecting the minihalo and phoenix in the Abell 85 cluster. *Mon. Not. R. Astron. Soc.* **2023**, *526*, L70–L76. [CrossRef]
108. Ramatsoku, M.; Murgia, M.; Vacca, V.; Serra, P.; Makhathini, S.; Govoni, F.; Smirnov, O.; Andati, L.A.L.; de Blok, E.; Józsa, G.I.G.; et al. Collimated synchrotron threads linking the radio lobes of ESO 137-006. *Astron. Astrophys.* **2020**, *636*, L1. [CrossRef]
109. Rudnick, L.; Cotton, W.; Knowles, K.; Kolokythas, K. One Source, Two Source(s): Ribs and Tethers. *Galaxies* **2021**, *9*, 81. [CrossRef]
110. Condon, J.J.; Cotton, W.; White, S.V.; Legodi, S.; Goedhart, S.; McAlpine, K.; Ratcliffe, S.; Camilo, F. Threads, Ribbons, and Rings in the Radio Galaxy IC 4296. *Astrophys. J.* **2021**, *917*, 18. [CrossRef]
111. Brienza, M.; Shimwell, T.W.; de Gasperin, F.; Bikmaev, I.; Bonafede, A.; Botteon, A.; Brüggen, M.; Brunetti, G.; Burenin, R.; Capetti, A.; et al. A snapshot of the oldest active galactic nuclei feedback phases. *Nat. Astron.* **2021**, *5*, 1261–1267. [CrossRef]
112. Rudnick, L.; Brüggen, M.; Brunetti, G.; Cotton, W.D.; Forman, W.; Jones, T.W.; Nolting, C.; Schellenberger, G.; van Weeren, R. Intracluster Magnetic Filaments and an Encounter with a Radio Jet. *Astrophys. J.* **2022**, *935*, 168. [CrossRef]
113. Jaffe, W.J. Origin and transport of electrons in the halo radio source in the Coma cluster. *Astrophys. J.* **1977**, *212*, 1–7. [CrossRef]
114. Venturi, T.; Bardelli, S.; Dallacasa, D.; Brunetti, G.; Giacintucci, S.; Hunstead, R.W.; Morganti, R. The radio halo in the merging cluster A3562. *Astron. Astrophys.* **2003**, *402*, 913–920. [CrossRef]
115. Brunetti, G.; Setti, G.; Comastri, A. Inverse Compton X-rays from strong FRII radio-galaxies. *Astron. Astrophys.* **1997**, *325*, 898–910. [CrossRef]
116. Blasi, P.; Colafrancesco, S. Cosmic rays, radio halos and nonthermal X-ray emission in clusters of galaxies. *Astropart. Phys.* **1999**, *122*, 169–183. [CrossRef]
117. Aleksić, J.; Antonelli, L.A.; Antoranz, P.; Backes, M.; Baixeras, C.; Balestra, S.; Barrio, J.A.; Bastieri, D.; Becerra González, J.; Bednarek, W.; et al. MAGIC Gamma-ray Telescope Observation of the Perseus Cluster of Galaxies: Implications for Cosmic Rays, Dark Matter, and NGC 1275. *Astrophys. J.* **2010**, *710*, 634–647. [CrossRef]
118. Ackermann, M.; Ajello, M.; Albert, A.; Allafort, A.; Atwood, W.B.; Baldini, L.; Ballet, J.; Barbiellini, G.; Bastieri, D.; Bechtol, K.; et al. Search for Cosmic-Ray-induced Gamma-Ray Emission in Galaxy Clusters. *Astrophys. J.* **2014**, *787*, 18. [CrossRef]
119. Ackermann, M.; Ajello, M.; Albert, A.; Atwood, W.; Baldini, L.; Ballet, J.; Barbiellini, G.; Bastieri, D.; Bechtol, K.; Bellazzini, R.; et al. Search for Gamma-Ray Emission from the Coma Cluster with Six Years of Fermi-LAT Data. *Astrophys. J.* **2016**, *819*, 149. [CrossRef]

120. Gupta, N.; Saro, A.; Mohr, J.J.; Benson, B.A.; Bocquet, S.; Capasso, R.; Carlstrom, J.E.; Chiu, I.; Crawford, T.M.; de Haan, T.; et al. High-frequency cluster radio galaxies: Luminosity functions and implications for SZE-selected cluster samples. *Mon. Not. R. Astron. Soc.* **2017**, *467*, 3737–3750. [CrossRef]
121. Böckmann, K.; Brüggen, M.; Koribalski, B.; Veronica, A.; Reiprich, T.H.; Bulbul, E.; Bahar, Y.E.; Balzer, F.; Comparat, J.; Garrel, C.; et al. Central radio galaxies in galaxy clusters: Joint surveys by eROSITA and ASKAP. *Astron. Astrophys.* **2023**, *677*, A188. [CrossRef]
122. Stimpson, M.; Hardcastle, M.J.; Krause, M.G.H. Numerical modelling of the lobes of radio galaxies-Paper V: Universal pressure profile cluster atmospheres. *Mon. Not. R. Astron. Soc.* **2023**, *526*, 3421–3440. [CrossRef]
123. Völk, H.J.; Atoyan, A.M. Clusters of galaxies: Magnetic fields and nonthermal emission. *Astropart. Phys.* **1999**, *11*, 73–82. [CrossRef]
124. Farber, R.; Ruszkowski, M.; Yang, H.Y.K.; Zweibel, E.G. Impact of Cosmic-Ray Transport on Galactic Winds. *Astrophys. J.* **2018**, *856*, 112. [CrossRef]
125. Brüggen, M.; Kaiser, C.R.; Churazov, E.; Enßlin, T.A. Simulation of radio plasma in clusters of galaxies. *Mon. Not. R. Astron. Soc.* **2002**, *331*, 545–555. [CrossRef]
126. Mendygral, P.J.; Jones, T.W.; Dolag, K. MHD Simulations of Active Galactic Nucleus Jets in a Dynamic Galaxy Cluster Medium. *Astrophys. J.* **2012**, *750*, 166. [CrossRef]
127. Turner, R.J.; Shabala, S.S. Dynamics of Powerful Radio Galaxies. *Galaxies* **2023**, *11*, 87. [CrossRef]
128. Yates-Jones, P.M.; Shabala, S.S.; Power, C.; Krause, M.G.H.; Hardcastle, M.J.; Mohd Noh Velastín, E.A.N.; Stewart, G.S.C. CosmoDRAGoN simulations—I. Dynamics and observable signatures of radio jets in cosmological environments. *Publ. Astron. Soc. Aust.* **2023**, *40*, e014. [CrossRef]
129. Shabala, S.; Yates-Jones, P.; Jerrim, L.; Turner, R.; Krause, M.; Norris, R.; Koribalski, B.; Filipovic, M.; Rudnick, L.; Power, C.; et al. Are Odd Radio Circles phoenixes of powerful radio galaxies? *arXiv* **2024**, arXiv:2402.09708.
130. Kang, H. Re-Acceleration of Fossil Electrons by Shocks Encountering Hot Bubbles in the Outskirts of Galaxy Clusters. *J. Korean Astron. Soc.* **2018**, *51*, 185–195. [CrossRef]
131. Kang, H.; Ryu, D. Curved Radio Spectra of Weak Cluster Shocks. *Astrophys. J.* **2015**, *809*, 186. [CrossRef]
132. Jones, T.W.; Nolting, C.; O'Neill, B.J.; Mendygral, P.J. Using collisions of AGN outflows with ICM shocks as dynamical probes. *Phys. Plasmas* **2017**, *24*, 041402. [CrossRef]
133. Nolting, C.; Jones, T.W.; O'Neill, B.J.; Mendygral, P. Interactions between Radio Galaxies and Cluster Shocks. I. Jet Axes Aligned with Shock Normals. *Astrophys. J.* **2019**, *876*, 154. [CrossRef]
134. Nolting, C.; Jones, T.W.; O'Neill, B.J.; Mendygral, P. Simulated Interactions between Radio Galaxies and Cluster Shocks. II. Jet Axes Orthogonal to Shock Normals. *Astrophys. J.* **2019**, *885*, 80. [CrossRef]
135. Nolting, C.; Ball, J.; Nguyen, T.M. Simulations of Precessing Jets and the Formation of X-shaped Radio Galaxies. *Astrophys. J.* **2023**, *948*, 25. [CrossRef]
136. ZuHone, J.; Ehlert, K.; Weinberger, R.; Pfrommer, C. Turning AGN Bubbles into Radio Relics with Sloshing: Modeling CR Transport with Realistic Physics. *Galaxies* **2021**, *9*, 91. [CrossRef]
137. Vazza, F.; Wittor, D.; Brunetti, G.; Brüggen, M. Simulating the transport of relativistic electrons and magnetic fields injected by radio galaxies in the intracluster medium. *Astron. Astrophys.* **2021**, *653*, A23. [CrossRef]
138. Vazza, F.; Wittor, D.; Di Federico, L.; Brüggen, M.; Brienza, M.; Brunetti, G.; Brighenti, F.; Pasini, T. Life cycle of cosmic-ray electrons in the intracluster medium. *Astron. Astrophys.* **2023**, *669*, A50. [CrossRef]
139. Vazza, F.; Wittor, D.; Brüggen, M.; Brunetti, G. Simulating the Enrichment of Fossil Radio Electrons by Multiple Radio Galaxies. *Galaxies* **2023**, *11*, 45. [CrossRef]
140. Inchingolo, G.; Wittor, D.; Rajpurohit, K.; Vazza, F. Radio relics radio emission from multishock scenario. *Mon. Not. R. Astron. Soc.* **2022**, *509*, 1160–1174. [CrossRef]
141. Smolinski, D.C.; Wittor, D.; Vazza, F.; Brüggen, M. A multishock scenario for the formation of radio relics. *Mon. Not. R. Astron. Soc.* **2023**, *526*, 4234–4244. [CrossRef]
142. Vazza, F.; Brüggen, M.; van Weeren, R.; Bonafede, A.; Dolag, K.; Brunetti, G. Why are central radio relics so rare? *Mon. Not. R. Astron. Soc.* **2012**, *421*, 1868–1873. [CrossRef]
143. Nuza, S.E.; Gelszinnis, J.; Hoeft, M.; Yepes, G. Can cluster merger shocks reproduce the luminosity and shape distribution of radio relics? *Mon. Not. R. Astron. Soc.* **2017**, *470*, 240–263. [CrossRef]
144. Hitomi Collaboration. The quiescent intracluster medium in the core of the Perseus cluster. *Nature* **2016**, *535*, 117–121. [CrossRef] [PubMed]

galaxies

MDPI

Review

Impact of Active Galactic Nuclei Feedback on the Dynamics of Gas: A Review Across Diverse Environments

Mojtaba Raouf [1,2,*], Mohammad Hossein Purabbas [2,3] and Fatemeh Fazel Hesar [1,4]

1 Leiden Observatory, Leiden University, P.O. Box 9513, 2300 RA Leiden, The Netherlands; fazel@strw.leidenuniv.nl
2 School of Astronomy, Institute for Research in Fundamental Sciences (IPM), Tehran 19538-33511, Iran; mohammad_hossein_purabbas@birjand.ac.ir
3 Department of Physics, Faculty of Sciences, University of Birjand, Shahid Aviny Street, Birjand 97174-34765, Iran
4 Eurospacehub Academy & LUNEX, ESA BIC, 2201 BB Noordwijk, The Netherlands
* Correspondence: raouf@strw.leidenuniv.nl

Abstract: This review examines the relationship between black hole activity and kinematic gas–star misalignment in brightest group galaxies (BGGs) with different merger rates. The formation history of galaxy groups is assessed through "age-dating" as an indicator of distinct major mergers involving the BGGs. BGGs within groups characterized by a higher frequency of major mergers are more likely to host active SMBHs. A consistent correlation is identified between the level of black hole activity, as indicated by the 1.4 GHz and 325 MHz radio emissions, and the degree of kinematic misalignment between the gas and stellar components in BGGs. In dynamically fossil groups, where black hole accretion rate is relatively ($\sim$1 dex) lower due to the lack of recent ($\leq$1 Gyr) major mergers, there is reduced ($\sim$30%) misalignment between the gas and stellar components of BGGs compared to non-fossil groups. Additionally, this study reveals that BGGs in non-fossil groups show higher levels of star formation rate and increased occurrence of mergers, contributing to observed color differences. Exploring the properties and dynamics of the gas disk influenced by mechanical AGN feedback through hydrodynamic simulations suggests that AGN wind-induced effects further lead to the persistent gas misalignment in the disk around the supermassive black hole.

Keywords: galaxy groups; black hole activity; AGN feedback; kinematic misalignment; mergers; group dynamics; radio luminosity; stellar populations; galaxy formation

1. Introduction

The presence of super massive black holes (SMBHs) at the centers of massive galaxies provides a mechanism for the existence of active galactic nuclei (AGN), which can suppress excessive star formation and regulate galaxy growth [1,2]. AGN feedback, through processes such as heating the surrounding gas with AGN jets, plays a crucial role in quenching massive galaxies and controlling the growth of SMBHs [3]. This feedback has a direct influence on the gas content of the galaxy, resulting in negative/positive feedback on star formation [4–8]. The role of AGN feedback in shaping galaxy properties is supported by numerous studies [9–15].

Brightest group galaxies (BGGs), at the core of galaxy groups, serve as windows into the intricate processes governing galaxy assembly, star formation, and the interplay between SMBHs and their host galaxies [16,17]. In a group, galaxy interactions shape galaxy evolution by leaving observable marks like tidal features and kinematic perturbations in stars and gas [18–20]. A remarkable perturbation is the stellar–gas misalignment, wherein stars and gas rotate in different (in the case of counter-rotation, opposite) directions due to their various (opposite) angular momentums [21]. A significant deviation exceeding 30° in the kinematic position angle (PA) between gas and stars can arise from various sources, both internal and external to the galaxy. These sources may include mergers between galaxies, gas accretion from neighboring galaxies, or the influence of black hole activity, such as AGN feedback [22–24].

Dynamically fossil groups, which lack recent group-scale mergers[1] and major galaxy mergers, are particularly suitable for such studies [25,26]. The dynamical state/age of galaxy groups can be characterized by indicators such as the luminosity gap and the offset between the BGG and the luminosity centroid, with the luminosity gap being a key factor [27]. A large luminosity gap suggests the absence of recent major mergers that could trigger cold mode accretion. In dynamically fossil groups, where the intergalactic medium (IGM) reaches its peak density at the bottom of the group/cluster potential well, an AGN is subject to hot gas accretion. In fact, the observational data are much clearer for non-fossil systems (with the low luminosity gap), showing that many of the most powerful radio galaxies in groups in the local universe are the BGGs of dynamically relaxed systems with cooling cores, fueled by gas cooling from the ICM, not from galaxy mergers. On one hand, Ref. [28] found that a combination of the low luminosity gap (an indicator for internal event and cold accretion) and the large BGG offset from the center (an indicator for external event and hot accretion) in a group, is the key driver behind the observed high AGN activity probed by the radio emission. The presence of nuclear gas that can be accreted by a central supermassive black hole is considered crucial for fueling and powering AGN [29]. On the other hand, the prevalence of misalignments in early-type galaxies (which are mostly located at the center of groups and clusters), along with the possibility of aligned stellar–gas kinematics resulting from external gas accretion depending on the interaction geometry and galaxy morphology, indicates a substantial contribution of external accretion to the total gas content in these galaxies [30], which in turn affects the availability of gas for black hole fueling. Based on the research conducted by [31], they considered the possibility that kinematic misalignments in clusters could be attributed to gas stripped from the galactic disk as the galaxy moves through the intra-cluster medium. This process may lead to a gradient in the gas velocity field in the direction of infall. However, it is important to note that, after a thorough examination of the same dataset, Ref. [30] found that only one galaxy, representing less than 1% of the sample, supported a misalignment origin due to ram-pressure stripping. These studies highlight the complexities involved in understanding the causes of kinematic misalignments in the cluster environment. While gas stripping through hydrodynamic effects can contribute to misalignments in some cases, the prevalence of such misalignments in clusters is not well-supported by the available evidence.

The alignment or misalignment of gas and stars in galaxies, along with the influence of external gas accretion, plays a crucial role in understanding the fueling of SMBHs and the AGN. Theoretical studies have proposed that the presence of counter-rotating or significantly misaligned structures facilitates gas inflow and potentially contributes to the fueling of supermassive black holes in galaxies of various types [32–36]. Recently, a hydrodynamic simulation of idealized gas disk around SMBHs suggested that the stability of the disk's PA across different radii can be attributed to the mechanical feedback from the AGN, indicating a direct influence of AGN activity on the kinematic properties of the molecular gas [37].

This comprehensive review primarily focuses on results derived from simulations, offering valuable insights into the impacts of black hole activity on the kinematics of gas during the evolution of galaxies. This paper is structured into key sections that cover essential aspects, starting with an examination of the dynamical state of galaxy groups, followed by an exploration of the complex interplay between black hole evolution and their actions, and finally delving into the dramatic narratives of major merger history. In addition to the simulation-based findings, this review incorporates observational studies that further enhance our understanding of the subject matter. These studies shed light on various important aspects, including the variations in radio emissions from BGGs, the influence of galaxy groups' dynamical states on their stellar populations, and the intriguing phenomenon of kinematic gas–star misalignment. Furthermore, by employing hydrodynamic simulations, we present the fascinating phenomena of gas misalignment in AGN-dominated galaxies, specifically focusing on an idealized gas disk around the supermassive black hole. Each segment of this review contributes a valuable piece to the puzzle, enriching our understanding of the complex interplay between black hole activity, kinematic misalignment, and the environments of galaxies. By combining simulation results and observational insights, we aim to provide a comprehensive and nuanced perspective on this captivating field of research.

2. Perspectives from Cosmological Simulations

2.1. Age Dating Galaxy Groups

Fossil galaxy groups refer to systems that have undergone relatively few recent mergers or interactions [38]. These groups are characterized by a more stable and virialized state, where the member galaxies and dark matter halos have had sufficient time to settle into a dynamically old state/age configuration. In fossil groups, galaxies are more centrally concentrated, which suggests that a greater number of member galaxies are located near the center rather than evenly distributed throughout the group [27]. Additionally, the X-ray emitting hot gas in these groups exhibits a smoother and more symmetric distribution [39]. In contrast, non-fossil galaxy groups are systems that have experienced recent merger events or interactions with other groups or galaxies. Due to ongoing interactions, the member galaxies in non-fossil groups may exhibit more chaotic motions, and the spatial distribution of galaxies can be more dispersed [40]. The X-ray emission from the hot gas in non-fossil groups may show signs of substructures or asymmetries [39]. Such differences in the dynamical state of galaxy groups significantly impact the stellar populations and gas kinematics of the BGGs, with non-fossil groups showing significantly bluer NUV-r colors and higher star formation rates compared to their fossil counterparts [17,41]. The advantage of the study [27] is its use of optical data to distinguish between these fossil and non-fossil groups. By analyzing various measurable parameters derived from optical observations of the galaxy groups, the study aims to distinguish between groups with different merger histories [42]. Several studies explore the accuracy of photometric, spectroscopic, and X-ray proxies in assessing the dynamical state of galaxy groups [39,43]. In the current era of extensive surveys, having a reliable and cost-effective method to characterize the dynamical state of halos would greatly benefit statistical studies focused on understanding the influence of the environment on galaxy properties. By utilizing the luminosity gap and BGG offset approach, Ref. [44] shows that such an alternative approach mitigates the need to rely exclusively on X-ray data or other observational techniques, which may be limited by sample size or coverage.

The review study of fossil groups of galaxies offers a distinct perspective on the processes influencing galaxy evolution, providing a transitional picture [40]. Fossil groups are characterized by a large luminosity gap ($\Delta M_{12} > 2$ mag) between the BGG and the other group members along with relatively high X-ray luminosity ($L_{X,bol} \approx 10^{42}\ \mathrm{h}_{50}^{-2}$ erg s^{-1}) [25,26], indicating a long period of dynamical relaxation [45,46]. Moreover, by adding the offset from the group luminosity centroid, we can considerably improve the accuracy and precision of *age-dating* techniques for the optical (without X-ray criterion) selection of fossil groups of galaxies [27]. In this review, we refer to the term "fossil" to describe an "evolved" group that is more commonly observed and represents an "older" system in simulations, indicating its earlier formation. The non-fossil group, in contrast, represents the "evolving" groups, which are younger and formed at a later stage.

2.2. Black Hole Activity and Formation History

The BGGs in fossil groups typically harbor more massive black holes, underscoring a correlation between the historical mass assembly of group halos and the growth of SMBHs. Figure 1 provides a visual representation of the variance in black hole accretion rates and mass within BGGs hosted by fossil and non-fossil groups using cosmological simulations [47]. The BGGs in the dynamically fossil galaxy groups display a lower accretion rate (left panel) compared to the BGGs in dynamically non-fossil groups. At the same time, the black hole mass (right panel) in the BGGs dominating the dynamically fossil groups is larger than in the BGGs of dynamically non-fossil or young groups. This means that the mass assembly history of the host group (halos) has a significant impact on the activity and growth of BGGs' SMBHs. The BGGs of fossil systems seem to be very efficient in black hole growth by consuming the gas which could have been generally found with a higher density in early stages of the halo formation [47].

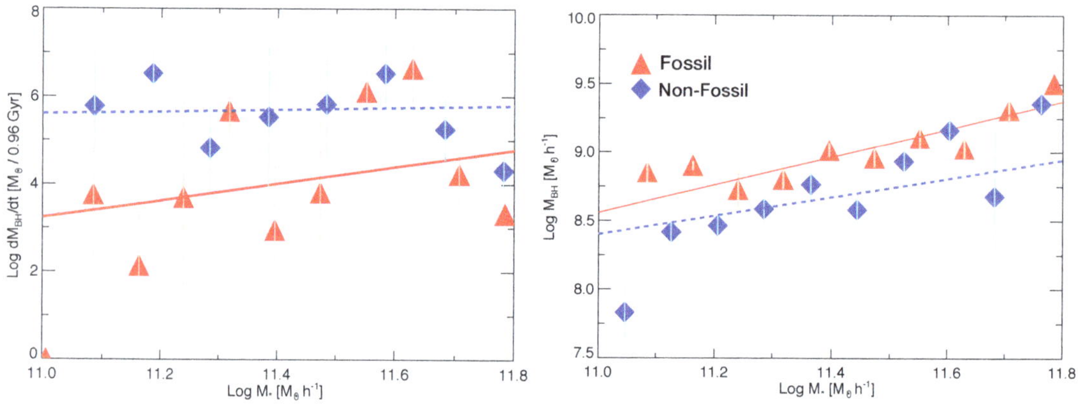

Figure 1. (**Left**): The black hole accretion rate ($\dot{M}_{BH}$) as function of stellar mass for BGGs, comparing dynamically fossil groups (red) against BGGs in dynamically non-fossil groups (blue). The red line and blue dashed line show the linear regressions to the fossil and non-fossil systems, respectively. (**Right**): The black hole mass (M_{BH}) as function of stellar mass for BGGs in fossil and non-fossil groups. The BGGs in fossil group generally possess larger black holes than those in non-fossil groups, reflecting the influence of the group's dynamical history on black hole growth. The error bars in all panels are based on the standard deviation over the mean.

Through the growth history of fossil and non-fossil halos, we found a significant time difference ($\sim$2 Gyr) in their peak merging activity [42]. As shown in the left panel of Figure 2, BGGs in non-fossil groups had their merger peak about 2 billion years (Gyr) later than BGGs in fossil groups. As expected, galaxies in the two types of groups have different merger histories. Specifically, our analysis reveals that non-fossil groups, on average, have experienced the last major mergers with a frequency approximately $\sim$2 times higher within the past 1 Gyr compared to the BGGs in fossil groups (middle panel). It is important to note that these last major mergers refer to the overall history of the groups and may not necessarily correspond to immediate major mergers involving their galaxies within the specified 1 Gyr time frame. This relationship is further supported by the increased radio brightness ($L_{1.4\ \mathrm{GHz}}/M_{star}$) illustrated in the right panel of Figure 2 utilizing a radio semi-analytic galaxy evolution model (Radio-SAGE; [48,49]). In contrast, BGGs in fossil groups exhibit an earlier peak in merging activity and demonstrate lower levels of radio brightness, even when considering the same duration since the last major merger [42].

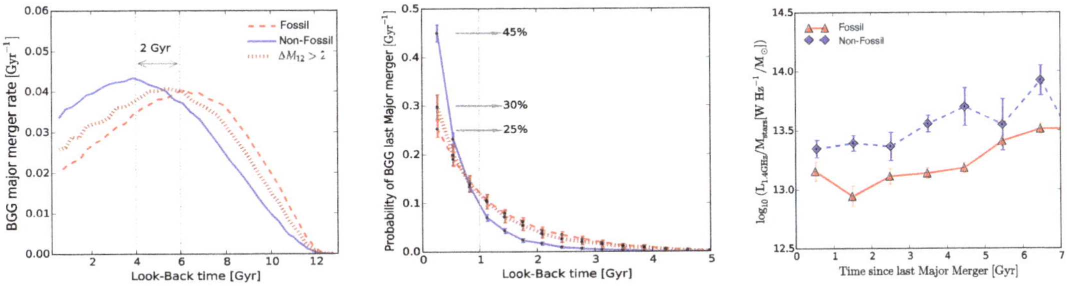

Figure 2. (**Left**): The major merger rates of the BGGs in three categories of groups using Radio-SAGE galaxy formation model: dynamically fossil (red dashed line) and non-fossil (blue line) groups of galaxies along the fossil ($\Delta M_{12} > 2$) groups (brown dotted line), for halo masses above $10^{13} M_{\odot}$. Major mergers are characterized by mergers with $m_1/m_2 < 3$. As seen, the BGGs of fossil systems show a higher probability for a major merger at earlier epochs compared with non-fossil systems (about 2 Gyr in peak). (**Middle**): Comparison of the distributions of time since the last major merger between the three group categories. The error bars represent Poisson uncertainties. The most recent major merger event for BGGs in non-fossil groups occurred more recently compared to the BGGs in high-gap-only and fossil groups, respectively. (**Right**): Radio loudness (luminosity at 1.4 GHz divided by stellar mass) of BGGs at z = 0 versus the time since the last major merger suffered by the BGG, in sub-samples of groups defined by the epoch of their last major merger. The red triangles and blue diamonds show the medians and standard deviation over mean error for BGGs of fossil and non-fossil groups, respectively.

3. Perspectives from Observational Surveys

3.1. Radio Emissions of BGGs

Several observational studies consistently demonstrate a notable increase in AGN activity within merging galaxies [50,51]. As shown in Figure 3 for 1.4 GHz (left panel) and 325 MHz (right panel) radio emission, the study by [28] indicates that, in the dynamically fossil environments, BGGs exhibit a diminished level of radio activity (~1 dex), indicating a relatively quiet phase of AGN activity. Our findings reveal that the proportion of radio-loud brightest group galaxies in observed dynamically non-fossil groups is approximately twice that found in dynamically fossil groups. Incorporating the Radio-SAGE[2] [48] model's findings into the discussion on radio emissions in BGGs, it becomes evident that the refined modeling of AGN feedback significantly contributes to our understanding of AGN activity and its various impacts. The model's advanced approach to simulating AGN jets and cooling processes under different dynamical states of galaxy groups complements the observed radio emission patterns in BGGs. This shows the complex connection between the dynamical state of the galaxy group, the behavior of the central black hole, and the resultant AGN feedback, elucidating the varied radio emission properties of BGGs in different group environments. The successful reproduction of this observation has been achieved through the utilization of a Radio-SAGE model. This model enables us to investigate the different factors contributing to these findings. The consistent agreement between observations and Radio-SAGE model predictions indicates a higher rate of accretion[3] for the central black hole hosted by the BGG in dynamically non-fossil galaxy groups compared to those hosted by dynamically fossil galaxy groups at a given stellar mass. Ref. [52] also emphasizes the influence of group dynamics on AGN feedback mechanisms of BGGs through the study of kinematic properties and scaling relations of radio galaxies in galaxy groups. Ref. [53] utilized a small sample of fossil groups to investigate the radio emission properties of the most luminous galaxies in these dynamically fossil halos. Their findings indicated that these galaxies exhibit a lower level of radio emissions at 610 MHz and 1.4 GHz. The findings of these studies suggest that mergers, which are the primary events driving the formation of a large luminosity gap, are likely the main factor contributing to the differences observed in the radio properties.

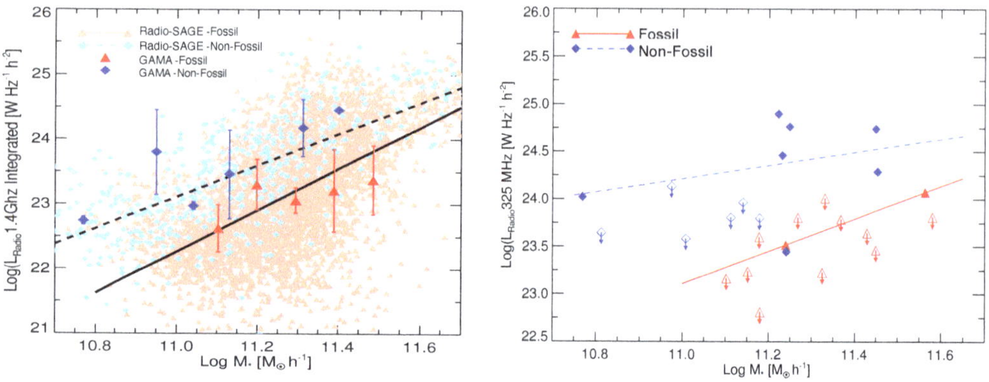

Figure 3. (**Left**): The radio emission at 1.4 GHz for the BGG in fossil groups (red) and non-fossil groups (blue) as function of stellar mass. The radio luminosity refers to the integrated flux densities obtained from the VLA FIRST catalog. The bold symbols indicate the average value over the bin. We overlay the central galaxies in fossil (light red triangles) and unrelaxed (sky blue diamonds) galaxy groups corresponding to different stellar masses as a function of the 1.4 GHz radio luminosity predicted by Radio-SAGE galaxy formations model. The solid and dashed lines represent linear fits to the model data points for the BGGs in fossil and non-fossil galaxy groups, respectively. The majority of BGGs in dynamically non-fossil groups are radio-loud (i.e., $L_{1.4\ \mathrm{GHz}} > 10^{23}$ W Hz^{-1}h^{-2}). (**Right**): 325 MHz radio power of the BGGs in fossil (filled red) and non-fossil (filled blue) groups using GMRT observations. An upper limit is given for the undetected fossil (open red) and non-fossil (open blue) BGGs.

3.2. Stellar Population of BGGs

Figure 4 shows the metallicity and specific star formation rate (sSFR) derived from SED fitting as a function of stellar mass for a sample of galaxy and mass assembly (GAMA) observations [41]. As shown in the right panel, BGGs in non-fossil groups tend to show higher levels of star formation rate (SFR) compared to those in fossil groups. Also, the BGGs in fossil groups tend to have a higher percentage of red NUV-r colors (with a higher percentage of NUV-r > 4.5) considering a fixed Sérsic index and dust mass, again indicating a lower star formation rate. This suggests that the increased occurrence of mergers in non-fossil groups contributes to the observed NUV-r color differences. In the left panel, BGGs in non-fossil groups tend to have higher stellar metallicities ($\sim$0.05 dex) compared to those in fossil groups. This could be attributed to the mass and metal enrichment history of the building blocks from which the groups formed.

3.3. The Kinematics of BGG

Figure 5 (b_1 and c_1) shows the fraction of misaligned BGGs at a given stellar mass and Sérsic index for sample of BGGs in fossil versus non-fossil groups using the Sydney–Australian–Astronomical–Observatory Multi-object Integral-Field Spectrograph (SAMI) galaxy survey [17]. In the left panel, for a given stellar mass, the fraction of misaligned[4] gas–star systems is primarily lower ($\sim$30% less in all masses) in the fossil group's BGGs compared to the BGGs in the non-fossil sample and the median value across all galaxies. In the right, the non-fossil groups tend to have a higher fraction of misaligned BGGs at a fixed Sersic index, again verifying the role of group dynamical state on the BGG dynamics. These differences are likely attributed to the extended period of tranquility experienced by BGGs in fossil groups, allowing them to settle and relax following earlier mergers and interactions. BGGs in fossil groups, which are characterized by a more stable and settled environment, exhibit ∼35% more regular rotation fields and around 10% more higher fractions of slow rotators compare to BGG in unrelaxed groups. The authors of [33] employed cosmological zoom-in simulations to carefully investigate the influence of mergers on the misalignment of gas discs in early-type galaxies. The research demonstrated that a significant merger could give rise to the formation of a new gas disk that is misaligned with the stellar rotation of the galaxy and persists for several gigayears.

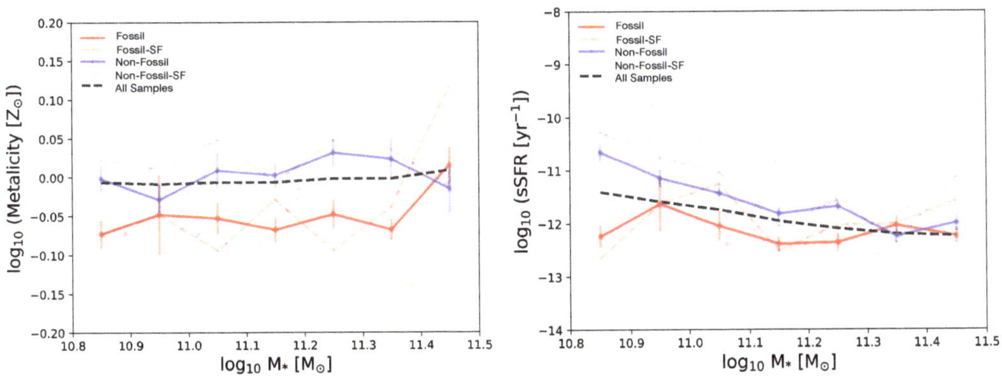

Figure 4. The variation in metallicity (**left**) and sSFR (**right**) of BGGs in relation to their stellar mass with the medians and standard deviation over mean errors, for fossil and non-fossil groups with specified all and star-forming (SF, light colors) BGGs selected by BPT diagram.The dashed black line in each panel is the metallicity and sSFR for the full sample. BGGs in fossil groups (red line) generally have lower metallicity than those in non-fossil groups (blue line) while the median sSFR is higher for BGGs in non-fossil groups compared to those in fossil groups.

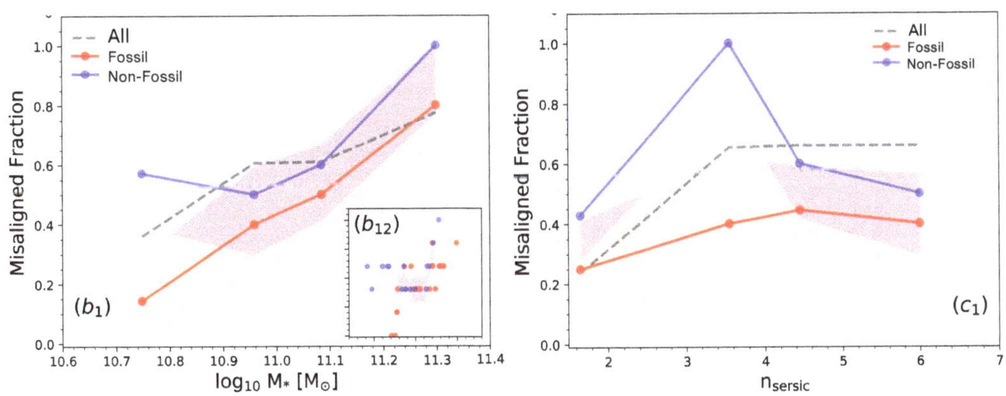

Figure 5. (**Left**): The fraction of gas–star misalignment angel ($\delta PA > 30$ degree) as a function of stellar mass for BGGs hosted by relaxed (red) and non-fossil (blue) groups. Dashed lines show the median trend for all BGGs in our sample (b_1). The inset figure (b_{12}) shows the median fraction of regular-rotation BGGs as a function of stellar mass using the moving average method wherein the axes' range is the same as in the main plot. (**Right**): The fraction of gas–star misalignment angel as a function of BGGs Sérsic index (c_1), n_{Sersic}, with the same description as given in the left panel. In each panel, the errors (color-shaded lines) are 1σ confidence intervals on the fractions calculated using the bootstrap method.

4. Gas Misalignment in AGN-Dominated Galaxies: Insights from Hydrodynamic Simulations

Molecular observations of the gas enveloping AGNs have uncovered a intricate scenario in which the observational signatures of mechanical and radiative feedback are confounded by the probable existence of starburst activity [54–57]. Galaxies exhibiting both an AGN and robust circumnuclear star formation are particularly intriguing, as they encompass a intricate amalgamation of energetic processes, such as gas accretion and external infall, which contribute to a complex geometry and kinematics [58–60]. These combined factors give rise to a multifaceted and dynamic system. The ISM gas within the circumnuclear disk (CND) of an AGN-dominated galaxy, particularly a Seyfert-2 type galaxy like NGC 1068[5], was extensively investigated through hydrodynamic simulations referred to as HDGAS[6] [37]. The study utilized these simulations to explore the properties and dynamics of the gas in the CND influenced by mechanical AGN feedback. The study incorporated mechanical feedback from the AGN and employed the CHIMES non-equilibrium chemistry network to compare models with and without AGN feedback. The results reveal that the presence of an AGN significantly enhances CO formation in dense, clumpy regions surrounding supermassive black holes. This phenomenon is further elucidated by the examination of the counter-rotating gas disk in NGC 1068, as evidenced by kinematic maps of CO line emission. Furthermore, in the models incorporating AGN feedback, specific regions of CND show higher velocity dispersion, represented by the parameter σ, which can be attributed to the superposition of velocity components associated with the outflow phenomenon.

Figure 6 illustrates the variation in the disk's PA at various CND radii (<50 pc and <100 pc around SMBH) for four distinct AGN feedback models (high/low wind velocity and mass loading, see Table 1 in [37]), each differing in energy and momentum loading compared to the model without AGN feedback. The momentum and energy loading driven by the wind's velocity play a crucial role in maintaining the disk's misalignment at a scale of 100 pc over extended periods. The wind-induced effects lead to a persistent misalignment of the disk, characterized by a high offset in PA on the 100 pc scale, demonstrating a continued state of misalignment (top-left, AGN-BH3V10B1). However, in the bottom panels, when the wind velocity is lower (AGN-BH3V5B1, AGN-BH3V5B6), the disk appears to realign within a relatively short period of time. Increasing the mass loading results in the smoothing out of the disk, following an initial substantial offset (top-right panel, AGN-BH3V10B6). Notably, in the top-right panel, during earlier time frames (around 3 to 4 million years), when both mass loading and wind velocity are high, the dynamics are predominantly influenced by the inflow phase. In contrast, the NoAGN model does not exhibit a significant PA offset at any time or radius. We note that our study, conducted through hydrodynamic simulations, was not limited to specific environments and encompassed AGN-dominated galaxies across diverse conditions. The main focus of our investigation was to explore the effects of mechanical AGN feedback. Specifically, we examined the impacts of different jet velocities and momentum-loading outflows on the dynamics of gas within an idealized disk surrounding a supermassive black hole. To gain insights into the role of AGN feedback, we compared the results with a model that did not incorporate any AGN feedback mechanisms.

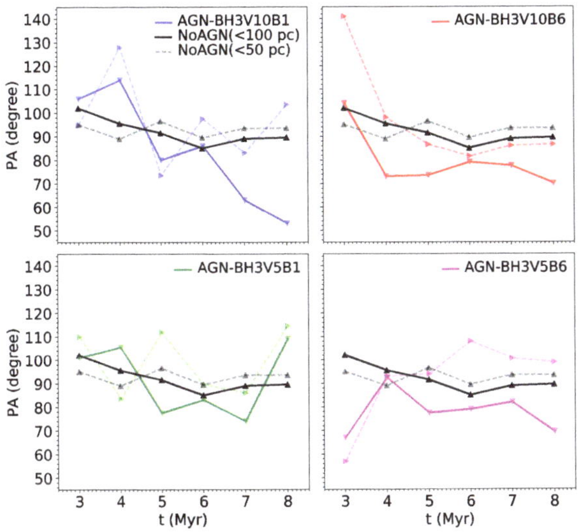

Figure 6. Kinematic position angles (PAs) of CO(1-0) moment-1 maps within the central 50 (dashed lines) and 100 (solid lines) pc of the disk for AGN models with different energy and momentum-loading factors and the model NoAGN (black lines) measured at times between 3 and 8 Myr. A large proportion of such PA offsets occurs within the central r = 50 pc of the disk surrounding the SMBH in low mass-loading wind models (BH3V10B1, BH3V5B1), where the change in PA occurs within a shorter time frame. There is a significant early offset (at 3–4 Myr) in models with a high mass-loading factor (BH3V10B6, BH3V5B6), followed by the steady PA trend to the later times. Moreover, different radii scales of AGN models with low velocity winds (BH3V5B6) exhibit a higher PA offset. The different radii reveal similar trends for all models except for the model with high wind velocity (i.e., BH3V10B1), which exhibits a large PA offset at a later time (t∼8 Myr) compared to earlier times.

5. Discussion

In this review, we have gained valuable insights into the impacts of black hole activity on the kinematics of gas in galaxies with different merger histories. By utilizing state-of-the-art cosmological simulations and observational surveys, we have explored various dimensions of this complex interplay. We start by examining the dynamics of galaxy groups and their formation states. It has been found that the BGGs in fossil systems, which are characterized by stable and settled environments, exhibit highly efficient black hole growth. The consumption of gas during the early stages of halo formation significantly contributes to the growth of black holes in these fossil systems (Figure 1). Furthermore, we have investigated the growth history of both fossil and non-fossil halos and observed a significant difference in the timing of their peak merging activities (Figure 2). In dynamically fossil environments, the BGGs exhibit lower levels of radio activity (see Figure 3), indicating a relatively quiet phase of AGN activity. On the other hand, BGGs in non-fossil groups, experiencing more dynamic and turbulent environments, tend to have higher levels of star formation rates (Figure 4). The occurrence of mergers in non-fossil groups contributes to the observed differences in the stellar populations and colors of BGGs. A notable finding from our study is that the gas–star misalignment fraction in the BGGs of fossil groups is predominantly lower compared to both the BGGs in the non-fossil sample and the median value of all galaxies (Figure 5). This suggests that the alignment between gas and stars in the BGGs is more preserved in fossil environments.

By examining idealized gas discs surrounding SMBHs, our study demonstrates that the influence of winds on galaxies results in a persistent misalignment of the galactic disk. The interaction between the galactic disk and the wind can disrupt the alignment of gas and stars, leading to changes in the kinematical major axis PA (Figure 6). It is important to note that the spatial orientation of a galactic disk is typically described by two angles: the PA and the inclination angle. The PA represents the angle of the major axis of the disk on the sky, while the inclination angle describes the tilt of the disk with respect to the line of sight. In the case of net circular gas motion, the PA can provide information about the line of nodes, which is the intersection between the galactic disk and the plane of the sky. However, in galaxies influenced by complex gas dynamics, such as those affected by winds, the PA may not solely reflect the line of nodes. Factors like radial gas motions induced by winds can cause deviations in the PA from the disk's overall orientation. These radial gas motions, although induced by winds, can still have an impact on the angular momentum distribution and result in shifts in the disk's orientation. Therefore, the PA turn induced by winds can be considered as a significant influence on the orientation of the galactic disk.

In line with recent studies, we have found that BGGs in fossil groups with lower black hole activity display less kinematic misalignment. This finding supports the notion that external accretion events, such as galaxy mergers or the accretion of gas from the immediate environment, play a significant role in disrupting gas kinematics [24]. Misalignment between ionized gas and stellar kinematic angles is more prevalent in galaxies with a higher fraction of active black holes. Studies have also demonstrated the stable misalignment of gas discs after mergers [33] and the influence of mechanical AGN feedback [37] on the gas misalignment. To further advance our understanding, future research should focus on exploring the sources and dynamics of gas accretion in BGGs, particularly the roles of hot and cold accretion processes and the influence of active galactic nuclei (AGN). Larger samples of BGGs and the integration of data on stellar kinematics will strengthen our understanding of misalignment patterns. Additionally, investigating the temporal dynamics of gas–star realignment within galaxies is crucial, as it may mirror the evolutionary timescales of galaxy groups. This synchronized pattern of internal and external galactic evolution warrants further exploration.

By shedding light on the factors influencing BGG kinematics, this review lays the groundwork for future inquiries into the intricate processes governing the life cycles of galaxies and their group environments. The insights gained from this research will contribute to a more comprehensive understanding of the interplay between black hole activity, kinematic misalignment, and the environments of galaxies.

Author Contributions: The other co-authors M.H.P. and F.F.H. made significant contributions to writing this paper and assisted the main author M.R. in preparing the visualisations and interpreting results. All authors have read and agreed to the published version of the manuscript.

Funding: This research received no external funding.

Data Availability Statement: Data from observation are publicly available in GAMA (http://www.gama-survey.org/) and SAMI (https://sami-survey.org/) galaxy surveyd. Data from the hydrodynamics simulations are based on the output from publicly available: (a) HDGAS simulation [37] following the GIZMO [61] code at the following repository: (https://bitbucket.org/phopkins/gizmo-public/src/master/) and (b) Illustris-TNG simulation (www.tng-project.org). Data from the semi-analytic model are based on the output from publicly availble model of (Radio-SAGE, https://github.com/mojtabaraouf/sage). The data underlying this article will be shared on reasonable request to the corresponding author .

Acknowledgments: We would like to express our sincere gratitude to the referees for their valuable comments and suggestions, which have greatly enhanced the quality of this review paper. An Advanced Research Grant from the European Union (833460) supported this study as part of the MOlecules as Probes of the Physics of EXternal (Moppex) galaxies project. FFH would like to express her heartfelt appreciation to EuroSpaceHub and LUNEX EuroMoonMars Earth Space Innovation for their generous funding and unwavering support.

Conflicts of Interest: The funders had no role in the design of the study; in the collection, analyses, or interpretation of data; in the writing of the manuscript; or in the decision to publish the results.

Notes

1 While group-group mergers can potentially lead to increased merging in the overall galaxy population over time, they do not necessarily guarantee that mergers involving the dominant galaxy will occur.

2 https://github.com/mojtabaraouf/sage.

3 The stable gravitational potential within fossil groups reduces the accretion rate onto the central supermassive black hole compared to more dynamically non-fossil environments.

4 In our analysis, we took into account a one-sigma uncertainty in the catalog, which corresponds to a maximum error of 30 degrees for the PAs. The PAs were measured in an anticlockwise direction, with 0 degrees representing North (y = 0) in the sky. It is important to note that PAs with uncertainty values close to 30 degrees indicate less reliable fits. However, we found that such cases occurred in less than 5% of all the SAMI samples, indicating that the majority of the fits were relatively robust. Therefore, we considered misalignment PAs greater than 30 degrees.

5 NGC 1068, discussed in this paper, is a spiral galaxy with a Seyfert nucleus. It is important to acknowledge that NGC 1068 is most representative of AGN activity in BGGs of relatively low-mass groups. The well-studied X-ray luminous groups, for which our knowledge is most complete, typically exhibit early-type BGGs with FR-I radio galaxies that are considerably more radio luminous than Seyferts. We also acknowledge the multifaceted nature of AGN activity and its various manifestations. However, we focus on the investigation of mechanical AGN feedback and its impact on the dynamics of gas within an idealized disk surrounding a supermassive black hole, using hydrodynamic simulations.

6 https://github.com/mojtabaraouf/HDGAS.

References

1. Silk, J.; Rees, M.J. Quasars and galaxy formation. *arXiv* **1998**, arXiv:astro-ph/9801013.
2. Binney, J.; Tabor, G. Evolving cooling flows. *Mon. Not. R. Astron. Soc.* **1995**, *276*, 663–678. [CrossRef]
3. Kondapally, R.; Best, P.N.; Raouf, M.; Thomas, N.L.; Davé, R.; Shabala, S.S.; Röttgering, H.J.A.; Hardcastle, M.J.; Bonato, M.; Cochrane, R.K.; et al. Cosmic evolution of radio-AGN feedback: Confronting models with data. *Mon. Not. R. Astron. Soc.* **2023**, *523*, 5292–5305. [CrossRef]
4. Silk, J. Ultraluminous starbursts from supermassive black hole-induced outflows. *Mon. Not. R. Astron. Soc.* **2005**, *364*, 1337–1342. [CrossRef]
5. Silk, J.; Begelman, M.C.; Norman, C.; Nusser, A.; Wyse, R.F.G. Which Came First: Supermassive Black Holes or Galaxies? Insights from JWST. *Astrophys. J. Lett.* **2024**, *961*, L39. [CrossRef]
6. Gaibler, V.; Khochfar, S.; Krause, M.; Silk, J. Jet-induced star formation in gas-rich galaxies. *Mon. Not. R. Astron. Soc.* **2012**, *425*, 438–449. [CrossRef]
7. Santini, P.; Rosario, D.; Shao, L.; Lutz, D.; Maiolino, R.; Alexander, D.; Altieri, B.; Andreani, P.; Aussel, H.; Bauer, F.; et al. Enhanced star formation rates in AGN hosts with respect to inactive galaxies from PEP-Herschel observations. *Astron. Astrophys.* **2012**, *540*, A109. [CrossRef]
8. Bieri, R.; Dubois, Y.; Silk, J.; Mamon, G.A. Playing with positive feedback: External pressure-triggering of a star-forming disk galaxy. *Astrophys. J. Lett.* **2015**, *812*, L36. [CrossRef]
9. Benson, A.; Hoyle, F.; Torres, F.; Vogeley, M.S. Galaxy voids in cold dark matter universes. *Mon. Not. R. Astron. Soc.* **2003**, *340*, 160–174. [CrossRef]
10. Di Matteo, T.; Springel, V.; Hernquist, L. Energy input from quasars regulates the growth and activity of black holes and their host galaxies. *Nature* **2005**, *433*, 604–607. [CrossRef] [PubMed]
11. Bower, R.G.; Benson, A.; Malbon, R.; Helly, J.; Frenk, C.; Baugh, C.; Cole, S.; Lacey, C.G. Breaking the hierarchy of galaxy formation. *Mon. Not. R. Astron. Soc.* **2006**, *370*, 645–655. [CrossRef]
12. Croton, D.J.; Springel, V.; White, S.D.; De Lucia, G.; Frenk, C.S.; Gao, L.; Jenkins, A.; Kauffmann, G.; Navarro, J.; Yoshida, N. The many lives of active galactic nuclei: Cooling flows, black holes and the luminosities and colours of galaxies. *Mon. Not. R. Astron. Soc.* **2006**, *365*, 11–28. [CrossRef]
13. Sijacki, D.; Springel, V.; Di Matteo, T.; Hernquist, L. A unified model for AGN feedback in cosmological simulations of structure formation. *Mon. Not. R. Astron. Soc.* **2007**, *380*, 877–900. [CrossRef]
14. Cattaneo, A.; Faber, S.; Binney, J.; Dekel, A.; Kormendy, J.; Mushotzky, R.; Babul, A.; Best, P.; Brüggen, M.; Fabian, A.; et al. The role of black holes in galaxy formation and evolution. *Nature* **2009**, *460*, 213–219. [CrossRef] [PubMed]
15. Fabian, A.C. Observational evidence of active galactic nuclei feedback. *Annu. Rev. Astron. Astrophys.* **2012**, *50*, 455–489. [CrossRef]
16. Einasto, M.; Einasto, J.; Tenjes, P.; Korhonen, S.; Kipper, R.; Tempel, E.; Liivamägi, L.J.; Heinämäki, P. Galaxy groups and clusters and their brightest galaxies within the cosmic web. *Astron. Astrophys.* **2024**, *681*, A91. [CrossRef]
17. Raouf, M.; Smith, R.; Khosroshahi, H.G.; Sande, J.v.d.; Bryant, J.J.; Cortese, L.; Brough, S.; Croom, S.M.; Hwang, H.S.; Driver, S.; et al. The SAMI Galaxy Survey: Kinematics of Stars and Gas in Brightest Group Galaxies—The Role of Group Dynamics. *Astrophys. J.* **2021**, *908*, 123. [CrossRef]
18. Haynes, M.P.; Giovanelli, R.; Chincarini, G.L. The influence of environment on the HI content of galaxies. *Annu. Rev. Astron. Astrophys.* **1984**, *22*, 445–470. [CrossRef]

19. Bertola, F. Gaseous Disks in Elliptical Galaxies. In Proceedings of the Morphological and Physical Classification of Galaxies: Proceedings of the Fifth International Workshop of the Osservatorio Astronomico di Capodimonte, Sant'Agata Sui Due Golfi, Italy, 3–7 September 1990; Springer: Dordrecht, The Netherlands, 1992, pp. 115–126.
20. Sancisi, R.; Fraternali, F.; Oosterloo, T.; Van Der Hulst, T. Cold gas accretion in galaxies. *Astron. Astrophys. Rev.* **2008**, *15*, 189–223. [CrossRef]
21. Kannappan, S.; Fabricant, D. A broad search for counterrotating gas and stars: Evidence for mergers and accretion. *Astron. J.* **2001**, *121*, 140. [CrossRef]
22. Bertola, F.; Bettoni, D.; Rusconi, L.; Sedmak, G. Stellar versus gaseous kinematics in E and SO galaxies. *Astron. J.* **1984**, *89*, 356–364. [CrossRef]
23. Davis, T.A.; Bureau, M. On the depletion and accretion time-scales of cold gas in local early-type galaxies. *Mon. Not. R. Astron. Soc.* **2016**, *457*, 272–280. [CrossRef]
24. Raimundo, S.I.; Malkan, M.; Vestergaard, M. An increase in black hole activity in galaxies with kinematically misaligned gas. *Nat. Astron.* **2023**, *7*, 463–472. [CrossRef]
25. Ponman, T.; Allan, D.; Jones, L.; Merrifield, M.; McHardy, I.; Lehto, H.; Luppino, G. A possible fossil galaxy group. *Nature* **1994**, *369*, 462–464. [CrossRef]
26. Jones, L.; Ponman, T.; Horton, A.; Babul, A.; Ebeling, H.; Burke, D. The nature and space density of fossil groups of galaxies. *Mon. Not. R. Astron. Soc.* **2003**, *343*, 627–638. [CrossRef]
27. Raouf, M.; Khosroshahi, H.G.; Ponman, T.J.; Dariush, A.A.; Molaeinezhad, A.; Tavasoli, S. Ultimate age-dating method for galaxy groups; clues from the Millennium Simulations. *Mon. Not. R. Astron. Soc.* **2014**, *442*, 1578–1585. [CrossRef]
28. Khosroshahi, H.G.; Raouf, M.; Miraghaei, H.; Brough, S.; Croton, D.J.; Driver, S.; Graham, A.; Baldry, I.; Brown, M.; Prescott, M.; et al. Galaxy And Mass Assembly (GAMA): A "No Smoking" Zone for Giant Elliptical Galaxies? *Astrophys. J.* **2017**, *842*, 81. [CrossRef]
29. Shlosman, I.; Begelman, M.C.; Frank, J. The fuelling of active galactic nuclei. *Nature* **1990**, *345*, 679–686. [CrossRef]
30. Ristea, A.; Cortese, L.; Fraser-McKelvie, A.; Brough, S.; Bryant, J.J.; Catinella, B.; Croom, S.M.; Groves, B.; Richards, S.N.; van de Sande, J.; et al. The SAMI Galaxy Survey: Physical drivers of stellar-gas kinematic misalignments in the nearby Universe. *Mon. Not. R. Astron. Soc.* **2022**, *517*, 2677–2696. [CrossRef]
31. Bryant, J.J.; Croom, S.M.; van de Sande, J.; Scott, N.; Fogarty, L.M.R.; Bland-Hawthorn, J.; Bloom, J.V.; Taylor, E.N.; Brough, S.; Robotham, A.; et al. The SAMI Galaxy Survey: Stellar and gas misalignments and the origin of gas in nearby galaxies. *Mon. Not. R. Astron. Soc.* **2019**, *483*, 458–479. [CrossRef]
32. Thakar, A.R.; Ryden, B.S. Smoothed particle hydrodynamics simulations of counterrotating disk formation in spiral galaxies. *Astrophys. J.* **1998**, *506*, 93. [CrossRef]
33. van de Voort, F.; Davis, T.A.; Kereš, D.; Quataert, E.; Faucher-Giguère, C.A.; Hopkins, P.F. The creation and persistence of a misaligned gas disc in a simulated early-type galaxy. *Mon. Not. R. Astron. Soc.* **2015**, *451*, 3269–3277. [CrossRef]
34. Negri, A. Hydrodynamical Simulations of Early-Type Galaxies: Effects of Galaxy Shape and Stellar Dynamics on Hot Coronae. Ph.D. Thesis, Università di Bologna, Bologna, Italy, 2014.
35. Capelo, P.R.; Dotti, M.; Volonteri, M.; Mayer, L.; Bellovary, J.M.; Shen, S. A survey of dual active galactic nuclei in simulations of galaxy mergers: Frequency and properties. *Mon. Not. R. Astron. Soc.* **2017**, *469*, 4437–4454. [CrossRef]
36. Taylor, P.; Federrath, C.; Kobayashi, C. The origin of kinematically distinct cores and misaligned gas discs in galaxies from cosmological simulations. *Mon. Not. R. Astron. Soc.* **2018**, *479*, 141–152. [CrossRef]
37. Raouf, M.; Viti, S.; García-Burillo, S.; Richings, A.J.; Schaye, J.; Bemis, A.; Nobels, F.S.J.; Guainazzi, M.; Huang, K.Y.; Schaller, M.; et al. Hydrodynamic simulations of the disc of gas around supermassive black holes (HDGAS)—I. Molecular gas dynamics. *Mon. Not. R. Astron. Soc.* **2023**, *524*, 786–800. [CrossRef]
38. Khosroshahi, H.G.; Maughan, B.J.; Ponman, T.J.; Jones, L.R. A fossil galaxy cluster: An X-ray and optical study of RX J1416.4+2315. *Mon. Not. R. Astron. Soc.* **2006**, *369*, 1211–1220. [CrossRef]
39. Khosroshahi, H.G.; Jones, L.R.; Ponman, T.J. An old galaxy group: Chandra X-ray observations of the nearby fossil group NGC 6482. *Mon. Not. R. Astron. Soc.* **2004**, *349*, 1240–1250. [CrossRef]
40. Aguerri, J.A.L.; Zarattini, S. Properties of Fossil Groups of Galaxies. *Universe* **2021**, *7*, 132. [CrossRef]
41. Raouf, M.; Smith, R.; Khosroshahi, H.G.; Dariush, A.A.; Driver, S.; Ko, J.; Hwang, H.S. The Impact of the Dynamical State of Galaxy Groups on the Stellar Populations of Central Galaxies. *Astrophys. J.* **2019**, *887*, 264. [CrossRef]
42. Raouf, M.; Khosroshahi, H.G.; Mamon, G.A.; Croton, D.J.; Hashemizadeh, A.; Dariush, A.A. Merger History of Central Galaxies in Semi-analytic Models of Galaxy Formation. *Astrophys. J.* **2018**, *863*, 40. [CrossRef]
43. Parekh, V.; van der Heyden, K.; Ferrari, C.; Angus, G.; Holwerda, B. Morphology parameters: Substructure identification in X-ray galaxy clusters. *Astron. Astrophys.* **2015**, *575*, A127. [CrossRef]
44. Zhoolideh Haghighi, M.H.; Raouf, M.; Khosroshahi, H.G.; Farhang, A.; Gozaliasl, G. On the Reliability of Photometric and Spectroscopic Tracers of Halo Relaxation. *Astrophys. J.* **2020**, *904*, 36. [CrossRef]
45. Khosroshahi, H.G.; Miraghaei, H.; Raouf, M. Fossil Systems; a Multi-wavelength Approach towards Understanding Galaxy Formation. *Galaxies* **2016**, *4*, 5. [CrossRef]
46. Farhang, A.; Khosroshahi, H.G.; Mamon, G.A.; Dariush, A.A.; Raouf, M. Evolution of Compact and Fossil Groups of Galaxies from Semi-analytical Models of Galaxy Formation. *Astrophys. J.* **2017**, *840*, 58. [CrossRef]

47. Raouf, M.; Khosroshahi, H.G.; Dariush, A. Evolution of Galaxy Groups in the Illustris Simulation. *Astrophys. J.* **2016**, *824*, 140. [CrossRef]
48. Raouf, M.; Shabala, S.S.; Croton, D.J.; Khosroshahi, H.G.; Bernyk, M. The many lives of active galactic nuclei-II: The formation and evolution of radio jets and their impact on galaxy evolution. *Mon. Not. R. Astron. Soc.* **2017**, *471*, 658–670. [CrossRef]
49. Raouf, M.; Silk, J.; Shabala, S.S.; Mamon, G.A.; Croton, D.J.; Khosroshahi, H.G.; Beckmann, R.S. Feedback by supermassive black holes in galaxy evolution: Impacts of accretion and outflows on the star formation rate. *Mon. Not. R. Astron. Soc.* **2019**, *486*, 1509–1522. [CrossRef]
50. Alonso, M.S.; Lambas, D.G.; Tissera, P.; Coldwell, G. Active galactic nuclei and galaxy interactions. *Mon. Not. R. Astron. Soc.* **2007**, *375*, 1017–1024. [CrossRef]
51. Weston, M.E.; McIntosh, D.H.; Brodwin, M.; Mann, J.; Cooper, A.; McConnell, A.; Nielsen, J.L. Incidence of WISE -selected obscured AGNs in major mergers and interactions from the SDSS. *Mon. Not. R. Astron. Soc.* **2017**, *464*, 3882–3906. [CrossRef]
52. Pasini, T.; Finoguenov, A.; Brüggen, M.; Gaspari, M.; de Gasperin, F.; Gozaliasl, G. Radio galaxies in galaxy groups: Kinematics, scaling relations, and AGN feedback. *Mon. Not. R. Astron. Soc.* **2021**, *505*, 2628–2637. [CrossRef]
53. Miraghaei, H.; Khosroshahi, H.G.; Klöckner, H.R.; Ponman, T.J.; Jetha, N.N.; Raychaudhury, S. IGM Heating and AGN activity in Fossil Galaxy Groups. *Proc. Int. Astron. Union Stud.* **2014**, *304*, 349–350. [CrossRef]
54. Aalto, S.; Garcia-Burillo, S.; Muller, S.; Winters, J.M.; van der Werf, P.; Henkel, C.; Costagliola, F.; Neri, R. Detection of HCN, HCO^+, and HNC in the Mrk 231 molecular outflow. Dense molecular gas in the AGN wind. *Astron. Astrophys.* **2012**, *537*, A44. [CrossRef]
55. Martin, C.L.; HIGGS Team. HIGGS: The Herschel Inner Galaxy Gas Survey, First Results. *Am. Astron. Soc. Meet. Abstr.* **2011**, *217*, 255.06.
56. Aladro, R.; Viti, S.; Bayet, E.; Riquelme, D.; Martín, S.; Mauersberger, R.; Martín-Pintado, J.; Requena-Torres, M.A.; Kramer, C.; Weiß, A. A λ = 3 mm molecular line survey of NGC 1068. Chemical signatures of an AGN environment. *Astron. Astrophys.* **2013**, *549*, A39. [CrossRef]
57. Watanabe, Y.; Sakai, N.; Sorai, K.; Yamamoto, S. Spectral Line Survey toward the Spiral Arm of M51 in the 3 and 2 mm Bands. *Astrophys. J.* **2014**, *788*, 4. [CrossRef]
58. Knapen, J.H.; Comerón, S.; Seidel, M.K. MUSE-AO view of the starburst-AGN connection: NGC 7130. *Astron. Astrophys.* **2019**, *621*, L5. [CrossRef]
59. Winkel, N.; Husemann, B.; Davis, T.A.; Smirnova-Pinchukova, I.; Bennert, V.N.; Combes, F.; Gaspari, M.; Jahnke, K.; Neumann, J.; O'Dea, C.P.; et al. The Close AGN Reference Survey (CARS). Tracing the circumnuclear star formation in the super-Eddington NLS1 Mrk 1044. *Astron. Astrophys.* **2022**, *663*, A104. [CrossRef]
60. Imanishi, M.; Nakanishi, K.; Izumi, T.; Wada, K. ALMA Reveals an Inhomogeneous Compact Rotating Dense Molecular Torus at the NGC 1068 Nucleus. *Astrophys. J. Lett.* **2018**, *853*, L25. [CrossRef]
61. Hopkins, P.F. A new class of accurate, mesh-free hydrodynamic simulation methods. *Mon. Not. R. Astron. Soc.* **2015**, *450*, 53–110. [CrossRef]

galaxies

Review

What Have We Learned about the Life Cycle of Radio Galaxies from New Radio Surveys

Raffaella Morganti [1,2]

[1] ASTRON- The Netherlands Institute for Radio Astronomy, Oude Hoogeveensedijk 4, 7991 PD Dwingeloo, The Netherlands; morganti@astron.nl

[2] Kapteyn Astronomical Institute, University of Groningen, Landleven 12, 9747 AD Groningen, The Netherlands

Abstract: The recurrent activity of radio AGN, with phases of activity alternating with periods of quiescence, has been known since the early studies of these objects. The full relevance of this cycle is emphasised by the requirement, from the AGN feedback scenario, of a recurrent impact of the energy released by the SMBH during the lifetime of the host galaxy: only in this way can AGN feedback influence galaxy evolution. Radio AGN in different evolutionary phases can be identified by their properties, like morphology and spectral indices. Dying/remnant and restarted sources have been the most elusive to select and characterise, but they are crucial to quantify the full life cycle. Thanks to the availability of new, large radio surveys (particularly at low frequencies), it is finally possible to make a more complete census of these rare sources and start building larger samples. This paper gives an overview of the recent work conducted using a variety of radio telescopes and surveys, highlighting some of the new results characterising the properties of dying/remnant and restarted radio sources and what has been learned about the life cycle of radio AGN. The comparison with the predictions from numerical simulations is also discussed. The results so far show that remnant and restarted radio AGN have a variety of properties which make these objects more complex than previously thought.

Keywords: survey–radio continuum; radio continuum; galaxies; active

1. Introduction

A fascinating aspect of radio active galactic nuclei (AGN) is that their evolutionary stage can be identified and timed. This is something quite unique for AGN and extremely valuable not only for understanding the processes that trigger or halt the activity but also for assessing the relevance of these phases in the context of AGN feedback. Recurring nuclear activity during the life of a galaxy is a requirement of the cosmological simulations in order to limit, by regulating accretion and the amount of gas present, the growth of super-massive black holes (SMBH) and the star formation in massive galaxies [1–3].

Different approaches have been taken to explore the life cycle of radio AGN. For example, some studies [4,5] have looked at the fraction of galaxies that host a radio AGN above a given luminosity limit. To the first order (and under the assumption that every massive galaxy becomes a radio source during its life), this fraction can be used as a proxy for the fraction of time a radio AGN is "on" (i.e., active). Figure 1 (taken from [5]) shows that the fraction of low-power radio sources is higher, suggesting that the most massive galaxies host always a radio AGN. This indicates a different and likely shorter duty cycle compared to the one of less massive galaxies [4,5]. Interestingly, the study of [6], while confirming these results, also finds radio sources hosted by massive galaxies which are in the remnant stage: this suggests that the actual evolution of these objects can be more complex and highlights the relevance of detailed studies. The presence of multiple phases of activity is also seen imprinted in the X-ray cavities observed in cool-core clusters, groups and in the halo of massive early-type galaxies (see, e.g., [7–9]). The X-ray cavities can show the impact of multiple episodes of radio emission, even when this emission has already faded away (i.e., ghost cavities, [1]). However, these studies require extremely deep X-ray observations available only for a limited group of objects.

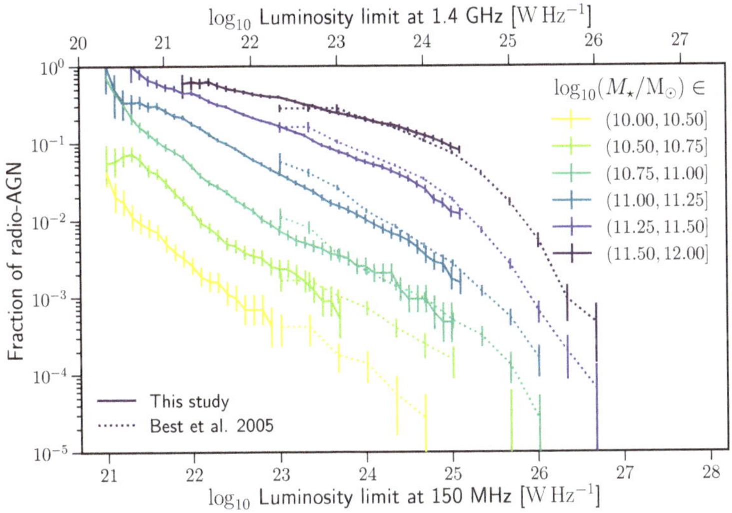

Figure 1. Fraction of galaxies hosting radio AGN brighter than a given radio luminosity, separated by their stellar mass (shown in different colours). Figure taken from [5]. Radio emission from AGN was separated from starformation using a number of criteria, see paper for details. The solid lines represent the results obtained at 150 MHz ([5]), and dotted lines show the results from [4] at 1.4 GHz (converted assuming a spectral index of 0.7). Reproduced with permission from Astronomy & Astrophysics, ©ESO.

These results show that the full characterisation of the life cycle of radio AGN requires to derive the physical properties of objects in all phases of evolution, i.e., young, evolved, dying (which will be indicated as remnant throughout this paper) and restarted sources. This has been historically performed using the spectral properties of the sources and, in some cases, using their morphology. A common way of estimating the ages of radio AGN is based on the theory of synchrotron (see [10] and references therein), the mechanism at the origin of the radio emission in these objects. The properties of the radio spectra can be used to select, for example, young sources, indicated by their peaked spectra due to synchrotron self-absorption or free–free absorption (e.g., [11–13]), or remnant radio sources, indicated by an ultra-steep spectrum of aged electrons due to energy losses (radiative and adiabatic expansion, see [10,14,15]). The morphological properties have been used to identify some types of restarted radio sources (i.e., double–double radio galaxies, DDRG [16]).

However, radio sources in the remnant phase (when the activity from the central SMBH stops or dims) and in the restarted phase (when the SMBH activity starts again) have been the most elusive to identify and it has been difficult until recently to build large samples and obtain a good census of their properties. This is a major limitation because without understanding more about these phases, we cannot time the full life cycle of radio AGN. Because the observations are so time consuming, the studies of these objects have been, until recently, mostly limited to single cases, often serendipitously discovered, or to small samples (see [15,17–24] to mention some examples). However, we have now entered a new era with new possibilities to select remnant and restarted radio sources over a larger area of the sky and using multiple selection criteria. This has provided a boost in the field with exciting new results as described in this short review. For interested readers, the overview in this paper can be complemented by a number of recent reviews on radio galaxies, including aspects of their life cycle, see [13,25–28].

2. Using Multiple Criteria to Select Samples of Remnants and Restarted Radio Galaxies

In the past years, new (or upgraded) radio telescopes have become available, and new surveys have been produced. They have made it possible to start building samples, larger than ever before, of rare objects, like remnant and restarted radio sources. In particular, we now have the following:

- Deep low-frequency ($\lesssim$200 MHz) surveys (i.e., the LOFAR Two-metre Sky Survey LoTSS, [29]; the Galactic and Extra-Galactic All-Sky MWA Survey (GLEAM), [30], and TIFR-GMRT Sky Survey (TGSS) [31]), tracing low surface-brightness radio structures. This emission is particularly sensitive to the location of old (remnant) electrons.
- Surveys at higher frequencies (e.g., at 1.4 GHz using the APERture Tile In Focus (Apertif) on the WSRT telescope [32] and the Australian Square Kilometre Array Pathfinder (ASKAP) [33]) and surveys at high spatial resolution (e.g., at 3 GHz, the Very Large Array Sky Survey (VLASS), [34,35]). The latter is particularly useful for characterising the spectral properties of the central regions (e.g., cores).
- Spectral indices[1] extended to low frequencies can be derived over large areas of the sky. Furthermore, by combining some of the surveys, spectral index images can be obtained.
- Availability of high-quality ancillary data (e.g., optical ID and redshift) for larger areas of the sky (e.g., [36–38]) as well as improved methods for deriving photometric redshifts for a large number of galaxies [37].

The selection of remnant and restarted radio sources was performed in the past, using, as mentioned earlier, a limited set of spectral and/or morphological features. In practice, this meant limiting the selection to groups of objects likely reflecting particular conditions or evolutionary path, e.g., DDRG for restarted and ultra-steep spectrum (USS, i.e., spectral index steeper than 1.2) sources for remnants. However, single object studies have already shown us that the situation is more complex and that radio AGN can follow different evolutionary paths. Cases like the remnant nicknamed *blob 1*, where the USS component is seen only at high frequencies [17], or restarted sources only identified thanks to the distribution of the spectral indices [18,39], show the variety of situations that can be encountered.

Thanks to the wealth of (radio) data available, this variety of properties can now be captured by using multiple criteria for selecting remnants and restarted candidates. This broader approach results in a more complete census of these objects. Thus, the studies that are presented below use, as much as possible, a combination of multiple criteria derived from the studies of single objects. The ones used to select remnant radio sources are (i) morphology (amorphous/low surface brightness, see also [40]); (ii) absence of core (or low core prominence, defined as CP = S_{core}/S_{total}, [41]); (iii) ultra-steep spectrum emission (USS, e.g., [15,42]); and (iv) spectral break (or high spectral curvature). In a similar way, the parameters that can be used to select restarted radio sources are (i) morphology (e.g., double–double); (ii) high core prominence combined with low surface brightness lobes, suggesting the presence of an active, newly started radio source surrounded by a remnant emission from the previous active phase; (iii) peaked or steep spectrum core, indicating synchrotron self-absorbed (or free–free absorption) emission typical of young radio sources [13]; and (iv) resolved spectral index showing two well-distinct populations (one of which is USS tracing a past phase of activity [18,39]).

3. Selecting Samples of Remnant and Restarted Radio Sources

A few studies aimed at selecting and characterising samples of remnant and restarted radio sources have been recently published. These projects are focused on famous, well-studied regions (i.e., relatively "small" areas between 5 and 20 sq deg) because of the wealth of ancillary data available. They represent a first step forward in the field, and below is a summary of some of their new results. For reference with respect to Figure 1, the sources included in the samples described below have low-frequency radio luminosity between a few $\times 10^{24}$ and a few $\times 10^{27}$ W Hz^{-1}.

3.1. Remnants Radio Sources

As mentioned above, the selection of remnant radio sources used to be mostly limited to USS, with a minority of studies looking for the lack of cores. The use of the morphology and, in particular, the presence of low surface brightness structures, has been pioneered by the work of [40], which, however, was limited to observations at 1.4 GHz. This selection can be more efficiently performed at low frequencies (i.e., frequencies well below 1 GHz), where the low-surface brightness structures can be more easily traced. LOFAR, providing a combination of relatively high spatial resolution and high sensitivity to low surface brightness structure, has given a unique contribution. An example of this is the study presented in [43], which has selected remnant radio sources in the Lockman Hole (LH) region imaged by LOFAR at 150 MHz [44]. For the selection, they used the morphology and the properties of the spectral indices. This study (and the follow up at 6 GHz with the VLA, [45]) identified 13 remnant sources in this area, corresponding to about 7% of the original sample made of sources larger than 60 arcsec. Interestingly, only 4.1% of the remnant sources are USS, confirming the relatively low fraction of USS even after extending the search to low frequencies. Thus, this work has shown that the amorphous, low surface brightness morphology can provide the larger contribution of selected remnants and, crucial for future searches, not all remnants sources are USS at low frequencies. While the selection was not based on the lack of cores, the candidate remnants all show a low CP. Surprisingly, Ref. [45] finds cases of remnants with USS total emission where the core is detected by the deeper 6 GHz observations (see Section 3.2).

The finding of a variety of spectral indices of the remnant structures is interesting. This result, confirmed by other studies (see below), suggests that remnant sources must be going through an evolution. This was further investigated and confirmed by some simple modelling; see Section 4. A follow-up study performed in the same region has derived resolved spectral indices and showed that in some of the remnant sources, the entire emission, which can extend a few hundred kpc, is USS, [28,46]. These are likely the older remnants still visible by our radio telescopes. Finally, unlike previous studies, only a minority of the selected remnant sources are in a cluster environment.

The study of [47] focused instead on the Herschel-ATLAS region and used the core-undetected method to identify remnant radio sources. This required follow-up VLA 6 GHz high-resolution observations of the objects pre-selected in the LOFAR image by [48]. They derived an upper limit on the fraction of remnants of about 9%, consistent with [43]. They also found that the remnant structures have a wide range of spectral indices ($0.5 < \alpha_{1400}^{150} < 1.5$). This confirms that the lobes of some remnants may be characterised by spectra indices at low frequencies similar to the one of active sources, suggesting that the steepening is occurring only at higher frequencies and has not yet reached and affected the low frequencies.

Quici et al. (2021) looked at the GAMA23 region using MWA, ASKAP and ATCA to identify sources which lack radio cores. These selection criteria resulted in 10 remnant candidates, corresponding to a fraction of remnants between 4% and 10%. One of their best cases in presented in Figure 2. Of the selected remnant sources, only three appear to be USS. This study also confirmed that cores can appear when deeper observations are available. Interestingly, they also found (like [45]) that objects with core can have aged spectra (USS). As in the case of [47], hot spots are seen in some objects: based on their spectra properties, the authors derived that these structures can persist 5–10 Myr after the emission from the core has switched off.

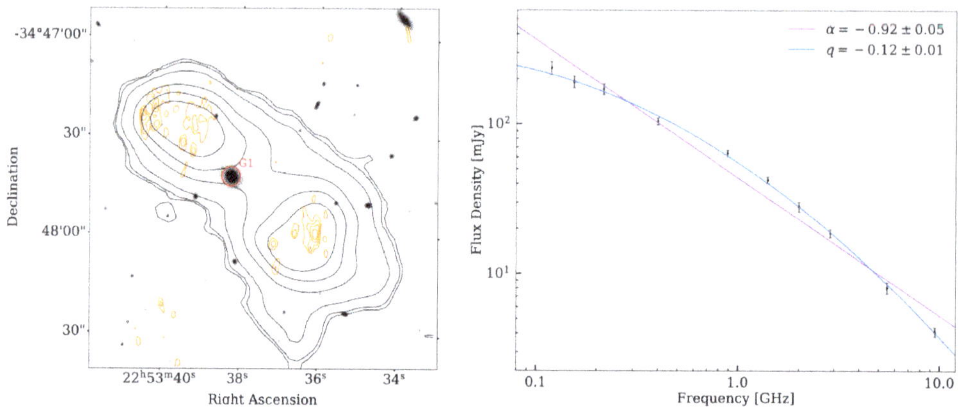

Figure 2. The continuum image (left) and integrated spectrum (right) of MIDAS J225543−344047 from [49]. Background image is a VIKING Ks band cutout set on a linear stretch. Three sets of contours are overlaid, representing the radio emission as seen by ASKAP EMU−ES (black contours, levels: [3, 4, 5, 7, 15, 30, 100] × σ), ATCA GLASS 5.5 GHz (orange contours, levels: [3, 5, 10, 20] × σ). Red marker indicates the positions of the host galaxy. Right: The radio continuum spectrum between 119 MHz and 9 GHz. A simple power-law and curved power-law model are fit to the spectrum, indicated by the purple and blue models, respectively. See [49] for details. Copyright © 2021, Cambridge University Press.

Finally, Ref. [50] searched for remnant sources in the XMM-LSS field (using images from GMRT, LOFAR and VLA for a subregion), covering an area between 5 and 12 sq deg. Using both the undetected radio cores method as well as the high spectral curvature and ultra-steep spectrum, they identified 21 remnant candidates (resulting in ∼8% of remnant sources). Like in [45], remnant sources were found to reside mostly in non-cluster environments. As noted also in other studies, the selected remnant sources exhibit a variety of properties in terms of morphology, spectral index (in the range of 0.75 to 1.71), and linear radio size (ranging from 242 kpc–1.3 Mpc). Similar to other studies, only ∼3.9% of the sources can be identified as USS sources.

3.2. The Presence of Cores in Candidates Remnant

One of the unexpected properties of the remnant sources is the finding, when deep enough observations are available, of faint core emission in (some of the) objects characterised by USS extended emission. This was already seen in the detailed studies of the *blob 1* and NGC 507 (see [17,51], respectively) but it is now found to be more common. A nice example obtained from the sample of remnants studied by [43,45] is the source shown in Figure 3. In this source, a core is detected at 6 GHz using A-array observations with the VLA while no large-scale jets are observed, and the extended emission has a steep spectral index, $\alpha_{1400}^{150} \sim 1.2$.

Explaining this finding provides an interesting puzzle, and two possibilities have been put forward. Either—at least in some cases—the cores never switch off completely but the emission only dims, reaching the very low CP characteristics of remnant sources, or, alternatively, these faint cores represent the restarting of the nuclear activity and a new phase for the radio source. If the faint cores represent a restarted activity in a remnant structure, this would suggest that in some objects, the "off" time is relatively short; see Section 4. This is also suggested by the properties of the spectral index of some sources (see also Section 3.3).

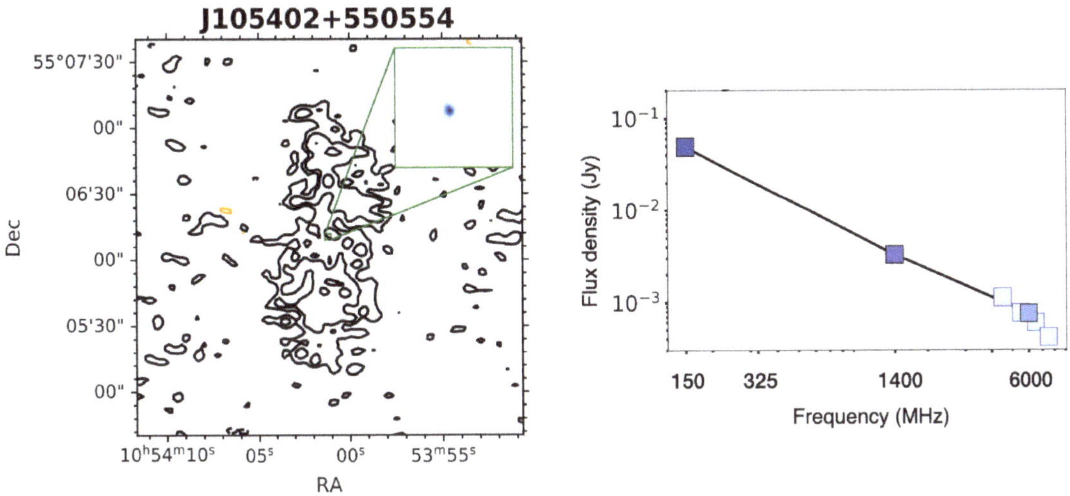

Figure 3. Radio image (left) of one of the remnants selected in [43] and followed up at higher frequencies and resolution by [45]. The green square represents the position of the optical host galaxy and show the detection of a corresponding nuclear emission at 6 cm using the VLA ($S_{core,6GHz} = 0.142$ mJy). The contours of the LOFAR 6 arcsec image are shown in black. The right panel shows the radio spectra of the total emission, derived by using fluxes from LoTSS, GMRT, NVSS and VLA (see [45] for details). Despite the presence of the core, the extended emission is USS indicative of a remnant emission. Taken from [45], see text for details. Reproduced with permission from Astronomy & Astrophysics, ©ESO.

Finally, it is interesting to note that when faint cores appear from deep, high-resolution radio observations, they help in confirming the identifications (made with other methods) of the host galaxy of the remnant radio structure. Because of their amorphous structure, it is usually difficult to find the optical identification of remnant radio sources and methods based on the barycenter of the emission used (see Ref. [45] for some examples and discussion). Thus, the confirmation a posteriori of the optical identification using the faint cores gives more confidence in this method.

3.3. Candidates Restarted Radio Sources

The selection of candidate restarted radio sources remains challenging and detailed studies are needed to test the criteria listed in Section 2. Despite these challenges, a number of interesting results have been recently reported.

The prototypical restarted radio AGN are considered to be the DDRG. They are identified by two (or in some rare cases, three [52,53]) pairs of radio lobes and characterised by an extremely stable axis of the ejecta (see [16,21,54]). The large-scale (many hundred kpc to Mpc) lobes are produced by past episode(s) of activity, while on the smaller scale are the more recently restarted lobes. The largest sample of DDRG is presented in [54]. These sources were selected from LOFAR LoTSS 150 MHz images of the HETDEX area (400 sq deg) and were followed up with higher-resolution VLA observations to confirm the structure of the inner lobes; see Figure 4 for some examples.

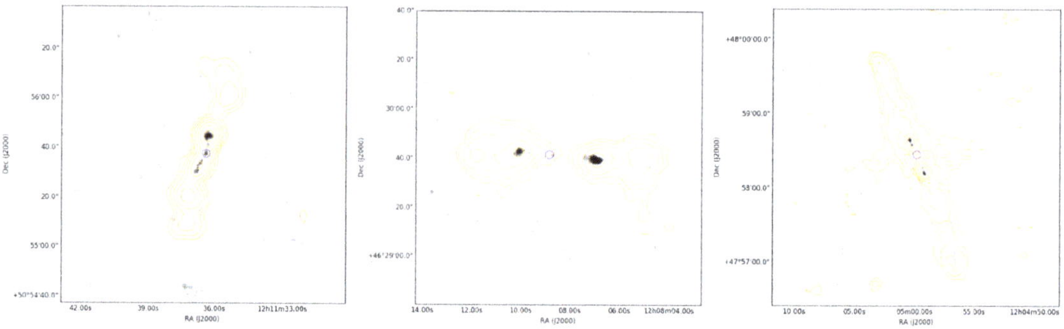

Figure 4. Examples of double–double radiogalaxies from the study of [54]. The yellow contours show the emission at 150 MHz from LOFAR while the grey scale represent the high frequency, high spatial resolution emission from follow-up VLA observations. The combination of the two highlights the presence of two sets of lobes. See [54] for details. Reproduced with permission from Astronomy & Astrophysics, ©ESO.

However, as mentioned in Section 2, the restarted activity can be identified also by other properties. Similar to what was performed for the remnant sources (and using the same reference sample of sources larger than 60 arcsec extracted from the LOFAR 150 MHz image of the LH region), the study by [55] selected, using multiple criteria, candidate restarted radio sources. Most of the restarted candidates were selected based on their high CP combined with low surface brightness of the extended emission (as described in Section 2). The restarted nature of these objects was later confirmed by a follow-up study which made use of the LOFAR International Baselines image of the LH [56], reaching a spatial resolution of 0.3 arcsec. This has allowed to characterise the central regions of these objects and to show the presence of small jets seen in the central region while being surrounded by low surface brightness and amorphous emission on large scales; see [57] for details. DDRG identified in the LH region by visual inspection were found to represent a minority of the group of candidate restarted sources. Taking all the candidates together, the study of [55,57] derived an incidence of about 13% of restarted radio sources.

Useful for identifying restarted radio sources are also the resolved spectral index images. These were produced in the LH region by combining 150 MHz LOFAR and 1400 MHz Apertif images (see [28,46]), the latter providing higher spatial resolution ($12'' \times 12''/sin\delta$) and higher sensitivity (noise of about 40 μJy beam^{-1}) than other surveys at the same frequency like the NVSS. A number of objects were found having a central active region (with $\alpha_{1400}^{150} \sim 0.7$) surrounded by a diffuse, low-surface brightness characterised by USS emission. Two examples are shown in Figure 5. In these objects, the old (remnant) phase of activity is still visible while a new phase has already re-started. Therefore, they are particularly useful for deriving the time scales of the life cycle of radio sources. Until recently, only one example of restarted radio source identified by the spectral properties was known (i.e., 3C 388, [18,39]), while more cases have now been found by building spectral index images of large areas (see [28,46,58] for a more recent example).

While the above samples of remnant and candidate restarted sources include only large sources (in order to resolve their morphology using the available surveys), observations of the young and small-size radio sources are also relevant for understanding the variety of time scales associated with the dying and restarting phases. These studies have shown the existence of not only a population of young dying sources (with ages $< 10^6$ yrs, [59–61]) but also cases of restarted sources in young remnant objects [60]. These findings are consistent with the presence, at least in some sources, of a rapid duty cycle, i.e., restarting activity before the emission of the remnant structure fades completely.

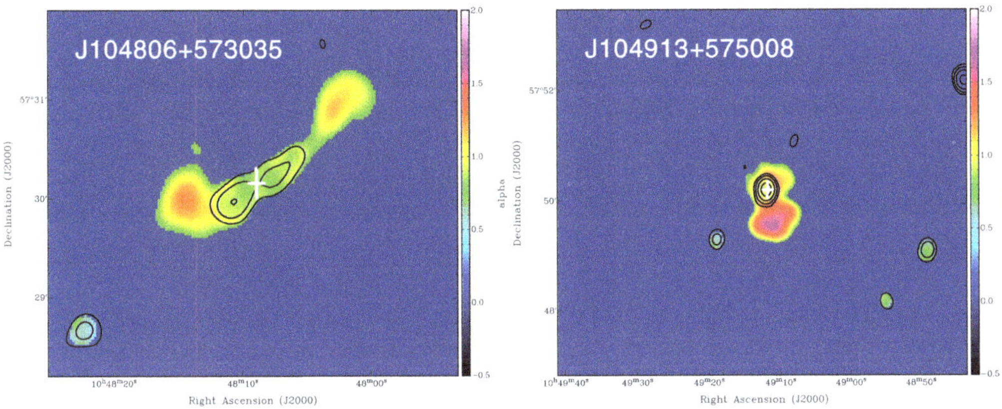

Figure 5. Examples taken from [46] of restarted radio galaxies identified thanks to the properties of their spectral index between 150 and 1400 MHz. The red regions represent spectral indices steeper than 1.2 (i.e., USS), yellow regions represent a spectral index around 0.8. The yellow edge around the source is an artefact due to the different sensitivity of the two observations (LOFAR and Apertif; see [46] for details). Reproduced with permission from Astronomy & Astrophysics, ©ESO.

Thus, the studies so far show that a variety of time scales characterise the restarting activity. Indeed, the time scales can be very short, resulting, in the extreme cases, in a "flickering" of the radio emission. An example of this flickering (on Myr time) emission is the case of Fornax A [23]. These extreme situations are difficult to trace by the available radio surveys, and instead detailed multi-wavelength observations are needed.

Finally, it is worth mentioning that the derived ages of the restarted jet can be affected by the presence of interaction between the radio plasma jet and the surrounding medium. In these cases, the flow in the jet can be temporarily disrupted by the rich medium in which they expand, and the derived ages should be taken with care (see Ref. [22] and the references therein). On the other hand, these cases are relevant for quantifying the impact that a new phase of radio AGN can have on the surrounding gas (see Section 5).

4. Implications for the Timing of the Life Cycle

The results described above, schematically summarised in the illustration in Figure 6, have already expanded our view of the life cycle of radio sources.

The fraction of remnants sources is consistently found in radio selected samples to be $\lesssim$10%. This relatively small fraction suggests that the remnant emission evolves quickly, and that fades rapidly once the core/jets turn off [47]. However, other findings suggest that the situation is likely more complicated than this. Somewhat surprisingly, the number of USS has not increased by observing at low frequencies: only about 4% of the remnants are USS. Instead, most of the studies indicate that remnants radio sources are characterised (at low frequencies) by a range of spectral indices. Although it may look surprising, this is consistent with the presence of remnants with a distribution of ages. This has been confirmed by the comparison of the results of the observations of the LH region (described above) with mock catalogues constructed to reproduce—in term of sensitivity—these observations (see [43,62]). Interestingly, the number of remnant sources observed can be reproduced by the mock catalogues only if in the dying phase adiabatic expansion losses are included in addition to radiative losses. This suggests that the radio lobes are still overpressured (with respect to the ambient medium) when reaching the off phase and implies that the expansion of the lobes continue during the remnant phase with a consequent (fast) evolution and dimming of the remnant emission (see the modelling presented in [43,62]). The distribution of the ages of the sources in the mock catalogue is shown in Figure 7 (see [43] for details). Table 5 in [43] shows the fraction of remnant radio sources in the mock catalogue: the models, including radiative and expansion losses predict that all remnant sources can be selected by the USS criteria only if the spectra can be built covering from 150 to, for example, 5000 MHz. For spectra extending only to 1400 MHz, half of the remnant sources are not USS; therefore, other criteria (like the morphology) are needed to identify them. Indeed, the histogram shows that some remnants have just switched off (light brown in the figure), while others are old remnants (dark brown). Most of the remnants are "young", i.e., in the phase shortly after the switching off (i.e., a few $\times 10^7$ yr). This means that the low-frequency spectrum is not yet ultra steep. Remnants selected based on their ultra-steep, low-frequency spectrum represent the older remnants in the sky (ages $> 10^8$ yrs). Indeed, the USS emission observed at low frequencies (that in some of the remnant sources can extend up to few $\times 100$ kpc) suggests that the older remnant sources have ages up to 160–300 Myr, if a frequency break between 600 and 150 MHz and the equipartition magnetic field Beq $= 3$ μG are assumed [28]. Because they are dependent on the magnetic field (notoriously difficult to derive) and from the expansion of the plasma, which is also difficult to quantify, these numbers should be taken with care (see also discussion in [25]).

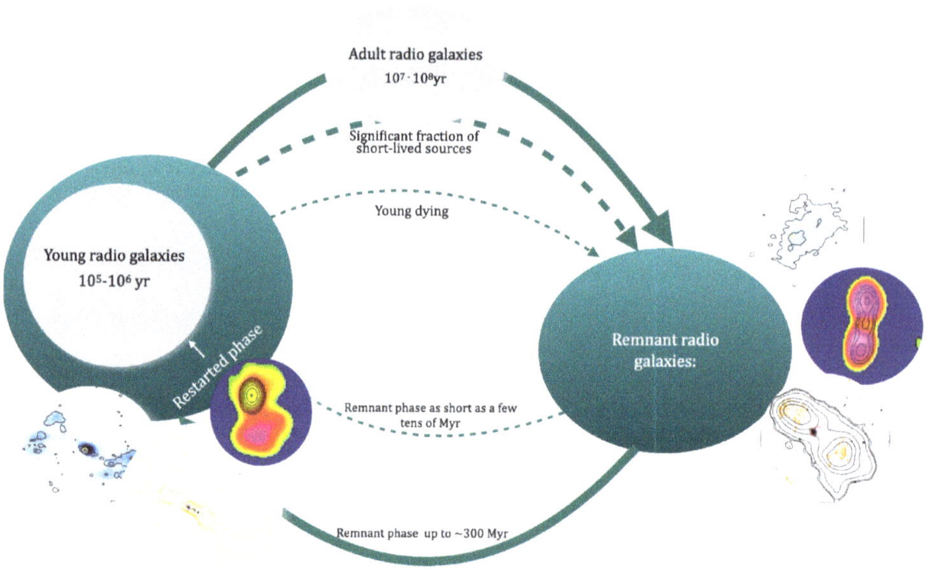

Figure 6. Illustration summarising the main phases of the life cycle of a radio source with some of the time scales added based on the studies discussed in this paper and modified from [28]. The existence of "young dying" has been discussed in [59,60] and the references therein. The significant fraction of short-lived sources is the result of the simulations of [63], see Section 4 for details. Examples of the objects (remnant and restarted radio sources) used to derive the time scales (some of them discussed in this paper) are also shown.

It is also interesting to consider the presence of faint cores in some of the bona fide remnant structures (i.e., in sources with USS extended emission) when deep enough high-resolution, high-frequency observations are obtained. As mentioned above, two hypotheses have been proposed to explain this: either some cores never really switch off (but only dim in flux), or we are seeing the beginning of a new phase of activity. The latter scenario has interesting implications for the timing of the restarted phase. All these properties can be explained if the "off" time of the radio AGN has not been longer than a few tens of Myr, i.e., the remnant emission had not enough time to fade before the new phase of activity started. It is worth noting that a relatively fast cycle of activity has been also suggested for the DDRG. The studies by [21,64] have shown that the "off" phase in these sources tends to be comparable or shorter than their on phase, ranging from a few Myr to a few tens of Myr. However, as a word of caution, we need to keep in mind that observations may favour the detection of DDRG with short off times, where the old lobes have not faded yet.

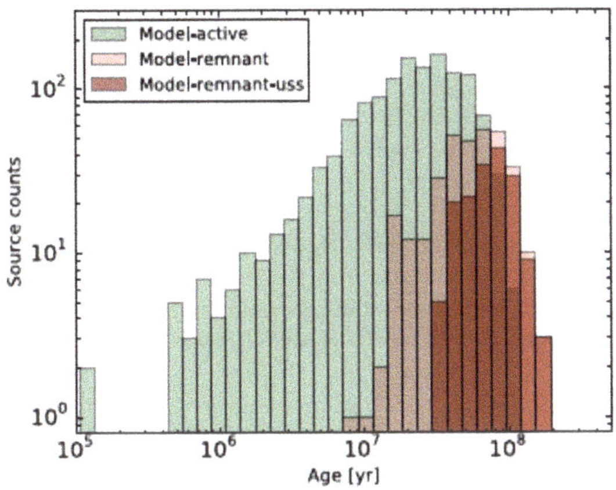

Figure 7. Distribution of ages which better describes the observations obtained from a mock catalogue, see details in [43]. Green marks active sources. Remnant sources (selected either from their morphology or their radio spectrum) are indicated in brown. Most of the remnants are "young" (light brown), i.e., in the phase shortly after the switching off (i.e., a few $\times 10^7$ yr) and their low-frequency spectrum is not yet ultra steep. Remnants selected based on their steep, low-frequency spectrum are the older remnants in the sky (ages $> 10^8$ yrs).

The new results from the observations of remnant and restarted radio AGN have stimulated new developments of the theoretical modelling of the evolution of radio sources (see [25,63,65,66]). For example, [63] have used RAiSE (Radio AGN in Semi-Analytic Environments, [67]) to compare the predictions of the simulations with the results from the observations obtained in the LH region and described above [43,45]. The strategy used has been to (i) use a sample of active radio galaxies [45] with host information to derive their physical parameters using RAiSE (i.e., sampling the distribution of physical properties of the radio galaxy population); (ii) employ outputs of galaxy formation models to quantify jet environments; and (iii) use these distributions to make predictions for remnants. The main result obtained from the comparison with the LH study is that the age distribution of radio galaxies most likely follows a power law: short-lived jets should exist to obtain enough remnant lobes which are still visible [63]. As mentioned above, other observational findings seem to confirm this result [60].

While these simulations describe the general occurrence of the remnant and restarted phases, numerical simulations are also in the process of being developed for describing in more detail the morphology and dynamics of the observed radio sources. Particularly challenging is the modelling of the restarted phase. The "off" time is a key parameter for the simulations in order to reproduce the conditions of the ISM in which the restarted jet will expand (e.g., [68,69]). Thus, the improved estimates of the length of the "on" and "off" phases derived from the new observations are providing key constraints to define the parameter space that needs to be explored by the new generation of simulations.

5. Life Cycle and Gas

Characterising the cycle of activity of radio AGN has broader implications, in particular, for the understanding of the feeding and feedback of AGN. How this cycle is regulated, i.e., the process of triggering and stopping the activity of the SMBH, and how the starting (or restarting) of a jet impacts the surrounding ISM/IGM in the host galaxy, i.e., feedback effect, are two key open questions.

Although various processes have been suggested to explain the triggering of AGN (see e.g., [70–72] for reviews), what actually originates the stop-and-start of the radio emission (and on which time scales) it is not fully understood. To explain the short-lived young radio sources, [73] suggested a radiation pressure instability scenario which may cause an intermittent activity of the central engine: outbursts of radio emission, with a duration of 10^{3-4} years, repeated regularly every 10^{5-6} years. To explain the longer cycle of activity, the availability of gas, perhaps connected to chaotic cold accretion [2], and the self-regulating feedback cycle, could be possible processes. Interestingly, some studies have suggested that young or restarted radio sources are more rich in gas, e.g., [13,74], while powerful radio galaxies seem to be more likely hosted by merger or interacting galaxies [70].

More progress have been made in the investigation of the impact of radio jets on the surrounding medium. While the main role of radio jets for feedback is usually considered to be preventing the cooling of the gas from the cluster scale [75], evidence has emerged that radio jets, like radiation and winds, can also originate gas outflows on galaxy scale. Simulations indicate that this can be more efficiently performed by a jet in its initial phase of evolution [76,77] (or in the restarted phase). Single object studies have also confirmed that this indeed can happen (see [13,74,78,79]). A statistical study specifically aimed at selecting young radio sources based on the peaked radio spectrum has further confirmed these results (see [80] and Figure 8). These results suggest that the young and restarted phases are those where the gas in the nuclear region of the host galaxies is mostly affected by the radio jet. These results do not exclude the effect of radiation/wind but show that when a young (or restarted) jet is present, the impact on the gas is visible. This impact may last for a short time (order of 10^6 yr) as seems to emerge from the actual size of the outflows (typically limited to the kpc scale).

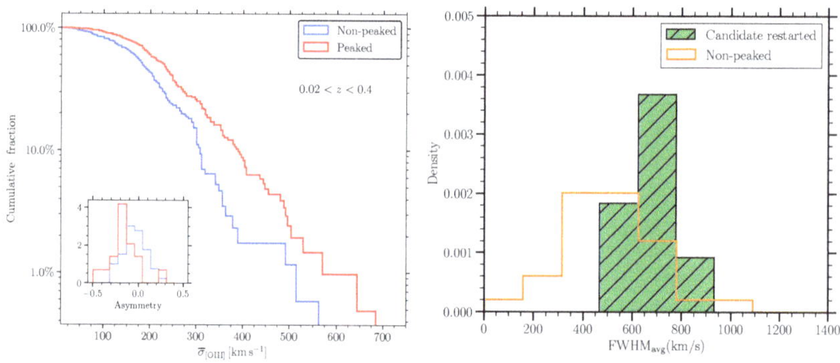

Figure 8. Left: Cumulative distribution of the [OIII] lines width (obtained from SDSS spectra) for peaked and non-peaked sources from the sample of Kukreti et al., in prep. Right: Average FWHM distributions of the [OIII] profiles of candidate restarted and the non-peaked radio AGN. Plots taken from [80]; see this paper for details.

6. Conclusions and Future Possibilities

This paper has briefly summarised some of the recent progresses made, thanks to the availability of new surveys, in the study of the life cycle of radio AGN. The view of the two most elusive phases, remnant (dying) and restarted, has significantly improved: these phases appear now richer and more complex than previously thought. Low radio-frequency surveys (i.e., using LOFAR, MWA and uGMRT) together with the expanded availability of ancillary data (i.e., optical identifications and redshifts) have opened up the possibility of selecting larger samples of sources in these elusive phases. Crucially, this can now be performed using multiple selection criteria: the results have demonstrated that this is essential for a more comprehensive census of remnant and restarted radio sources. The studies described in this paper provide the starting point and now can be expanded to larger areas, pushed to fainter targets. This will provide better statistics, covering a larger parameter space in terms of radio luminosity and stellar mass of the host galaxy, and better constraints for the models of radio galaxy evolution.

However, performing these larger selections provides a challenge in itself. With many million sources to classify, the development of advanced software tools for an automatic selection and classification of the objects is needed. This is a rapidly expanding field (e.g., [81,82]) which will also benefit from tools specifically tuned to identify rare objects like remnant radio source (see [83] and Brienza et al., in prep).

Many other new opportunities for improving our knowledge of the life cycle are already available (or will be soon). For example, these studies can be further expanded by using spectral index images, which can complement the analysis of the morphology and CP described above. As shown in this paper, surveys at GHz frequencies can be used together with the low-frequency ones. The results so far have also shown the importance of being able to achieve a good spatial resolution in order to separate the central and the extended emission. At low frequencies, LOFAR is also providing new possibilities for these studies. The LOFAR LBA Sky Survey (LoLSS) at 50 MHz (see [84]) will become available soon at full resolution. This will allow to build spectral index images extended to this very low frequency. A proof of concept has been performed in the Boötes region with spectral indices obtained at three frequencies (50, 150 and 1400 MHz) at about 15 arcsec resolution [58]. Other exciting opportunities are given by the high resolution that can be achieved by the LOFAR International Baselines Array. Thanks to the improved pipeline [56,85], images at 150 MHz with $\sim$0.3 arcsec resolution will become routinely available and will help identify and characterise the nuclear regions of radio AGN as shown for the LH area by [57].

The expanding possibilities for obtaining optical identifications for the radio sources and characterise the properties of the host galaxy are a major, key addition. The optical identification of the entire LoTSS DR2 has been recently published [38] and provides a great database of crucial ancillary information. These data also improve our knowledge of the environment, which is needed for a complete understanding of the parameters impacting the evolution of the radio sources. For tracing the presence and conditions of the gas, Integral Field Unit (IFU) instruments which will perform large surveys are also becoming available (e.g., WEAVE [86] and 4MOST [87]) and will play an important role.

In summary, a wealth of new data are opening up great possibilities for improving our understanding of the life cycle of radio AGN. Handling the data will provide major challenges as well as making sure theoretical models for interpreting the data are becoming available. Already from the results presented in this paper, it is clear that this effort will be rewarding in terms of scientific results, and it will also make sure the community is ready for the next big step ahead that will be made possible by the Square Kilometre Array (SKA).

Funding: This research received no external funding.

Data Availability Statement: Data from the surveys discussed in this paper (LOFAR, Apertif, MWA, etc.) are available mostly through the Virtual Observatory.

Acknowledgments: I would like to thank Marisa Brienza, Nika Jurlin, Stas Shabala and the LOFAR and Apertif Survey Teams for the help, comments and discussions. Some of the results presented in this overview have only been possible thanks to their contribution.

Conflicts of Interest: The author declares no conflicts of interest.

Abbreviations

The following abbreviations are used in this manuscript:

AGN	Active galactic nucleus
Apertif	APERture Tile In Focus
ASKAP	Australian Square Kilometre Array Pathfinder
ATCA	Australian Telescope Compact Array

CP	core prominence, S_{core}/S_{total}
DDRG	Double–double radio galaxy
FIRST	Faint Images of the Radio Sky at Twenty centimetres
GLEAM	Galactic and Extra-Galactic All-Sky MWA Survey
uGMRT	upgraded Giant Metrewave Radio Telescope
HBA	High-band Array
LBA	Low-Band Array
LH	Lockman Hole region
LoLSS	LOFAR LBA Sky Survey
LOFAR	LOw Frequency ARray
LoTSS	LOFAR Two-metre Sky Survey
MWA	Murchison Widefield Array
NVSS	The NRAO VLA Sky Survey
SMBH	Super Massive Black Hole
TGSS	TIFR-GMRT Sky Survey
USS	Ultra Steep Spectra
VLA	Karl G. Jansky Very Large Array
VLASS	Very Large Array Sky Survey
WSRT	Westerbork Synthesis Radio Telescope

Note

1 In this paper, the spectral index α is defined through $S \propto \nu^{-\alpha}$.

References

1. Fabian, A.C. Observational Evidence of Active Galactic Nuclei Feedback. *Annu. Rev. Astron. Astrophys.* **2012**, *50*, 455–489. [CrossRef]
2. Gaspari, M.; Temi, P.; Brighenti, F. Raining on black holes and massive galaxies: The top-down multiphase condensation model. *Mon. Not. R. Astron. Soc.* **2017**, *466*, 677–704. [CrossRef]
3. Silk, J.; Rees, M.J. Quasars and galaxy formation. *Astron. Astrophys.* **1998**, *331*, L1–L4. [CrossRef]
4. Best, P.N.; Kauffmann, G.; Heckman, T.M.; Brinchmann, J.; Charlot, S.; Ivezić, Z.; White, S.D.M. The host galaxies of radio-loud active galactic nuclei: Mass dependences, gas cooling and active galactic nuclei feedback. *Mon. Not. R. Astron. Soc.* **2005**, *362*, 25–40. [CrossRef]
5. Sabater, J.; Best, P.N.; Hardcastle, M.J.; Shimwell, T.W.; Tasse, C.; Williams, W. L.; Brüggen, M.; Cochrane, R.K.; Croston, J.H.; de Gasperin, F.; et al. The LoTSS view of radio AGN in the local Universe. The most massive galaxies are always switched on. *Astron. Astrophys.* **2019**, *622*, 17. [CrossRef]
6. Capetti, A.; Brienza, M.; Balmaverde, B.; Best, P.N.; Baldi, R.D.; Drabent, A.; Gürkan, G.; Rottgering, H.J.A.; Tasse, C.; Webster, B. The LOFAR view of giant, early-type galaxies: Radio emission from active nuclei and star formation. *Astron. Astrophys.* **2022**, *660*, 93. [CrossRef]
7. Biava, N.; Brienza, M.; Bonafede, A.; Gitti, M.; Bonnassieux, E.; Harwood, J.; Edge, A.C.; Riseley, C.J.; Vantyghem, A. Constraining the AGN duty cycle in the cool-core cluster MS 0735.6+7421 with LOFAR data. *Astron. Astrophys.* **2021**, *650*, 170. [CrossRef]
8. Randall, S.W.; Forman, W.R.; Giacintucci, S.; Nulsen, P.E.J.; Sun, M.; Jones, C.; Churazov, E.; David, L.P.; Kraft, R.; Donahue, M. et al. Shocks and Cavities from Multiple Outbursts in the Galaxy Group NGC 5813: A Window to Active Galactic Nucleus Feedback. *Astrophys. J.* **2011**, *726*, 86. [CrossRef]
9. Vantyghem, A.N.; McNamara, B.R.; Russell, H.; Main, R.; Nulsen, P.; Wise, M.W.; Hoekstra, H.; Gitti, M. Cycling of the powerful AGN in MS 0735.6+7421 and the duty cycle of radio AGN in clusters. *Mon. Not. R. Astron. Soc.* **2014**, *442*, 3192–3205. [CrossRef]
10. Harwood, J.J.; Hardcastle, M.J.; Croston, J.H.; Goodger, J.L. Spectral ageing in the lobes of FR-II radio galaxies: New methods of analysis for broad-band radio data. *Mon. Not. R. Astron. Soc.* **2013**, *435*, 3353–3375. [CrossRef]
11. Fanti, R.; Fanti, C.; Schilizzi, R.T.; Spencer, R.E.; Nan, R.; Parma, P.; van Breugel, W.J.M.; Venturi, T. On the nature of compact steep spectrum radio sources. *Astron. Astrophys.* **1990**, *231*, 333–346.
12. O'Dea, C.P. The Compact Steep-Spectrum and Gigahertz Peaked-Spectrum Radio Sources. *Publ. Astron. Soc. Pac.* **1998**, *110*, 493–532. [CrossRef]
13. O'Dea, C.P.; Saikia, D.J. Compact steep-spectrum and peaked-spectrum radio sources. *Astron. Astrophys. Rev.* **2021**, *29*, 3. [CrossRef]
14. Komissarov, S.S.; Gubanov, A.G. Relic radio galaxies: Evolution of synchrotron spectrum. *Astron. Astrophys.* **1994**, *285*, 27–43.
15. Murgia, M.; Parma, P.; Mack, K.-H.; De Ruiter, H.R.; Fanti, R.; Govoni, F.; Tarchi, A.; Giacintucci, S.; Markevitch, M. Dying radio galaxies in clusters. *Astron. Astrophys.* **2011**, *526*, A148. [CrossRef]

16. Schoenmakers, A.P.; De Bruyn, A.G.; Röttgering, H.J.A.; Van Der Laan, H.; Kaiser, C.R. Radio galaxies with a 'double–double morphology'—I. Analysis of the radio properties and evidence for interrupted activity in active galactic nuclei. *Mon. Not. R. Astron. Soc.* **2000**, *315*, 371–380. [CrossRef]
17. Brienza, M.; Godfrey, L.; Morganti, R.; Vilchez, N.; Maddox, N.; Murgia, M.; Orru, E.; Shulevski, A.; Best, P.N.; Brüggen, M.; et al. LOFAR discovery of a 700-kpc remnant radio galaxy at low redshift. *Astron. Astrophys.* **2015**, *585*, A29. [CrossRef]
18. Brienza, M.; Morganti, R.; Harwood, J.; Duchet, T.; Rajpurohit, K.; Shulevski, A.; Hardcastle, M.J.; Mahatma, V.; Godfrey, L.E.H.; Prandoni, I.; et al. Radio spectral properties and jet duty cycle in the restarted radio galaxy 3C388. *Astron. Astrophys.* **2020**, *638*, 29. [CrossRef]
19. Candini, S.; Brienza, M.; Bonafede, A.; Rajpurohit, K.; Biava, N.; Murgia, M.; Loi, F.; van Weeren, R.J.; Vazza, F. New filamentary remnant radio emission and duty cycle constraints in the radio galaxy NGC 6086. *Astron. Astrophys.* **2023**, *677*, 4. [CrossRef]
20. Cordey, R.A. IC 2476: A possible relic radio galaxy. *Mon. Not. R. Astron. Soc.* **1987**, *227*, 695–700. [CrossRef]
21. Konar, C.; Hardcastle, M.; Jamrozy, M.; Croston, J.H. Episodic radio galaxies J0116−4722 and J1158+2621: Can we constrain the quiescent phase of nuclear activity? *Mon. Not. R. Astron. Soc.* **2013**, *430*, 2137–2153. [CrossRef]
22. Kukreti, P.; Morganti, R.; Shimwell, T.W.; Morabito, L.K.; Beswick, R.J.; Brienza, M.; Hardcastle, M.J.; Sweijen, F.; Jackson, N.; Miley, G.K.; et al. Unmasking the history of 3C 293 with LOFAR sub-arcsecond imaging. *Astron. Astrophys.* **2022**, *658*, 6. [CrossRef]
23. Maccagni, F.M.; Murgia, M.; Serra, P.; Govoni, F.; Morokuma-Matsui, K.; Kleiner, D.; Buchner, S.; Józsa, G.I.G.; Kamphuis, P.; Makhathini, S.; et al. The flickering nuclear activity of Fornax A. *Astron. Astrophys.* **2020**, *634*, 9. [CrossRef]
24. Shulevski, A.; Morganti, R.; Harwood, J.J.; Barthel, P.D.; Jamrozy, M.; Brienza, M.; Brunetti, G.; Röttgering, H.J.A.; Murgia, M.; White, G.J.; et al. Radiative age mapping of the remnant radio galaxy B2 0924+30: The LOFAR perspective. *Astron. Astrophys.* **2017**, *600*, A65. [CrossRef]
25. Hardcastle, M.; Croston, J. Radio galaxies and feedback from AGN jets. *New Astron. Rev.* **2020**, *88*, 101539. . j.newar.2020.101539. [CrossRef]
26. Mahatma, V.H. The Dynamics and Energetics of Remnant and Restarting RLAGN. *Galaxies* **2023**, *11*, 74. [CrossRef]
27. Morganti, R. Archaeology of active galaxies across the electromagnetic spectrum. *Nat. Astron.* **2017**, *1*, 39–48. [CrossRef]
28. Morganti, R.; Jurlin, N.; Oosterloo, T.; Brienza, M.; Orrú, E.; Kutkin, A.; Prandoni, I.; Adams, E.A.K.; Dénes, H.; Hess, K.M.; et al. Combining LOFAR and Apertif Data for Understanding the Life Cycle of Radio Galaxies. *Galaxies* **2021a**, *9*, 88. [CrossRef]
29. Shimwell, T.W.; Hardcastle, M.J.; Tasse, C.; Best, P.N.; Röttgering, H.J.A.; Williams, W.L.; Botteon, A.; Drabent, A.; Mechev, A.; Shulevski, A.; et al. The LOFAR Two-metre Sky Survey. V. Second data release. *Astron. Astrophys.* **2022**, *659*, 1. [CrossRef]
30. Hurley-Walker, N.; Galvin, T.J.; Duchesne, S.W.; Zhang, X.; Morgan, J.; Hancock, P.J.; An, T.; Franzen, T.M.O.; Heald, G.; Ross, K.; et al. GaLactic and Extragalactic All-sky Murchison Widefield Array survey eXtended (GLEAM-X) I: Survey description and initial data release. *Publ. Astron. Soc. Aust.* **2022**, *39*, 35. [CrossRef]
31. Intema, H.T.; Jagannathan, P.; Mooley, K.P.; Frail, D.A. The GMRT 150 MHz all-sky radio survey. First alternative data release TGSS ADR1. *Astron. Astrophys.* **2017**, *598*, 78. . [CrossRef]
32. Adams, E.A.K.; Adebahr, B.; de Blok, W.J.G.; Dénes, H.; Hess, K. M.; van der Hulst, J.M.; Kutkin, A.; Lucero, D.M.; Morganti, R.; Moss, V.A.; et al. First release of Apertif imaging survey data. *Astron. Astrophys.* **2022**, *667*, 38. . [CrossRef]
33. McConnell, D.; Hale, C.L.; Lenc, E.; Banfield, J.K.; Heald, G.; Hotan, A.W.; Leung, J.K.; Moss, V.A.; Murphy, T.; O'Brien, A.; et al. The Rapid ASKAP Continuum Survey I: Design and first results. *Publ. Astron. Soc. Aust.* **2020**, *37*, 48. [CrossRef]
34. Gordon, Y.A.; Boyce, M.M.; O'Dea, C.P.; Rudnick, L.; Andernach, H.; Vantyghem, A.N.; Baum, S.A.; Bui, J.-P.; Dionyssiou, M.; Safi-Harb, S.; et al. A Quick Look at the 3 GHz Radio Sky. I. Source Statistics from the Very Large Array Sky Survey. *Astrophys. J. Suppl. Ser.* **2021**, *255*, 30 . [CrossRef]
35. Lacy, M.; Baum, S.A.; Chandler, C.J.; Chatterjee, S.; Clarke, T.E.; Deustua, S.; English, J.; Farnes, J.; Gaensler, B.M.; Gugliucci, N.; et al. The Karl G. Jansky Very Large Array Sky Survey (VLASS). Science Case and Survey Design. *Publ. Astron. Soc. Pac.* **2020**, *132*, 035001. [CrossRef]
36. DESI Collaboration; Adame, A.G.; Aguilar, J.; Ahlen, S.; Alam, S.; Aldering, G.; Alexander, D.M.; Alfarsy, R.; Allende Prieto, C.; Alvarez, M.; Alves, O.; et al. The Early Data Release of the Dark Energy Spectroscopic Instrument. *arXiv* **2023**, arXiv:2306.06308. https://doi.org/10.48550/arXiv.2306.06308.
37. Duncan, K.J.; Kondapally, R.; Brown, M.J.I.; Bonato, M.; Best, P. N.; Röttgering, H.J.A.; Bondi, M.; Bowler, R.A.A.; Cochrane, R.K.; Gürkan, G.; et al. The LOFAR Two-meter Sky Survey: Deep Fields Data Release 1. IV. Photometric redshifts and stellar masses. *Astron. Astrophys.* **2021**, *648*, 4. [CrossRef]
38. Hardcastle, M.J.; Horton, M.A.; Williams, W.L.; Duncan, K.J.; Alegre, L.; Barkus, B.; Croston, J.H.; Dickinson, H.; Osinga, E.; Röttgering, H.J.A.; et al. The LOFAR Two-Metre Sky Survey. VI. Optical identifications for the second data release. *Astron. Astrophys.* **2023**, *678*, 151. [CrossRef]
39. Roettiger, K.; Burns, J.O.; Clarke, D.A.; Christiansen, W.A. Relic Radio Emission in 3C 388. *Astrophys. J.* **1994**, *421*, L23. [CrossRef]
40. Saripalli, L.; Subrahmanyan, R.; Thorat, K.; Ekers, R.; Hunstead, R.W.; Johnston, H.; Sadler, E. ATLBS Extended Source Sample: The Evolution in Radio Source Morphology with Flux Density. *Astrophys. J. Suppl. Ser.* **2012**, *199*, 27. [CrossRef]
41. Giovannini, G.; Feretti, L.; Gregorini, L.; Parma, P. Radio nuclei in elliptical galaxies. *Astron. Astrophys.* **1988**, *199*, 73–84.
42. Parma, P.; Murgia, M.; de Ruiter, H.R.; Fanti, R.; Mack, K.-H.; Govoni, F. In search of dying radio sources in the local universe. *Astron. Astrophys.* **2007**, *470*, 875–888. .:20077592. [CrossRef]

43. Brienza, M.; Godfrey, L.; Morganti, R.; Prandoni, I.; Harwood, J.; Mahony, E.; Hardcastle, M.J.; Murgia, M.; Röttgering, H.J.A.; Shimwell, T.W.; et al. Search and modelling of remnant radio galaxies in the LOFAR Lockman Hole field. *Astron. Astrophys.* **2017**, *606*, A98. [CrossRef]
44. Tasse, C.; Shimwell, T.; Hardcastle, M.J.; O'Sullivan, S.P.; van Weeren, R.; Best, P.N.; Bester, L.; Hugo, B.; Smirnov, O.; Sabater, J.; et al. The LOFAR Two-meter Sky Survey: Deep Fields Data Release 1. I. Direction-dependent calibration and imaging. *Astron. Astrophys.* **2021**, *648*, 1. [CrossRef]
45. Jurlin, N.; Brienza, M.; Morganti, R.; Wadadekar, Y.; Ishwara- Chandra, C.H.; Maddox, N.; Mahatma, V. Multi-frequency characterisation of remnant radio galaxies in the Lockman Hole field. *Astron. Astrophys.* **2021**, *653*, 110. [CrossRef]
46. Morganti, R.; Oosterloo, T.A.; Brienza, M.; Jurlin, N.; Prandoni, I.; Orrù, E.; Shabala, S.S.; Adams, E.A.K.; Adebahr, B.; Best, P.N.; et al. The best of both worlds: Combining LOFAR and Apertif to derive resolved radio spectral index images. *Astron. Astrophys.* **2021**, *648*, 9. [CrossRef]
47. Mahatma, V.H.; Hardcastle, M.J.; Williams, W.L.; Brienza, M.; Brüggen, M.; Croston, J.H.; Gürkan, G.; Harwood, J.J.; Kunert-Bajraszewska, M.; Morganti, R.; et al. Remnant radio-loud AGN in the Herschel-ATLAS field. *Mon. Not. R. Astron. Soc.* **2018**, *475*, 4557–4578. [CrossRef]
48. Hardcastle, M.J.; Gürkan, G.; van Weeren, R.J.; Williams, W.L.; Best, P.N.; de Gasperin, F.; Rafferty, D.A.; Read, S.C.; Sabater, J.; Shimwell, T.W.; et al. LOFAR/H-ATLAS: A deep low-frequency survey of the Herschel-ATLAS North Galactic Pole field. *Mon. Not. R. Astron. Soc.* **2016**, *462*, 1910–1936. [CrossRef]
49. Quici, B.; Hurley-Walker, N.; Seymour, N.; Turner, R.J.; Shabala, S.S.; Huynh, M.; Andernach, H.; Kapińska, A.D.; Collier, J.D.; Johnston-Hollitt, M.; et al. Remnant radio galaxies discovered in a multi-frequency survey. *Publ. Astron. Soc. Aust.* **2021**, *38*, e008. [CrossRef]
50. Dutta, S.; Singh, V.; Chandra, C.H.I.; Wadadekar, Y.; Kayal, A.; Heywood, I. Search and Characterization of Remnant Radio Galaxies in the XMM-LSS Deep Field. *Astrophys. J.* **2023**, *944*, 176. https:/doi.org/10.3847/1538-4357/acaf01. [CrossRef]
51. Brienza, M.; Lovisari, L.; Rajpurohit, K.; Bonafede, A.; Gastaldello, F.; Murgia, M.; Vazza, F.; Bonnassieux, E.; Botteon, A.; Brunetti, G.; et al. The galaxy group NGC 507: Newly detected AGN remnant plasma transported by sloshing. *Astron. Astrophys.* **2022**, *661*, 92. [CrossRef]
52. Brocksopp, C.; Kaiser, C.R.; Schoenmakers, A.P.; de Bruyn, A.G. Three episodes of jet activity in the Fanaroff-Riley type II radio galaxy B0925+420. *Mon. Not. R. Astron. Soc.* **2007**, *382*, 1019–1028. [CrossRef]
53. Chavan, K.; Dabhade, P.; Saikia, D.J. A giant radio galaxy with three cycles of episodic jet activity from LoTSS DR2. *Mon. Not. R. Astron. Soc.* **2023**, *525*, L87–L92. [CrossRef]
54. Mahatma, V.H.; Hardcastle, M.J.; Williams, W.L.; Best, P.N.; Croston, J.H.; Duncan, K.; Mingo, B.; Morganti, R.; Brienza, M.; Cochrane, R.K.; et al. LoTSS DR1: Double-double radio galaxies in the HETDEX field. *Astron. Astrophys.* **2019**, *622*, 13. [CrossRef]
55. Jurlin, N.; Morganti, R.; Brienza, M.; Mandal, S.; Maddox, N.; Duncan, K.J.; Shabala, S.S.; Hardcastle, M.J.; Prandoni, I.; Röttgering, H.J.A.; et al. The life cycle of radio galaxies in the LOFAR Lockman Hole field. *Astron. Astrophys.* **2020**, *638*, 34. [CrossRef]
56. Sweijen, F.; van Weeren, R.J.; Röttgering, H.J.A.; Morabito, L.K.; Jackson, N.; Offringa, A.R.; van der Tol, S.; Veenboer, B.; Oonk, J.B.R.; Best, P.N.; et al. Deep sub-arcsecond wide-field imaging of the Lockman Hole field at 144 MHz. *Nat. Astron.* **2022**, *6*, 350–356. [CrossRef]
57. Jurlin, N.; Morganti, R.; Sweijen, F.; Morabito, L.K.; Brienza, M.; Barthel, P.; Miley, G.K. Nuclear regions as seen with LOFAR international baselines: A high-resolution study of the recurrent activity. *Astronomy and Astrophysics* **2024**, *682*, 118. arXiv:2311.12114. https://doi.org/10.48550/arXiv.2311.12114.
58. Kutkin, A.M.; Oosterloo, T.A.; Morganti, R.; Offringa, A.R.; Adams, E.A.K.; Adebahr, B.; Dénes, H.; Hess, K.M.; van der Hulst, J.M.; de Blok, W.J.G.; et al. Apertif 1.4 GHz continuum observations of the Boötes field and their combined view with LOFAR. *Astron. Astrophys.* **2023**, *676*, 37. [CrossRef]
59. Kunert-Bajraszewska, M.; Marecki, A.; Thomasson, P. FIRST-based survey of compact steep spectrum sources. *Astron. Astrophys.* **2006**, *450*, 945–958. .:20054428. [CrossRef]
60. Orienti, M.; Murgia, M.; Dallacasa, D.; Migliori, G.; D'Ammando, F. Young but fading radio sources: Searching for remnants among compact steep-spectrum radio sources. *Mon. Not. R. Astron. Soc.* **2023**, *522*, 3877–3889. [CrossRef]
61. Webster, B.; Croston, J.H.; Harwood, J.J.; Baldi, R.D.; Hardcastle, M.J.; Mingo, B.; Röttgering, H.J.A. Investigating the spectra and physical nature of galaxy scale jets. *Mon. Not. R. Astron. Soc.* **2021**, *508*, 5972–5990. [CrossRef]
62. Godfrey, L.E.H.; Morganti, R.; Brienza, M. On the population of remnant Fanaroff–Riley type II radio galaxies and implications for radio source dynamics. *Mon. Not. R. Astron. Soc.* **2017**, *471*, 891–907. [CrossRef]
63. Shabala, S.S.; Jurlin, N.; Morganti, R.; Brienza, M.; Hardcastle, M.J.; Godfrey, L.E.H.; Krause, M.G.H.; Turner, R. The duty cycle of radio galaxies revealed by LOFAR: Remnant and restarted radio source populations in the Lockman Hole. *Mon. Not. R. Astron. Soc.* **2020**, *496*, 1706–1717. [CrossRef]
64. Orrù, E.; van Velzen, S.; Pizzo, R.F.; Yatawatta, S.; Paladino, R.; Iacobelli, M.; Murgia, M.; Falcke, H.; Morganti, R.; de Bruyn, A.G.; et al. Wide-field LOFAR imaging of the field around the double-double radio galaxy B1834+620. A fresh view on a restarted AGN and doubeltjes. *Astron. Astrophys.* **2015**, *584*, 112. [CrossRef]
65. Hardcastle, M.J. A simulation-based analytic model of radio galaxies. *Mon. Not. R. Astron. Soc.* **2018**, *475*, 2768–2786. [CrossRef]
66. Turner, R.J.; Shabala, S.S. Dynamics of Powerful Radio Galaxies. *Galaxies* **2023**, *11*, 87. [CrossRef]
67. Turner, R.J.; Shabala, S.S. Energetics and Lifetimes of Local Radio Active Galactic Nuclei. *Astrophys. J.* **2015**, *806*, 59. [CrossRef]

68. Safouris, V.; Subrahmanyan, R.; Bicknell, G.V.; Saripalli, L. PKS B1545-321: Bow shocks of a relativistic jet? *Mon. Not. R. Astron. Soc.* **2008**, *385*, 2117–2135. [CrossRef]
69. Walg, S.; Achterberg, A.; Markoff, S.; Keppens, R.; Porth, O. Relativistic AGN jets - II. Jet properties and mixing effects for episodic jet activity. *Mon. Not. R. Astron. Soc.* **2014**, *439*, 3969–3985. [CrossRef]
70. Ramos Almeida, C.; Bessiere, P.S.; Tadhunter, C.N.; Pérez-González, P.G.; Barro, G.; Inskip, K.J.; Morganti, R.; Holt, J.; Dicken, D. Are luminous radio-loud active galactic nuclei triggered by galaxy interactions? *Mon. Not. R. Astron. Soc.* **2012**, *419*, 687–705. [CrossRef]
71. Combes, F. Fueling Processes on (Sub-)kpc Scales. *Galaxies* **2023**, *11*, 120. [CrossRef]
72. Storchi-Bergmann, T.; Schnorr-Müller, A. Observational constraints on the feeding of supermassive black holes. *Nat. Astron.* **2019**, *3*, 48–61. [CrossRef]
73. Czerny, B.; Siemiginowska, A.; Janiuk, A.; Nikiel-Wroczyński, B.; Stawarz, Ł. Accretion Disk Model of Short-Timescale Intermittent Activity in Young Radio Sources. *Astrophys. J.* **2009**, *698*, 840–851. [CrossRef]
74. Holt, J.; Tadhunter, C.N.; Morganti, R. Fast outflows in compact radio sources: Evidence for AGN-induced feedback in the early stages of radio source evolution. *Mon. Not. R. Astron. Soc.* **2008**, *387*, 639–659. [CrossRef]
75. McNamara, B.R.; Nulsen, P.E.J. Mechanical feedback from active galactic nuclei in galaxies, groups and clusters. *New J. Phys.* **2012**, *14*, 055023. [CrossRef]
76. Mukherjee, D.; Bicknell, G.V.; Wagner, A.Y.; Sutherl, R.S.; Silk, J. Relativistic jet feedback—III. Feedback on gas discs. *Mon. Not. R. Astron. Soc.* **2018**, *479*, 5544–5566. [CrossRef]
77. Sutherl, R.S.; Bicknell, G.V. Interactions of a Light Hypersonic Jet with a Nonuniform Interstellar Medium. *Astrophys. J. Suppl. Ser.* **2007**, *173*, 37–69. [CrossRef]
78. Molyneux, S.J.; Harrison, C.M.; Jarvis, M.E. Extreme ionised outflows are more common when the radio emission is compact in AGN host galaxies. *Astron. Astrophys.* **2019**, *631*, A132. [CrossRef]
79. Shih, H.-Y.; Stockton, A.; Kewley, L. Ionized Outflows from Compact Steep Spectrum Sources. *Astrophys. J.* **2013**, *772*, 138. [CrossRef]
80. Kukreti, P.; Morganti, R.; Tadhunter, C.; Santoro, F. Ionised gas outflows over the radio AGN life cycle. *Astron. Astrophys.* **2023**, *674*, 198. [CrossRef]
81. Galvin, T.J.; Huynh, M.T.; Norris, R.P.; Wang, X.R.; Hopkins, E.; Polsterer, K.; Ralph, N.O.; O'Brien, A.N.; Heald, G.H. Cataloguing the radio-sky with unsupervised machine learning: A new approach for the SKA era. *Mon. Not. R. Astron. Soc.* **2020**, *497*, 2730–2758. [CrossRef]
82. Ndung'u, S.; Grobler, T.; Wijnholds, S.J.; Karastoyanova, D.; Azzopardi, G. Advances on the morphological classification of radio galaxies: A review. *New Astron. Rev.* **2023**, *97*, 101685. [CrossRef]
83. Mostert, R.I.J.; Morganti, R.; Brienza, M.; Duncan, K.J.; Oei, M.S.S.L.; Röttgering, H.J.A.; Alegre, L.; Hardcastle, M.J.; Jurlin, N. Finding AGN remnant candidates based on radio morphology with machine learning. *Astron. Astrophys.* **2023**, *674*, 208. [CrossRef]
84. de Gasperin, F.; Edler, H.W.; Williams, W.L.; Callingham, J.R.; Asabere, B.; Brüggen, M.; Brunetti, G.; Dijkema, T.J.; Hardcastle, M.J.; Iacobelli, M.; et al. The LOFAR LBA Sky Survey. II. First data release. *Astron. Astrophys.* **2023**, *673*, 165. [CrossRef]
85. Morabito, L.K.; Jackson, N.J.; Mooney, S.; Sweijen, F.; Badole, S.; Kukreti, P.; Venkattu, D.; Groeneveld, C.; Kappes, A.; Bonnassieux, E.; et al. Sub-arcsecond imaging with the International LOFAR Telescope. I. Foundational calibration strategy and pipeline. *Astron. Astrophys.* **2022**, *658*, 1. [CrossRef]
86. Jin, S.; Trager, S.C.; Dalton, G.B.; Aguerri, J.A.L.; Drew, J.E.; Falcón-Barroso, J.; Gänsicke, B.T.; Hill, V.; Iovino, A.; Pieri, M.M.; et al. The wide-field, multiplexed, spectroscopic facility WEAVE: Survey design, overview, and simulated implementation. *Mon. Not. R. Astron. Soc.* **2023**, stad557. [CrossRef]
87. de Jong, R.S.; Bellido-Tirado, O.; Brynnel, J.G.; Ezzati de Jong, R.S.; Bellido-Tirado, O.; Brynnel, J.G.; Ezzati Amini, A.; Frey, S.; Füßlein, C.; Gäbler, M.; Giannone, D.; Johl, D.; Kuba, S.; et al. 4MOST: The 4-metre multi-object spectroscopic telescope project in the assembly, integration, and test phase. In Proceedings of the Ground-Based and Airborne Instrumentation for Astronomy IX, Montréal, QC, Canada, 17–23 July 2022; Volume 12184. [CrossRef]

MDPI AG
Grosspeteranlage 5
4052 Basel
Switzerland
Tel.: +41 61 683 77 34

Galaxies Editorial Office
E-mail: galaxies@mdpi.com
www.mdpi.com/journal/galaxies

MDPI

www.ingramcontent.com/pod-product-compliance
Lightning Source LLC
Chambersburg PA
CBHW062051220926
51850CB00010B/2729
* 9 7 8 3 7 2 5 8 7 0 2 6 4 *